D1562157

VOLUME FOUR HUNDRED AND NINETY-FOUR

METHODS IN ENZYMOLOGY

Methods in Methane Metabolism, Part A

Methanogenesis

METHODS IN ENZYMOLOGY

VOLUME FOUR HUNDRED AND NINETY-FOUR

METHODS IN ENZYMOLOGY

Methods in Methane Metabolism, Part A

Methanogenesis

EDITED BY

AMY C. ROSENZWEIG
Departments of Biochemistry, Molecular Biology, and Cell Biology and of Chemistry
Northwestern University
Illinois, USA

STEPHEN W. RAGSDALE
Department of Biological Chemistry
University of Michigan Medical School
Michigan, USA

AMSTERDAM • BOSTON • HEIDELBERG • LONDON
NEW YORK • OXFORD • PARIS • SAN DIEGO
SAN FRANCISCO • SINGAPORE • SYDNEY • TOKYO
Academic Press is an imprint of Elsevier

Academic Press is an imprint of Elsevier
525 B Street, Suite 1900, San Diego, CA 92101-4495, USA
30 Corporate Drive, Suite 400, Burlington, MA 01803, USA
32 Jamestown Road, London NW1 7BY, UK

First edition 2011

ISBN: 978-0-12-385112-3
ISSN: 0076-6879

Printed and bound in United States of America
11 12 13 14 10 9 8 7 6 5 4 3 2 1

Contents

Contributors

Irini Angelidaki
Department of Environmental Engineering, Technical University of Denmark, Lyngby, Denmark

Damien J. Batstone
Advanced Water Management Centre, The University of Queensland, Brisbane, Queensland, Australia

Heather Brewer
Environmental and Molecular Sciences Laboratory, Pacific Northwest National Laboratory, Richland, Washington, USA

Charlene Brungess
Department of Chemistry and Biochemistry, Auburn University, Alabama, USA

Nicole Buan
Department of Biochemistry and the Redox Biology Center, University of Nebraska-Lincoln, N200 Beadle Center, Lincoln, Nebraska, USA

Lea Constan
Department of Microbiology and Immunology, University of British Columbia, Vancouver, British Columbia, Canada

Uwe Deppenmeier
Institute of Microbiology and Biotechnology, University of Bonn, Bonn, Germany

Evert C. Duin
Department of Chemistry and Biochemistry, Auburn University, Alabama, USA

Donald J. Ferguson, Jr.
Department of Microbiology, University of Miami, Oxford, Ohio, USA

James G. Ferry
Department of Biochemistry and Molecular Biology, Pennsylvania State University, University Park, Pennsylvania, USA

Gordon Fox
Department of Integrative Biology, University of South Florida, Tampa, Florida, USA

Peter R. Girguis
Department of Organismal and Evolutionary Biology, Harvard University, Cambridge, Massachusetts, USA

David E. Graham
Oak Ridge National Laboratory, Biosciences Division and University of Tennessee Knoxville, Microbiology Department, Oak Ridge, Tennessee, USA

David A. Grahame
Department of Biochemistry and Molecular Biology, Uniformed Services University of the Health Sciences, Bethesda, Maryland, USA

Steven J. Hallam
Department of Microbiology and Immunology, and Graduate Program in Bioinformatics, University of British Columbia, Vancouver, British Columbia, Canada

Dimitar Karakashev
Department of Environmental Engineering, Technical University of Denmark, Lyngby, Denmark

Joseph A. Krzycki
Department of Microbiology and the OSU Biochemistry Program, Ohio State University, Columbus, Ohio, USA

Gargi Kulkarni
Department of Microbiology, University of Illinois, Urbana-Champaign, B103 Chemical and Life Sciences Laboratory, Urbana, Illinois, USA

John A. Leigh
Department of Microbiology, University of Washington, Seattle, Washington, USA

David G. Longstaff
Department of Microbiology, Ohio State University, Columbus, Ohio, USA

William Metcalf
Department of Microbiology, University of Illinois, Urbana-Champaign, B103 Chemical and Life Sciences Laboratory, Urbana, Illinois, USA

Volker Müller
Molecular Microbiology & Bioenergetics, Institute of Molecular Biosciences, Goethe-University Frankfurt, Frankfurt, Germany

Angela D. Norbeck
Pacific Northwest National Laboratory, Richland, Washington, USA

Antoine P. Pagé
Department of Microbiology and Immunology, University of British Columbia, Vancouver, British Columbia, Canada

Ljiljana Pasa-Tolic
Environmental and Molecular Sciences Laboratory, Pacific Northwest National Laboratory, Richland, Washington, USA

Caroline M. Plugge
Laboratory of Microbiology, Wageningen University, Wageningen, The Netherlands

Divya Prakash
Department of Chemistry and Biochemistry, Auburn University, Alabama, USA

Michael Rother
Institut für Molekulare Biowissenschaften, Molekulare Mikrobiologie & Bioenergetik, Johann Wolfgang Goethe-Universität, Frankfurt am Main, Germany

Felipe Sarmiento B.
Department of Microbiology, University of Georgia, Athens, Georgia, USA

Christian Sattler
Institut für Molekulare Biowissenschaften, Molekulare Mikrobiologie & Bioenergetik, Johann Wolfgang Goethe-Universität, Frankfurt am Main, Germany

Michael Schick
Max Planck Institute for Terrestrial Microbiology, Marburg, Germany

Katharina Schlegel
Molecular Microbiology & Bioenergetics, Institute of Molecular Biosciences, Goethe-University Frankfurt, Frankfurt, Germany

Kathleen M. Scott
Department of Integrative Biology, University of South Florida, Tampa, Florida, USA

Seigo Shima
Max Planck Institute for Terrestrial Microbiology, Marburg, Germany, and PRESTO, Japan Science and Technology Agency (JST), Honcho, Kawaguchi, Saitama, Japan

Young C. Song
Graduate Program in Bioinformatics, University of British Columbia, Vancouver, British Columbia, Canada

Alfons J. M. Stams
Laboratory of Microbiology, Wageningen University, Wageningen, The Netherlands

Tilmann Stock
Institut für Molekulare Biowissenschaften, Molekulare Mikrobiologie & Bioenergetik, Johann Wolfgang Goethe-Universität, Frankfurt am Main, Germany

Haruka Tamura
Max Planck Institute for Terrestrial Microbiology, Marburg, Germany

Cornelia Welte
Institute of Microbiology and Biotechnology, University of Bonn, Bonn, Germany

William B. Whitman
Department of Microbiology, University of Georgia, Athens, Georgia, USA

Ralph S. Wolfe
Department of Microbiology, University of Illinois, Urbana, Illinois, USA

Preface

The production and consumption of methane by microorganisms is central to the global carbon cycle. It has been 2 decades since a *Methods in Enzymology* volume focused explicitly on this field (Volume 188, *Hydrocarbons and Methylotrophy*). During that time, interest in methane metabolism has steadily increased in the context of dwindling petroleum reserves, increased greenhouse gas emissions, and environmental hydrocarbon pollution. A field previously dominated by microbiology and protein biochemistry has exploded in multiple directions. In particular, the advent of genomic and proteomic techniques has transformed the way methane metabolic pathways are studied. In these volumes, we cover both the generation (Part A) and utilization (Part B) of methane.

Part A describes recent developments that enable a wide variety of experiments with methanogenic Archaea, which seemed intractable two decades ago to all but those initiates who "grew up" studying anaerobes. Methods are presented to readily culture and to perform genetic experiments on these oxygen-sensitive microbes as well as to characterize the remarkable enzymes and respiratory proteins that allow methanogens to generate energy for growth and produce as a byproduct a very important fuel that may be adopted as the "fuel of the future." Included is the state of the art in "biomethanation," the biotechnological use of methanogens to produce this important energy source. Finally, approaches are described for the deployment of "omics" technologies to understand how methanogens regulate metabolism.

The first methanotroph genome sequence was completed in 2004 and research has become progressively more "omic" in the past few years. Part B combines traditional approaches to methanotroph isolation and enzyme chemistry with state-of-the-art genomic and proteomic techniques. Novel methods have been developed and used to address challenging problems such as linking metagenomic data with environmental function or generating mutant forms of the methane monooxygenase enzymes. Moreover, whole new areas of research, such as the study of the copper chelator methanobactin, have been discovered. Taken together, we hope that the two volumes capture the excitement that pervades this rapidly developing field.

We thank an outstanding group of colleagues with diverse points of view for providing ideas on content and ultimately contributing a series of excellent chapters. The high level of enthusiasm in the community resulted in the unexpected final production of two, rather than one, volumes. The methods and novel approaches described here should inspire and guide future research in the field as well as provide a central resource for researchers interested in methane metabolism.

Amy C. Rosenzweig and Stephen W. Ragsdale

Methods in Enzymology

Volume I. Preparation and Assay of Enzymes
Edited by Sidney P. Colowick and Nathan O. Kaplan

Volume II. Preparation and Assay of Enzymes
Edited by Sidney P. Colowick and Nathan O. Kaplan

Volume III. Preparation and Assay of Substrates
Edited by Sidney P. Colowick and Nathan O. Kaplan

Volume IV. Special Techniques for the Enzymologist
Edited by Sidney P. Colowick and Nathan O. Kaplan

Volume V. Preparation and Assay of Enzymes
Edited by Sidney P. Colowick and Nathan O. Kaplan

Volume VI. Preparation and Assay of Enzymes *(Continued)*
Preparation and Assay of Substrates
Special Techniques
Edited by Sidney P. Colowick and Nathan O. Kaplan

Volume VII. Cumulative Subject Index
Edited by Sidney P. Colowick and Nathan O. Kaplan

Volume VIII. Complex Carbohydrates
Edited by Elizabeth F. Neufeld and Victor Ginsburg

Volume IX. Carbohydrate Metabolism
Edited by Willis A. Wood

Volume X. Oxidation and Phosphorylation
Edited by Ronald W. Estabrook and Maynard E. Pullman

Volume XI. Enzyme Structure
Edited by C. H. W. Hirs

Volume XII. Nucleic Acids (Parts A and B)
Edited by Lawrence Grossman and Kivie Moldave

Volume XIII. Citric Acid Cycle
Edited by J. M. Lowenstein

Volume XIV. Lipids
Edited by J. M. Lowenstein

Volume XV. Steroids and Terpenoids
Edited by Raymond B. Clayton

VOLUME XXXV. Lipids (Part B)
Edited by JOHN M. LOWENSTEIN

VOLUME XXXVI. Hormone Action (Part A: Steroid Hormones)
Edited by BERT W. O'MALLEY AND JOEL G. HARDMAN

VOLUME XXXVII. Hormone Action (Part B: Peptide Hormones)
Edited by BERT W. O'MALLEY AND JOEL G. HARDMAN

VOLUME XXXVIII. Hormone Action (Part C: Cyclic Nucleotides)
Edited by JOEL G. HARDMAN AND BERT W. O'MALLEY

VOLUME XXXIX. Hormone Action (Part D: Isolated Cells, Tissues, and Organ Systems)
Edited by JOEL G. HARDMAN AND BERT W. O'MALLEY

VOLUME XL. Hormone Action (Part E: Nuclear Structure and Function)
Edited by BERT W. O'MALLEY AND JOEL G. HARDMAN

VOLUME XLI. Carbohydrate Metabolism (Part B)
Edited by W. A. WOOD

VOLUME XLII. Carbohydrate Metabolism (Part C)
Edited by W. A. WOOD

VOLUME XLIII. Antibiotics
Edited by JOHN H. HASH

VOLUME XLIV. Immobilized Enzymes
Edited by KLAUS MOSBACH

VOLUME XLV. Proteolytic Enzymes (Part B)
Edited by LASZLO LORAND

VOLUME XLVI. Affinity Labeling
Edited by WILLIAM B. JAKOBY AND MEIR WILCHEK

VOLUME XLVII. Enzyme Structure (Part E)
Edited by C. H. W. HIRS AND SERGE N. TIMASHEFF

VOLUME XLVIII. Enzyme Structure (Part F)
Edited by C. H. W. HIRS AND SERGE N. TIMASHEFF

VOLUME XLIX. Enzyme Structure (Part G)
Edited by C. H. W. HIRS AND SERGE N. TIMASHEFF

VOLUME L. Complex Carbohydrates (Part C)
Edited by VICTOR GINSBURG

VOLUME LI. Purine and Pyrimidine Nucleotide Metabolism
Edited by PATRICIA A. HOFFEE AND MARY ELLEN JONES

VOLUME LII. Biomembranes (Part C: Biological Oxidations)
Edited by SIDNEY FLEISCHER AND LESTER PACKER

Volume LIII. Biomembranes (Part D: Biological Oxidations)
Edited by Sidney Fleischer and Lester Packer

Volume LIV. Biomembranes (Part E: Biological Oxidations)
Edited by Sidney Fleischer and Lester Packer

Volume LV. Biomembranes (Part F: Bioenergetics)
Edited by Sidney Fleischer and Lester Packer

Volume LVI. Biomembranes (Part G: Bioenergetics)
Edited by Sidney Fleischer and Lester Packer

Volume LVII. Bioluminescence and Chemiluminescence
Edited by Marlene A. DeLuca

Volume LVIII. Cell Culture
Edited by William B. Jakoby and Ira Pastan

Volume LIX. Nucleic Acids and Protein Synthesis (Part G)
Edited by Kivie Moldave and Lawrence Grossman

Volume LX. Nucleic Acids and Protein Synthesis (Part H)
Edited by Kivie Moldave and Lawrence Grossman

Volume 61. Enzyme Structure (Part H)
Edited by C. H. W. Hirs and Serge N. Timasheff

Volume 62. Vitamins and Coenzymes (Part D)
Edited by Donald B. McCormick and Lemuel D. Wright

Volume 63. Enzyme Kinetics and Mechanism (Part A: Initial Rate and Inhibitor Methods)
Edited by Daniel L. Purich

Volume 64. Enzyme Kinetics and Mechanism
(Part B: Isotopic Probes and Complex Enzyme Systems)
Edited by Daniel L. Purich

Volume 65. Nucleic Acids (Part I)
Edited by Lawrence Grossman and Kivie Moldave

Volume 66. Vitamins and Coenzymes (Part E)
Edited by Donald B. McCormick and Lemuel D. Wright

Volume 67. Vitamins and Coenzymes (Part F)
Edited by Donald B. McCormick and Lemuel D. Wright

Volume 68. Recombinant DNA
Edited by Ray Wu

Volume 69. Photosynthesis and Nitrogen Fixation (Part C)
Edited by Anthony San Pietro

Volume 70. Immunochemical Techniques (Part A)
Edited by Helen Van Vunakis and John J. Langone

Volume 89. Carbohydrate Metabolism (Part D)
Edited by Willis A. Wood

Volume 90. Carbohydrate Metabolism (Part E)
Edited by Willis A. Wood

Volume 91. Enzyme Structure (Part I)
Edited by C. H. W. Hirs and Serge N. Timasheff

Volume 92. Immunochemical Techniques (Part E: Monoclonal Antibodies and General Immunoassay Methods)
Edited by John J. Langone and Helen Van Vunakis

Volume 93. Immunochemical Techniques (Part F: Conventional Antibodies, Fc Receptors, and Cytotoxicity)
Edited by John J. Langone and Helen Van Vunakis

Volume 94. Polyamines
Edited by Herbert Tabor and Celia White Tabor

Volume 95. Cumulative Subject Index Volumes 61–74, 76–80
Edited by Edward A. Dennis and Martha G. Dennis

Volume 96. Biomembranes [Part J: Membrane Biogenesis: Assembly and Targeting (General Methods; Eukaryotes)]
Edited by Sidney Fleischer and Becca Fleischer

Volume 97. Biomembranes [Part K: Membrane Biogenesis: Assembly and Targeting (Prokaryotes, Mitochondria, and Chloroplasts)]
Edited by Sidney Fleischer and Becca Fleischer

Volume 98. Biomembranes (Part L: Membrane Biogenesis: Processing and Recycling)
Edited by Sidney Fleischer and Becca Fleischer

Volume 99. Hormone Action (Part F: Protein Kinases)
Edited by Jackie D. Corbin and Joel G. Hardman

Volume 100. Recombinant DNA (Part B)
Edited by Ray Wu, Lawrence Grossman, and Kivie Moldave

Volume 101. Recombinant DNA (Part C)
Edited by Ray Wu, Lawrence Grossman, and Kivie Moldave

Volume 102. Hormone Action (Part G: Calmodulin and Calcium-Binding Proteins)
Edited by Anthony R. Means and Bert W. O'Malley

Volume 103. Hormone Action (Part H: Neuroendocrine Peptides)
Edited by P. Michael Conn

Volume 104. Enzyme Purification and Related Techniques (Part C)
Edited by William B. Jakoby

Volume 123. Vitamins and Coenzymes (Part H)
Edited by Frank Chytil and Donald B. McCormick

Volume 124. Hormone Action (Part J: Neuroendocrine Peptides)
Edited by P. Michael Conn

Volume 125. Biomembranes (Part M: Transport in Bacteria, Mitochondria, and Chloroplasts: General Approaches and Transport Systems)
Edited by Sidney Fleischer and Becca Fleischer

Volume 126. Biomembranes (Part N: Transport in Bacteria, Mitochondria, and Chloroplasts: Protonmotive Force)
Edited by Sidney Fleischer and Becca Fleischer

Volume 127. Biomembranes (Part O: Protons and Water: Structure and Translocation)
Edited by Lester Packer

Volume 128. Plasma Lipoproteins (Part A: Preparation, Structure, and Molecular Biology)
Edited by Jere P. Segrest and John J. Albers

Volume 129. Plasma Lipoproteins (Part B: Characterization, Cell Biology, and Metabolism)
Edited by John J. Albers and Jere P. Segrest

Volume 130. Enzyme Structure (Part K)
Edited by C. H. W. Hirs and Serge N. Timasheff

Volume 131. Enzyme Structure (Part L)
Edited by C. H. W. Hirs and Serge N. Timasheff

Volume 132. Immunochemical Techniques (Part J: Phagocytosis and Cell-Mediated Cytotoxicity)
Edited by Giovanni Di Sabato and Johannes Everse

Volume 133. Bioluminescence and Chemiluminescence (Part B)
Edited by Marlene DeLuca and William D. McElroy

Volume 134. Structural and Contractile Proteins (Part C: The Contractile Apparatus and the Cytoskeleton)
Edited by Richard B. Vallee

Volume 135. Immobilized Enzymes and Cells (Part B)
Edited by Klaus Mosbach

Volume 136. Immobilized Enzymes and Cells (Part C)
Edited by Klaus Mosbach

Volume 137. Immobilized Enzymes and Cells (Part D)
Edited by Klaus Mosbach

Volume 138. Complex Carbohydrates (Part E)
Edited by Victor Ginsburg

Volume 139. Cellular Regulators (Part A: Calcium- and Calmodulin-Binding Proteins)
Edited by Anthony R. Means and P. Michael Conn

Volume 140. Cumulative Subject Index Volumes 102–119, 121–134

Volume 141. Cellular Regulators (Part B: Calcium and Lipids)
Edited by P. Michael Conn and Anthony R. Means

Volume 142. Metabolism of Aromatic Amino Acids and Amines
Edited by Seymour Kaufman

Volume 143. Sulfur and Sulfur Amino Acids
Edited by William B. Jakoby and Owen Griffith

Volume 144. Structural and Contractile Proteins (Part D: Extracellular Matrix)
Edited by Leon W. Cunningham

Volume 145. Structural and Contractile Proteins (Part E: Extracellular Matrix)
Edited by Leon W. Cunningham

Volume 146. Peptide Growth Factors (Part A)
Edited by David Barnes and David A. Sirbasku

Volume 147. Peptide Growth Factors (Part B)
Edited by David Barnes and David A. Sirbasku

Volume 148. Plant Cell Membranes
Edited by Lester Packer and Roland Douce

Volume 149. Drug and Enzyme Targeting (Part B)
Edited by Ralph Green and Kenneth J. Widder

Volume 150. Immunochemical Techniques (Part K: *In Vitro* Models of B and T Cell Functions and Lymphoid Cell Receptors)
Edited by Giovanni Di Sabato

Volume 151. Molecular Genetics of Mammalian Cells
Edited by Michael M. Gottesman

Volume 152. Guide to Molecular Cloning Techniques
Edited by Shelby L. Berger and Alan R. Kimmel

Volume 153. Recombinant DNA (Part D)
Edited by Ray Wu and Lawrence Grossman

Volume 154. Recombinant DNA (Part E)
Edited by Ray Wu and Lawrence Grossman

Volume 155. Recombinant DNA (Part F)
Edited by Ray Wu

Volume 156. Biomembranes (Part P: ATP-Driven Pumps and Related Transport: The Na, K-Pump)
Edited by Sidney Fleischer and Becca Fleischer

Volume 157. Biomembranes (Part Q: ATP-Driven Pumps and Related Transport: Calcium, Proton, and Potassium Pumps)
Edited by Sidney Fleischer and Becca Fleischer

Volume 158. Metalloproteins (Part A)
Edited by James F. Riordan and Bert L. Vallee

Volume 159. Initiation and Termination of Cyclic Nucleotide Action
Edited by Jackie D. Corbin and Roger A. Johnson

Volume 160. Biomass (Part A: Cellulose and Hemicellulose)
Edited by Willis A. Wood and Scott T. Kellogg

Volume 161. Biomass (Part B: Lignin, Pectin, and Chitin)
Edited by Willis A. Wood and Scott T. Kellogg

Volume 162. Immunochemical Techniques (Part L: Chemotaxis and Inflammation)
Edited by Giovanni Di Sabato

Volume 163. Immunochemical Techniques (Part M: Chemotaxis and Inflammation)
Edited by Giovanni Di Sabato

Volume 164. Ribosomes
Edited by Harry F. Noller, Jr., and Kivie Moldave

Volume 165. Microbial Toxins: Tools for Enzymology
Edited by Sidney Harshman

Volume 166. Branched-Chain Amino Acids
Edited by Robert Harris and John R. Sokatch

Volume 167. Cyanobacteria
Edited by Lester Packer and Alexander N. Glazer

Volume 168. Hormone Action (Part K: Neuroendocrine Peptides)
Edited by P. Michael Conn

Volume 169. Platelets: Receptors, Adhesion, Secretion (Part A)
Edited by Jacek Hawiger

Volume 170. Nucleosomes
Edited by Paul M. Wassarman and Roger D. Kornberg

Volume 171. Biomembranes (Part R: Transport Theory: Cells and Model Membranes)
Edited by Sidney Fleischer and Becca Fleischer

Volume 172. Biomembranes (Part S: Transport: Membrane Isolation and Characterization)
Edited by Sidney Fleischer and Becca Fleischer

Volume 173. Biomembranes [Part T: Cellular and Subcellular Transport: Eukaryotic (Nonepithelial) Cells]
Edited by Sidney Fleischer and Becca Fleischer

Volume 174. Biomembranes [Part U: Cellular and Subcellular Transport: Eukaryotic (Nonepithelial) Cells]
Edited by Sidney Fleischer and Becca Fleischer

Volume 175. Cumulative Subject Index Volumes 135–139, 141–167

Volume 176. Nuclear Magnetic Resonance (Part A: Spectral Techniques and Dynamics)
Edited by Norman J. Oppenheimer and Thomas L. James

Volume 177. Nuclear Magnetic Resonance (Part B: Structure and Mechanism)
Edited by Norman J. Oppenheimer and Thomas L. James

Volume 178. Antibodies, Antigens, and Molecular Mimicry
Edited by John J. Langone

Volume 179. Complex Carbohydrates (Part F)
Edited by Victor Ginsburg

Volume 180. RNA Processing (Part A: General Methods)
Edited by James E. Dahlberg and John N. Abelson

Volume 181. RNA Processing (Part B: Specific Methods)
Edited by James E. Dahlberg and John N. Abelson

Volume 182. Guide to Protein Purification
Edited by Murray P. Deutscher

Volume 183. Molecular Evolution: Computer Analysis of Protein and Nucleic Acid Sequences
Edited by Russell F. Doolittle

Volume 184. Avidin-Biotin Technology
Edited by Meir Wilchek and Edward A. Bayer

Volume 185. Gene Expression Technology
Edited by David V. Goeddel

Volume 186. Oxygen Radicals in Biological Systems (Part B: Oxygen Radicals and Antioxidants)
Edited by Lester Packer and Alexander N. Glazer

Volume 187. Arachidonate Related Lipid Mediators
Edited by Robert C. Murphy and Frank A. Fitzpatrick

Volume 188. Hydrocarbons and Methylotrophy
Edited by Mary E. Lidstrom

Volume 189. Retinoids (Part A: Molecular and Metabolic Aspects)
Edited by Lester Packer

Volume 190. Retinoids (Part B: Cell Differentiation and Clinical Applications)
Edited by Lester Packer

Volume 191. Biomembranes (Part V: Cellular and Subcellular Transport: Epithelial Cells)
Edited by Sidney Fleischer and Becca Fleischer

Volume 192. Biomembranes (Part W: Cellular and Subcellular Transport: Epithelial Cells)
Edited by Sidney Fleischer and Becca Fleischer

Volume 193. Mass Spectrometry
Edited by James A. McCloskey

Volume 194. Guide to Yeast Genetics and Molecular Biology
Edited by Christine Guthrie and Gerald R. Fink

Volume 195. Adenylyl Cyclase, G Proteins, and Guanylyl Cyclase
Edited by Roger A. Johnson and Jackie D. Corbin

Volume 196. Molecular Motors and the Cytoskeleton
Edited by Richard B. Vallee

Volume 197. Phospholipases
Edited by Edward A. Dennis

Volume 198. Peptide Growth Factors (Part C)
Edited by David Barnes, J. P. Mather, and Gordon H. Sato

Volume 199. Cumulative Subject Index Volumes 168–174, 176–194

Volume 200. Protein Phosphorylation (Part A: Protein Kinases: Assays, Purification, Antibodies, Functional Analysis, Cloning, and Expression)
Edited by Tony Hunter and Bartholomew M. Sefton

Volume 201. Protein Phosphorylation (Part B: Analysis of Protein Phosphorylation, Protein Kinase Inhibitors, and Protein Phosphatases)
Edited by Tony Hunter and Bartholomew M. Sefton

Volume 202. Molecular Design and Modeling: Concepts and Applications (Part A: Proteins, Peptides, and Enzymes)
Edited by John J. Langone

Volume 203. Molecular Design and Modeling: Concepts and Applications (Part B: Antibodies and Antigens, Nucleic Acids, Polysaccharides, and Drugs)
Edited by John J. Langone

Volume 204. Bacterial Genetic Systems
Edited by Jeffrey H. Miller

Volume 205. Metallobiochemistry (Part B: Metallothionein and Related Molecules)
Edited by James F. Riordan and Bert L. Vallee

Volume 206. Cytochrome P450
Edited by Michael R. Waterman and Eric F. Johnson

Volume 207. Ion Channels
Edited by Bernardo Rudy and Linda E. Iverson

Volume 208. Protein–DNA Interactions
Edited by Robert T. Sauer

Volume 209. Phospholipid Biosynthesis
Edited by Edward A. Dennis and Dennis E. Vance

Volume 210. Numerical Computer Methods
Edited by Ludwig Brand and Michael L. Johnson

Volume 211. DNA Structures (Part A: Synthesis and Physical Analysis of DNA)
Edited by David M. J. Lilley and James E. Dahlberg

Volume 212. DNA Structures (Part B: Chemical and Electrophoretic Analysis of DNA)
Edited by David M. J. Lilley and James E. Dahlberg

Volume 213. Carotenoids (Part A: Chemistry, Separation, Quantitation, and Antioxidation)
Edited by Lester Packer

Volume 214. Carotenoids (Part B: Metabolism, Genetics, and Biosynthesis)
Edited by Lester Packer

Volume 215. Platelets: Receptors, Adhesion, Secretion (Part B)
Edited by Jacek J. Hawiger

Volume 216. Recombinant DNA (Part G)
Edited by Ray Wu

Volume 217. Recombinant DNA (Part H)
Edited by Ray Wu

Volume 218. Recombinant DNA (Part I)
Edited by Ray Wu

Volume 219. Reconstitution of Intracellular Transport
Edited by James E. Rothman

Volume 220. Membrane Fusion Techniques (Part A)
Edited by Nejat Düzgüneş

Volume 221. Membrane Fusion Techniques (Part B)
Edited by Nejat Düzgüneş

Volume 222. Proteolytic Enzymes in Coagulation, Fibrinolysis, and Complement Activation (Part A: Mammalian Blood Coagulation Factors and Inhibitors)
Edited by Laszlo Lorand and Kenneth G. Mann

VOLUME 293. Ion Channels (Part B)
Edited by P. MICHAEL CONN

VOLUME 294. Ion Channels (Part C)
Edited by P. MICHAEL CONN

VOLUME 295. Energetics of Biological Macromolecules (Part B)
Edited by GARY K. ACKERS AND MICHAEL L. JOHNSON

VOLUME 296. Neurotransmitter Transporters
Edited by SUSAN G. AMARA

VOLUME 297. Photosynthesis: Molecular Biology of Energy Capture
Edited by LEE MCINTOSH

VOLUME 298. Molecular Motors and the Cytoskeleton (Part B)
Edited by RICHARD B. VALLEE

VOLUME 299. Oxidants and Antioxidants (Part A)
Edited by LESTER PACKER

VOLUME 300. Oxidants and Antioxidants (Part B)
Edited by LESTER PACKER

VOLUME 301. Nitric Oxide: Biological and Antioxidant Activities (Part C)
Edited by LESTER PACKER

VOLUME 302. Green Fluorescent Protein
Edited by P. MICHAEL CONN

VOLUME 303. cDNA Preparation and Display
Edited by SHERMAN M. WEISSMAN

VOLUME 304. Chromatin
Edited by PAUL M. WASSARMAN AND ALAN P. WOLFFE

VOLUME 305. Bioluminescence and Chemiluminescence (Part C)
Edited by THOMAS O. BALDWIN AND MIRIAM M. ZIEGLER

VOLUME 306. Expression of Recombinant Genes in Eukaryotic Systems
Edited by JOSEPH C. GLORIOSO AND MARTIN C. SCHMIDT

VOLUME 307. Confocal Microscopy
Edited by P. MICHAEL CONN

VOLUME 308. Enzyme Kinetics and Mechanism (Part E: Energetics of Enzyme Catalysis)
Edited by DANIEL L. PURICH AND VERN L. SCHRAMM

VOLUME 309. Amyloid, Prions, and Other Protein Aggregates
Edited by RONALD WETZEL

VOLUME 310. Biofilms
Edited by RON J. DOYLE

VOLUME 328. Applications of Chimeric Genes and Hybrid Proteins (Part C: Protein–Protein Interactions and Genomics)
Edited by JEREMY THORNER, SCOTT D. EMR, AND JOHN N. ABELSON

VOLUME 329. Regulators and Effectors of Small GTPases (Part E: GTPases Involved in Vesicular Traffic)
Edited by W. E. BALCH, CHANNING J. DER, AND ALAN HALL

VOLUME 330. Hyperthermophilic Enzymes (Part A)
Edited by MICHAEL W. W. ADAMS AND ROBERT M. KELLY

VOLUME 331. Hyperthermophilic Enzymes (Part B)
Edited by MICHAEL W. W. ADAMS AND ROBERT M. KELLY

VOLUME 332. Regulators and Effectors of Small GTPases (Part F: Ras Family I)
Edited by W. E. BALCH, CHANNING J. DER, AND ALAN HALL

VOLUME 333. Regulators and Effectors of Small GTPases (Part G: Ras Family II)
Edited by W. E. BALCH, CHANNING J. DER, AND ALAN HALL

VOLUME 334. Hyperthermophilic Enzymes (Part C)
Edited by MICHAEL W. W. ADAMS AND ROBERT M. KELLY

VOLUME 335. Flavonoids and Other Polyphenols
Edited by LESTER PACKER

VOLUME 336. Microbial Growth in Biofilms (Part A: Developmental and Molecular Biological Aspects)
Edited by RON J. DOYLE

VOLUME 337. Microbial Growth in Biofilms (Part B: Special Environments and Physicochemical Aspects)
Edited by RON J. DOYLE

VOLUME 338. Nuclear Magnetic Resonance of Biological Macromolecules (Part A)
Edited by THOMAS L. JAMES, VOLKER DÖTSCH, AND ULI SCHMITZ

VOLUME 339. Nuclear Magnetic Resonance of Biological Macromolecules (Part B)
Edited by THOMAS L. JAMES, VOLKER DÖTSCH, AND ULI SCHMITZ

VOLUME 340. Drug–Nucleic Acid Interactions
Edited by JONATHAN B. CHAIRES AND MICHAEL J. WARING

VOLUME 341. Ribonucleases (Part A)
Edited by ALLEN W. NICHOLSON

VOLUME 342. Ribonucleases (Part B)
Edited by ALLEN W. NICHOLSON

VOLUME 343. G Protein Pathways (Part A: Receptors)
Edited by RAVI IYENGAR AND JOHN D. HILDEBRANDT

VOLUME 344. G Protein Pathways (Part B: G Proteins and Their Regulators)
Edited by RAVI IYENGAR AND JOHN D. HILDEBRANDT

Volume 362. Recognition of Carbohydrates in Biological Systems (Part A)
Edited by Yuan C. Lee and Reiko T. Lee

Volume 363. Recognition of Carbohydrates in Biological Systems (Part B)
Edited by Yuan C. Lee and Reiko T. Lee

Volume 364. Nuclear Receptors
Edited by David W. Russell and David J. Mangelsdorf

Volume 365. Differentiation of Embryonic Stem Cells
Edited by Paul M. Wassauman and Gordon M. Keller

Volume 366. Protein Phosphatases
Edited by Susanne Klumpp and Josef Krieglstein

Volume 367. Liposomes (Part A)
Edited by Nejat Düzgüneş

Volume 368. Macromolecular Crystallography (Part C)
Edited by Charles W. Carter, Jr., and Robert M. Sweet

Volume 369. Combinational Chemistry (Part B)
Edited by Guillermo A. Morales and Barry A. Bunin

Volume 370. RNA Polymerases and Associated Factors (Part C)
Edited by Sankar L. Adhya and Susan Garges

Volume 371. RNA Polymerases and Associated Factors (Part D)
Edited by Sankar L. Adhya and Susan Garges

Volume 372. Liposomes (Part B)
Edited by Nejat Düzgüneş

Volume 373. Liposomes (Part C)
Edited by Nejat Düzgüneş

Volume 374. Macromolecular Crystallography (Part D)
Edited by Charles W. Carter, Jr., and Robert W. Sweet

Volume 375. Chromatin and Chromatin Remodeling Enzymes (Part A)
Edited by C. David Allis and Carl Wu

Volume 376. Chromatin and Chromatin Remodeling Enzymes (Part B)
Edited by C. David Allis and Carl Wu

Volume 377. Chromatin and Chromatin Remodeling Enzymes (Part C)
Edited by C. David Allis and Carl Wu

Volume 378. Quinones and Quinone Enzymes (Part A)
Edited by Helmut Sies and Lester Packer

Volume 379. Energetics of Biological Macromolecules (Part D)
Edited by Jo M. Holt, Michael L. Johnson, and Gary K. Ackers

Volume 380. Energetics of Biological Macromolecules (Part E)
Edited by Jo M. Holt, Michael L. Johnson, and Gary K. Ackers

Volume 399. Ubiquitin and Protein Degradation (Part B)
Edited by Raymond J. Deshaies

Volume 400. Phase II Conjugation Enzymes and Transport Systems
Edited by Helmut Sies and Lester Packer

Volume 401. Glutathione Transferases and Gamma Glutamyl Transpeptidases
Edited by Helmut Sies and Lester Packer

Volume 402. Biological Mass Spectrometry
Edited by A. L. Burlingame

Volume 403. GTPases Regulating Membrane Targeting and Fusion
Edited by William E. Balch, Channing J. Der, and Alan Hall

Volume 404. GTPases Regulating Membrane Dynamics
Edited by William E. Balch, Channing J. Der, and Alan Hall

Volume 405. Mass Spectrometry: Modified Proteins and Glycoconjugates
Edited by A. L. Burlingame

Volume 406. Regulators and Effectors of Small GTPases: Rho Family
Edited by William E. Balch, Channing J. Der, and Alan Hall

Volume 407. Regulators and Effectors of Small GTPases: Ras Family
Edited by William E. Balch, Channing J. Der, and Alan Hall

Volume 408. DNA Repair (Part A)
Edited by Judith L. Campbell and Paul Modrich

Volume 409. DNA Repair (Part B)
Edited by Judith L. Campbell and Paul Modrich

Volume 410. DNA Microarrays (Part A: Array Platforms and Web-Bench Protocols)
Edited by Alan Kimmel and Brian Oliver

Volume 411. DNA Microarrays (Part B: Databases and Statistics)
Edited by Alan Kimmel and Brian Oliver

Volume 412. Amyloid, Prions, and Other Protein Aggregates (Part B)
Edited by Indu Kheterpal and Ronald Wetzel

Volume 413. Amyloid, Prions, and Other Protein Aggregates (Part C)
Edited by Indu Kheterpal and Ronald Wetzel

Volume 414. Measuring Biological Responses with Automated Microscopy
Edited by James Inglese

Volume 415. Glycobiology
Edited by Minoru Fukuda

Volume 416. Glycomics
Edited by Minoru Fukuda

Volume 433. Lipidomics and Bioactive Lipids: Specialized Analytical Methods and Lipids in Disease
Edited by H. Alex Brown

Volume 434. Lipidomics and Bioactive Lipids: Lipids and Cell Signaling
Edited by H. Alex Brown

Volume 435. Oxygen Biology and Hypoxia
Edited by Helmut Sies and Bernhard Brüne

Volume 436. Globins and Other Nitric Oxide-Reactive Protiens (Part A)
Edited by Robert K. Poole

Volume 437. Globins and Other Nitric Oxide-Reactive Protiens (Part B)
Edited by Robert K. Poole

Volume 438. Small GTPases in Disease (Part A)
Edited by William E. Balch, Channing J. Der, and Alan Hall

Volume 439. Small GTPases in Disease (Part B)
Edited by William E. Balch, Channing J. Der, and Alan Hall

Volume 440. Nitric Oxide, Part F Oxidative and Nitrosative Stress in Redox Regulation of Cell Signaling
Edited by Enrique Cadenas and Lester Packer

Volume 441. Nitric Oxide, Part G Oxidative and Nitrosative Stress in Redox Regulation of Cell Signaling
Edited by Enrique Cadenas and Lester Packer

Volume 442. Programmed Cell Death, General Principles for Studying Cell Death (Part A)
Edited by Roya Khosravi-Far, Zahra Zakeri, Richard A. Lockshin, and Mauro Piacentini

Volume 443. Angiogenesis: *In Vitro* Systems
Edited by David A. Cheresh

Volume 444. Angiogenesis: *In Vivo* Systems (Part A)
Edited by David A. Cheresh

Volume 445. Angiogenesis: *In Vivo* Systems (Part B)
Edited by David A. Cheresh

Volume 446. Programmed Cell Death, The Biology and Therapeutic Implications of Cell Death (Part B)
Edited by Roya Khosravi-Far, Zahra Zakeri, Richard A. Lockshin, and Mauro Piacentini

Volume 447. RNA Turnover in Bacteria, Archaea and Organelles
Edited by Lynne E. Maquat and Cecilia M. Arraiano

Volume 464. Liposomes, Part F
Edited by Nejat Düzgüneş

Volume 465. Liposomes, Part G
Edited by Nejat Düzgüneş

Volume 466. Biothermodynamics, Part B
Edited by Michael L. Johnson, Gary K. Ackers, and Jo M. Holt

Volume 467. Computer Methods Part B
Edited by Michael L. Johnson and Ludwig Brand

Volume 468. Biophysical, Chemical, and Functional Probes of RNA Structure, Interactions and Folding: Part A
Edited by Daniel Herschlag

Volume 469. Biophysical, Chemical, and Functional Probes of RNA Structure, Interactions and Folding: Part B
Edited by Daniel Herschlag

Volume 470. Guide to Yeast Genetics: Functional Genomics, Proteomics, and Other Systems Analysis, 2nd Edition
Edited by Gerald Fink, Jonathan Weissman, and Christine Guthrie

Volume 471. Two-Component Signaling Systems, Part C
Edited by Melvin I. Simon, Brian R. Crane, and Alexandrine Crane

Volume 472. Single Molecule Tools, Part A: Fluorescence Based Approaches
Edited by Nils G. Walter

Volume 473. Thiol Redox Transitions in Cell Signaling, Part A Chemistry and Biochemistry of Low Molecular Weight and Protein Thiols
Edited by Enrique Cadenas and Lester Packer

Volume 474. Thiol Redox Transitions in Cell Signaling, Part B Cellular Localization and Signaling
Edited by Enrique Cadenas and Lester Packer

Volume 475. Single Molecule Tools, Part B: Super-Resolution, Particle Tracking, Multiparameter, and Force Based Methods
Edited by Nils G. Walter

Volume 476. Guide to Techniques in Mouse Development, Part A Mice, Embryos, and Cells, 2nd Edition
Edited by Paul M. Wassarman and Philippe M. Soriano

Volume 477. Guide to Techniques in Mouse Development, Part B Mouse Molecular Genetics, 2nd Edition
Edited by Paul M. Wassarman and Philippe M. Soriano

Volume 478. Glycomics
Edited by Minoru Fukuda

CHAPTER ONE

Techniques for Cultivating Methanogens

Ralph S. Wolfe

Contents

Abstract

Basic techniques for the cultivation of methanogenic archaea in anoxic media, where the O/R potential is maintained below (–) 330 mV under a pressurized atmosphere of 20% carbon dioxide, are described.

1. Introduction

Although techniques for growing methanogens vary in detail among laboratories, these procedures follow the legacy of Hungate who perfected the preparation of prereduced media (Bryant, 1972; Hungate, 1950, 1969). He developed methods for the exclusion of oxygen in the preparation and sterilization of anoxic media as well as methods for the aseptic inoculation and transfer of anaerobic microbes in media where an O/R potential below

Department of Microbiology, University of Illinois, Urbana, Illinois, USA

Methods in Enzymology, Volume 494
ISSN 0076-6879, DOI: 10.1016/B978-0-12-385112-3.00001-9

(−) 330 mV was maintained. Three essential steps persist today: (a) use of nature's buffer of carbon dioxide–bicarbonate–carbonate to maintain a pH near neutrality, (b) use of cysteine and sodium sulfide as reducing agents, and (c) use of resazurin as an O/R indicator that is reddish, when oxidized, and colorless, when reduced at (−) 330 mV. To aseptically remove a stopper from a culture tube in which a methanogen had generated a negative pressure required exceptional skill for the operator to prevent contamination or the entrance of oxygen into the tube. The demands for perfection in use of these techniques were so high that only a few methanogens had been isolated in pure culture prior to 1974. Development of procedures where methanogens could be cultured in a pressurized atmosphere produced a paradigm shift in the isolation and culture of these organisms; the chances for contamination or loss of reducing potential were essentially eliminated (Balch and Wolfe, 1976; Balch *et al.*, 1979). For additional details on the cultivation of methanogens, the article by Sowers and Noll (1995) should be consulted. The following presentation is designed to encourage the study of methanogens in laboratories where funds for equipment may be limiting or for the cultivation of methanogens in teaching laboratories. Use of an anoxic chamber is then discussed.

2. Methanogens on a Budget

2.1. Preparation of oxygen scrubbers

1. Essential to the study of methanogens is a mechanism for scrubbing traces of oxygen from commercially available compressed gases. Figure 1.1 presents a system for routing N_2, or mixtures of 80% N_2 and 20% CO_2 (N_2:CO_2) or 80% H_2 and 20% CO_2 (H_2:CO_2) through a heated cylinder (Fig. 1.1(1)) that contains reduced-copper filings or packed copper turnings, an inexpensive oxygen scrubber. Gas pressures up to 2 atm may be used, so it is essential that a sturdy gas-tight system with copper or stainless-steel (ss) tubing and connecters (Swagelok-type, Whitey Co.), be constructed. Figure 1.2A shows a simple scrubber that employs a section of copper tubing 23 × 3 cm that contains copper filings. Solid copper caps have been soldered to each end with high temperature (essential!) silver solder. The top cap is penetrated by two pieces of copper tubing, 3 mm (dia) and of desired length, soldered in place; one piece, the intake, extends to within 2 cm of the bottom, and the exit tube extends through the cap about 1 cm.
2. Figure 1.2B shows a method of heating the cylinder with electric heating tape connected to a variable transformer. Only heating tape specifically designed for use on metal can be used safely! (Fisher Scientific). For insulation, the cylinder should be wrapped in a thick layer of glass wool.

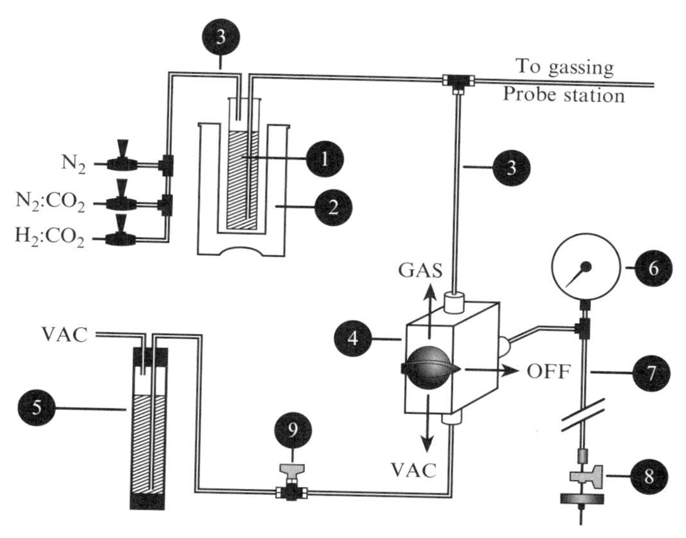

Figure 1.1 Gassing station for dispensing anoxic gases (not drawn to scale). (1) Welded stainless-steel (ss) column 3 × 25 cm containing reduced-copper filings; (2) ceramic furnace at 340 °C; (3) ss 3 mm tubing; (4) ss three-way valve with a center-off position and Swagelok fittings Whitey Co.; (5) moisture trap (5 × 40 cm) containing $CaCl_2$ pellets (not necessary for oil-less vacuum pumps); (6) vacuum pump pressure gauge, range (−) 100 kPa (−30 in.Hg) to 200 kPa (30 psi); (7) flexible 3 mm PVA perfluoroalkyloxy tubing (Swagelok Co.); (8) gas-dispensing attachment; (9) vacuum line on/off vent valve. Reproduced with permission, from Wolfe and Metcalf (2010).

Figure 1.2C shows a welded 316 ss-cylinder that is filled with copper filings and inserted in an electrically heated ceramic furnace (formerly distributed by a company named Sargent Welch, but is now difficult to find). Such equipment has functioned flawlessly for decades in many laboratories.

3. The copper filings are initially reduced by passing H_2 or H_2:CO_2 slowly through the hot filings. This reduction releases significant amounts of heat and should be performed carefully so as not to cause the copper filings to clump. After reduction, cool the filings by passing N_2 through the cylinder; then place a high temperature thermometer in contact with the cylinder. Turn on the transformer and gradually increase the setting until the column temperature approaches 340 °C. At higher temperatures, significant reduction of CO_2 to CO may occur. In active laboratories, the transformer remains on continuously. After continued use, oxidized copper is regenerated by passing H_2:CO_2 through the column for a few minutes.
4. From the scrubber (Fig. 1.1(1)) the gas line branches at a T-connecter, one line going to a gassing station with three probes (Fig. 1.3) and one to a gassing attachment (Fig. 1.1(8)). A probe is easily constructed (Fig. 1.3D) from a 3-cc glass Leur Lok syringe filled with cotton.

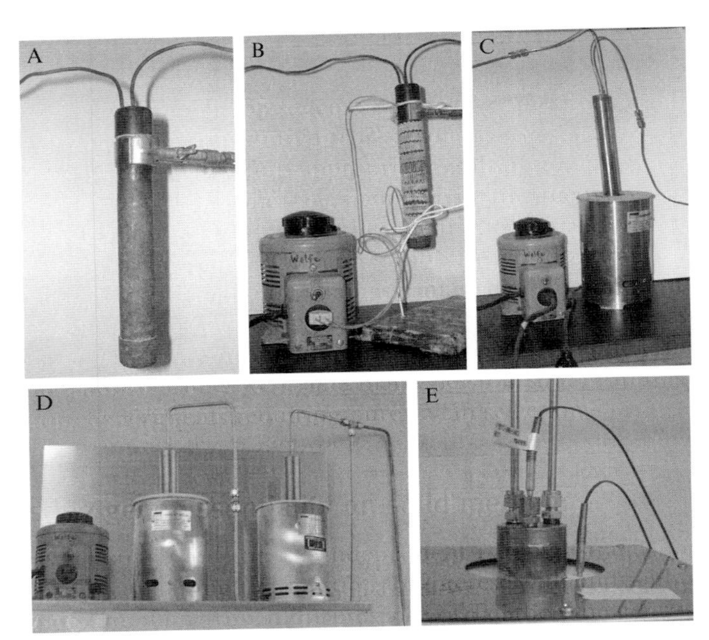

Figure 1.2 Heated reduced-copper oxygen scrubber. (A) Copper tube filled with copper filings; (B) simple method for heating cylinder with a variable transformer and metal-safe heating tape (Fisher); (C) welded ss-cylinder filled with copper filings and inserted in a ceramic furnace; (D) two ss-cylinders connected in series for scrubbing gases routed to an anoxic chamber (Section 3); (E) top of a larger ceramic furnace fitted with a custom-made ss-cylinder and a temperature probe, for use with an anoxic chamber.

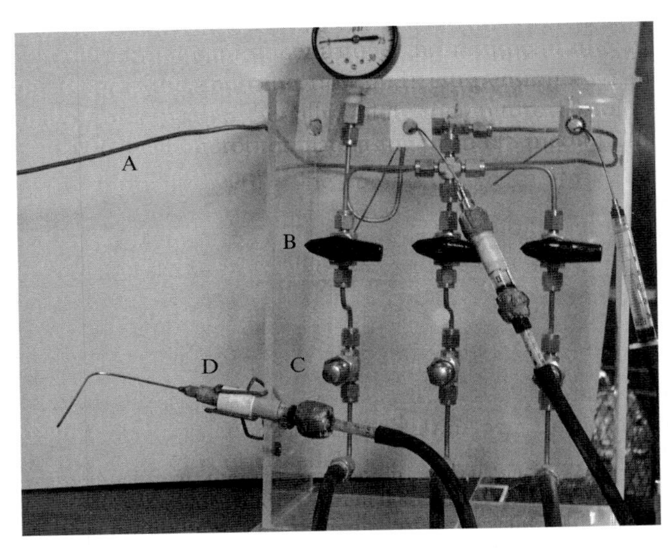

Figure 1.3 Gassing station for dispensing anoxic gases. (A) 3 mm copper tubing from an oxygen scrubber; (B) on/off valve; (C) adjustable metering valve; (D) gassing probe, a 3-mL glass Leur Lock syringe filled with cotton, fitted with a long, bent 16G needle (blunt end), and held in place for easy removal.

To the syringe add a bent 16-gauge (G) needle 12 cm long from which the point has been removed. The complete probe can be placed in a test tube holder, as shown. Two other gassing probes each with a 12-cm 18G needle are also constructed. Each probe is attached to the gas source with black-rubber tubing (OD 9 mm with a 3 mm-thick wall) by inserting the tubing firmly into the syringe. To initiate gas flow to a probe, turn on the gas flow valve (Fig. 1.3B); to regulate the amount of gas flowing to a probe, turn the metering valve (Fig. 1.3C).

2.2. Preparation of 200 mL of anoxic medium by boiling

Many methanogens grow in a simple medium where 4 mol of H_2 are oxidized and 1 mol of CO_2 is reduced to methane, resulting in a negative pressure within the culture tube or vial. The atmosphere of H_2:CO_2 is repressurized during growth of the methanogen.

1. Here is an example of such a medium (g/L): NH_4Cl, 1; NaCl, 0.6; $NaHCO_3$, 5; KH_2PO_4, 0.3; K_2HPO_4, 0.3; $MgCl_2 \cdot 6H_2O$, 0.16; $CaCl_2 \cdot 2H_2O$, 0.009; resazurin 0.1% solution, 1 mL; cysteine•HCl and $Na_2S \cdot 9H_2O$ are added separately (see item 4 below).
2. Prepare a solution of the following vitamins (10 mg of each/L) and add 10 mL of the solution to a liter of medium: *p*-aminobenzoic acid, nicotinic acid, calcium pantothenate, pyridoxine, riboflavin, thiamine, and 5 mg each of biotin, folic acid, α-lipoic acid, and B_{12}.
3. Prepare a solution of trace minerals (g/L) and add 10 mL of the solution to 1 L of medium: trisodium nitrilotriacetic acid, 1.5; $Fe(NH_4)_2(SO_4)_2$, 0.8; $NaSeO_3$, 0.2; $CoCl_2 \cdot 6H_2O$, 0.1; $MnSO_4 \cdot H_2O$, 0.1; $Na_2MoO_4 \cdot 2H_2O$, 0.1; $NaWO_4 \cdot 2H_2O$, 0.1; $ZnSO_4 \cdot 7H_2O$, 0.1; $NiCl_2 \cdot 6H_2O$, 0.1; H_3BO_3, 0.01; $CuSO_4 \cdot 5H_2O$, 0.01.
4. Reducing agents are most conveniently added as anoxic solutions. Prepare a 0.2-*M* solution of cysteine HCl by adding the powder to a volume of hot distilled water that is about one-third the volume of the container vial. For example, 50 mL in a 158-mL vial (listed as 125 mL, Wheaton). Flush out the atmosphere above the liquid with N_2 from a gassing probe; Hungate seal and crimp a Balch stopper in place (Fig. 1.4E–H). Likewise, prepare 0.2 *M* solution of $Na_2S \cdot 9H_2O$. (Sulfite on the surface of a crystal of sulfide is toxic to methanogens; so briefly rinse the surface of the crystal, held in tweezers, under flowing water and dry the crystal with a paper towel before weighing.) Unfortunately, the strong reducing agent, sodium dithionite, is toxic to methanogens and cannot be used.
5. Open the plastic stopcock on the gassing attachment (Figs. 1.1(8) and 1.5) so that N_2 is flowing through the 21G needle. Insert the needle through the Balch stopper on the vial of reducing solution and adjust the pressure of N_2 in each vial to about 35 kPa (5 psi) so that, when desired, a sample of

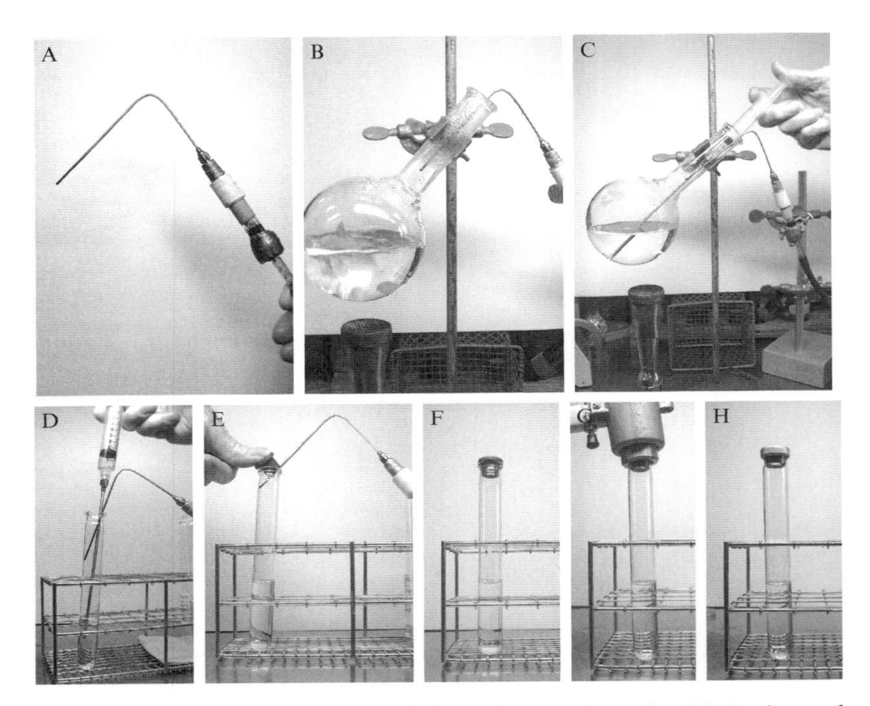

Figure 1.4 Procedure for preparing and dispensing anoxic media. (A) Gassing probe; (B) a boiling medium being flushed with anoxic N_2:CO_2 gas from a probe; (C) 10 mL plastic syringe with long, blunt 16G needle being flushed out with anoxic N_2:CO_2 and filled with anoxic medium; (D) medium being dispensed into the anoxic atmosphere of a tube; (E, F) Hungate sealing a tube with a Balch stopper; (G, H) crimping an aluminum seal in place.

liquid can be removed easily through a 21G needle on a syringe that has been flushed with anoxic gas. Reducing agents are added last, when preparing a medium, to avoid excessive loss of volatile sulfide.

6. To prepare the anoxic medium add 200 mL of medium to a 500-mL flask and place the flask over a Bunsen burner (Fig. 1.4B). As the medium approaches boiling, insert a gassing probe with an N_2:CO_2 flow rate of at least 1 L per min and inject 3 mL of the 0.2 *M* cysteine HCl solution. Allow the medium to boil for about 1 min; then turn off the burner. When boiling has stopped, inject 0.8 mL of the 0.2 *M* $Na_2S \cdot 9H_2O$ solution and mix by swirling the flask.
7. Assemble 20 aluminum seal anaerobic culture tubes or 20, 40 mL serum vials with stoppers (Balch aluminum-crimp stoppers), aluminum seals, and a crimper (Bellco, Vineland, NJ). Place a gassing probe in each of two empty culture tubes and start the flow of N_2:CO_2. Place a rubber bulb-type safety pipet-filler or a Scienceware type pipet pump filler/dispenser

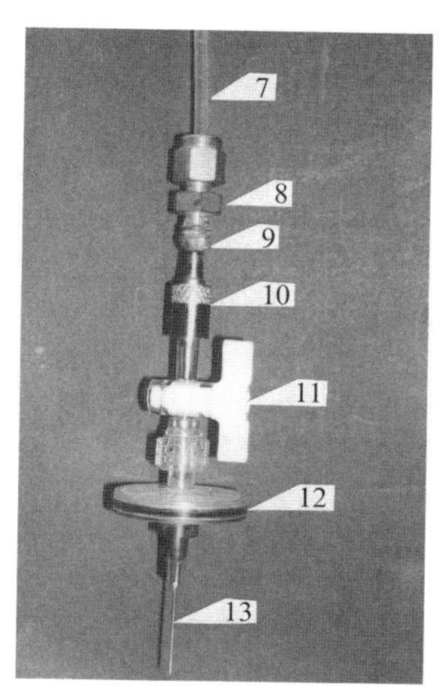

Figure 1.5 Anoxic gas-dispensing attachment. (7) PVA tubing; (8) Swagelok 3 mm brass union; (9) solder; (10) Leur Lok fitting with male tubing adapter, Beckton-Dickenson 3083; (11) plastic stopcock with Leur Lok inlet–outlet fittings; (12) 0.2 μm sterile membrane filter, preferably with Leur Lok inlet–outlet fittings; (13) 21G needle. Reproduced with permission, from Wolfe and Metcalf (2010).

on the end of a 10-mL pipet, or use a 15-cm long 16G needle (blunt end) on a 10cc plastic syringe. Place the pipet or 16G needle into the neck of the flask (Fig. 1.4C). Remove a syringe or pipet full of gas and expel it outside the flask; repeat twice, then lower the needle or pipet into the medium and remove 10 mL. Place the needle or pipet into a tube that is being flushed out and inject the medium into the tube (Fig. 1.4D). Place the syringe or pipet back into the flask. Perform a Hungate seal (Fig. 1.4E and F) by pressing a Balch stopper into the mouth of the culture tube as the gassing probe is withdrawn in one smooth motion. Place the gassing probe into an empty tube. Likewise add the anoxic medium to each tube. Crimp an aluminum seal over each stopper (Fig. 1.4G and H).

8. Now the gas atmosphere in each tube can be adjusted to the needs of the experiment. The medium is designed to have a neutral pH under pressure (173 kPa, 25 psi) of a gas mixture containing 20% CO_2. To exchange gas in a tube, use the gassing attachment (Fig. 1.5). Most methanogens can use hydrogen as substrate and reduce CO_2 to methane; so set the regulator of the gas mixture of H_2:CO_2 (80:20) at 173 kPa (25 psi). When using gas atmospheres at negative or positive pressures, wear plastic goggles and work with the tubes or bottles behind a Plexiglas barrier. Close the N_2:CO_2 tank (Fig. 1.1) open the H_2:CO_2 tank, and turn on the vacuum pump. Close vacuum-release valve (9).

(Before turning off the vacuum pump, when finished, open valve (9) to release the negative pressure in the line.)

9. To evacuate the N_2:CO_2 in a tube turn the three-way valve (Fig. 1.1(4)) to VAC, insert the 21G needle (Fig. 1.5(13)) of the gassing attachment through the Balch stopper, and open the plastic valve (Fig. 1.5(11)). After 10 s, turn the three-way valve to GAS and after 10 s turn the plastic valve off and insert the needle in the stopper of another tube. Repeat this procedure to add H_2:CO_2 to each tube. Sterilize the medium at 121 °C for 20 min, then fast exhaust the autoclave.

2.3. The vacuum–vortex method

Recently, Wolfe and Metcalf (2010) described a quick method for the preparation of anoxic solutions of nonvolatile compounds in culture tubes or in small vials. This method also was used to prepare small volumes of an anoxic medium in which a methanogen grew as well as in media prepared by boiling. This procedure avoids a boiling step and involves the use of the equipment described in Sections 2.1 and 2.2 with the addition of a common laboratory vortex mixer such as a Fisher Vortex/Genie 2. In this procedure, 10 mL of a solution in a culture tube or small vial can be rendered anoxic in 90 s by three alternate cycles of strong vortexing under high negative pressure followed by addition of N_2, (or with N_2:CO_2 for culture media). The gassing attachment is used as described in Section 2.2 (5) above, with the precaution that after each vacuum-vortexing period a positive gas pressure is passed through the attachment to clear any liquid from the needle prior the next vacuum–vortex cycle; any liquid that enters the 0.2 μm membrane may seal the filter, preventing gas from passing through the filter. If this happens, replace the filter.

1. Turn the three-way valve (Fig. 1.1(4)) to the OFF position. Turn on the vacuum pump and close valve 9 (Fig. 1.1). Adjust the gas pressure of N_2 in the line to 100 kPa (15 psi). Set the speed of the vortex mixer to near maximum and to vibrate on contact.
2. Turn the three-way valve to the GAS position. Open the plastic stopcock on the gassing attachment, and with gas flowing from the 21G needle, insert the needle through the Balch stopper of a tube that contains the aerobic solution that is to be made anoxic. Now, turn the three-way valve to the VAC position for 10 s. Turn the plastic stopcock on the gassing attachment off and turn the three-way valve to OFF.
3. Hold the tube firmly (Fig. 1.6A) so that it does not rotate as the tube is touched to the vortex pad for 10 s (Fig. 1.6B).
4. Turn the three-way valve to GAS and open the plastic stopcock on the gassing attachment. This procedure ensures that any liquid in the needle is removed and adds gas to mix with any oxygenic gas that has been removed from the solution.

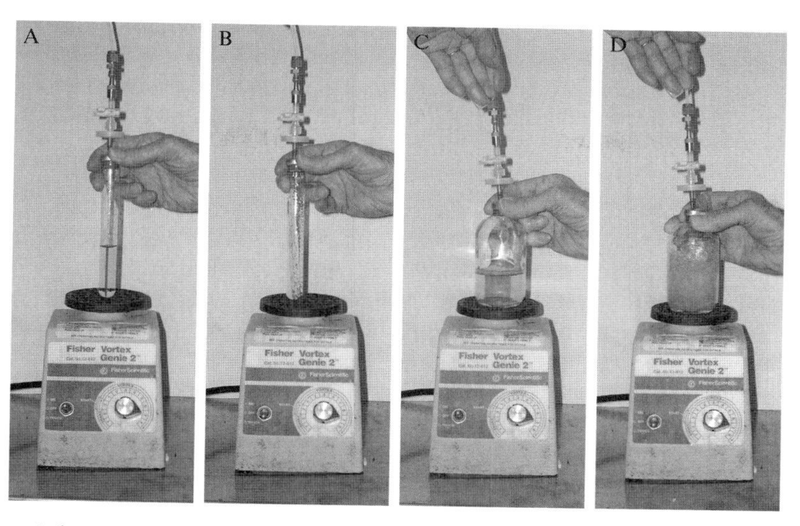

Figure 1.6 The vacuum–vortex procedure for preparing anoxic solutions. (A) Tube is held firmly with gassing attachment in place; (B) with its atmosphere under negative pressure tube contacts vortex mixer top; (C) 50 mL of liquid in a serum bottle; (D) serum bottle with its atmosphere under negative pressure contacts top of vortex mixer.

5. Turn the three-way valve to VAC and repeat the procedure (2–4) above two more times. After the final step, pressurize the atmosphere in the tube to 173 kPa, 25 psi. Sterilize the anoxic liquid at 121 °C for 20 min; then fast exhaust the autoclave.
6. When this procedure is used for 50 mL amounts of liquid (Fig. 1.6C and D) the evacuation time of the atmosphere in the bottle should be increased to 40 s and the vortex step to 30 s.
7. For culture media, prepare a medium aerobically without reducing agents, add it to tubes or vials, and by use of N_2:CO_2 instead of N_2, render the medium anoxic as described in Section 2.3 (1–5) above. If H_2 is to be the substrate, evacuate the N_2:CO_2 and replace it with H_2:CO_2. Now inject the appropriate amount of cysteine and sodium sulfide from anoxic solutions of these reducing agents to complete the medium; sterilize the medium and fast exhaust the autoclave. For more details, consult Wolfe and Metcalf (2010).

2.4. Anoxic, aseptic use of a syringe

1. Prepare a gassing probe (Fig. 1.3D) for aseptic use by passing the 16G needle through the flame of a gas burner. Heat exchange with the metal is so efficient that only a brief exposure to the flame is necessary. Replace the probe in its holder.

2. Add a sterile needle to a sterile 1 mL plastic syringe, and with anoxic gas flowing through the probe, insert the 21G needle into the 16G needle and pump the plunger a few times; then fill the syringe with sterile anoxic gas. If the gas is H_2:CO_2, hold the syringe upright to retain H_2 in the syringe.
3. Insert the needle into the flamed, rubber stopper of a tube or bottle as the plunger of the syringe is held firmly in place to avoid its being dislodged by the pressurized gas atmosphere. Invert the vessel, expel the gas from the syringe, and fill it with sterile fluid while holding the piston firmly. When removing the syringe needle from a rubber stopper, hold the plastic base of the syringe needle firmly to prevent separation of the needle from the syringe body. Transfer the liquid through the flamed, sterile stopper of the receiving vessel.
4. To economize, plastic syringes with needles may be reused many times. Simply wash out the syringe by pumping the plunger as the needle is held in running demineralized water. Then separate the plunger from the syringe body to prevent its rubber tip from being compressed during sterilization, and hang the syringe body and plunger in an 18 mm (dia) test tube. Add a 21-mm (inside dia) plastic cap to the tube, or cover the end of the tube with aluminum foil, and sterilize; then fast exhaust the autoclave.

2.5. Growth of methanogens on agar medium, the bottle plate

To isolate a methanogen in pure culture use of the Petri plate in an anoxic chamber is the method of choice (Section 3). Methods for obtaining isolated colonies on agar at the lab bench are described in this section and in Section 2.6. Use of an improved bottle plate, Bellco, (Olson, 1992) to obtain isolated colonies of a methanogen is the most simple method for use of agar (Herman *et al.*, 1986).

1. To the desired amount of medium (Section 2.2) in a 500-mL flask, add agar to yield a final concentration of 1.5%. Add about 300 mL of water to a 1-L metal beaker that has been placed on a ring stand above a gas burner. Clamp the flask of medium so that its bottom is immersed in the boiling water, and add a gassing probe from which N_2:CO_2 is flowing. Heat the medium with occasional swirling of the flask until the agar is completely melted, then add solutions of cysteine and sodium sulfide. Turn off the burner. Follow the procedures in Section 2.2 to dispense the anoxic agar medium, 10 mL per bottle. Hungate seal each bottle with a Balch stopper, add a screw cap, and pressurize the atmosphere with N_2:CO_2 to 105 kPa (15 psi).
2. After sterilization, cool the bottle of medium in a water bath to about 50 °C; then manipulate the agar into the plate area of the bottle, place the bottle in a horizontal position so that the bottom reservoir is devoid

of agar, and allow the agar to solidify. Store the bottle in an upright position (Fig. 1.8B) so that water of condensation moves to the bottom, allowing the agar surface to remain free of water.

3. To inoculate the agar, insert the sterile needle of a gassing attachment with N_2:CO_2 flowing through the bottom Balch stopper (Fig. 1.8C). Aseptically, remove the top Balch stopper, add a drop of inoculum to the lower region of the agar, and streak the inoculum over the agar surface by using a streaking probe made by melting the tip of a Pasteur pipet into a small glass ball. Dip the pipet into alcohol and touch it briefly to a flame before use. Hungate seal the stopper in place, replace the screw cap, and change the atmosphere in the bottle to H_2:CO_2. Incubate the bottle in an upright position.
4. After incubation, repeat the procedures to add N_2:CO_2 to the bottom part and open the top stopper. An isolated colony may be picked with a platinum or ss-inoculation loop, transferred quickly to the anoxic agar surface of another bottle, and streaked. After incubation, if all "second plate" colonies are identical, one colony may be considered to represent a pure culture.

2.6. Growth of methanogens in a Hungate roll tube

The Hungate roll tube (Bryant, 1972; Hungate 1950, 1969; Sowers and Noll, 1995) is the most economical method to obtain isolated colonies on agar. The inoculated agar medium in a culture tube is rotated in flowing cold water to solidify the medium on the inside surfaces of the tube, creating a cavity within the tube which can be pressurized with H_2:CO_2. Colonies which grow in the agar can easily be picked by use of aseptic, anoxic procedures. There are many variations of these procedures, including methods for streaking a roll tube. Holdeman *et al.* (1977) popularized an adaptation of the roll tube method for use in clinical diagnostic laboratories. Balch-stoppered roll tubes are readily adapted to these procedures for growth of methanogens.

The procedures presented below are intended as a simple introduction to the use of a roll tube.

1. Prepare 200 mL of anoxic medium as described in Section 2.2 and add agar as described in Section 2.5. Add 9 mL to each culture tube; use an anoxic gassing probes with N_2:CO_2. After sterilization, equilibrate the tubes in a water bath at 45 °C.
2. By aseptic, anoxic use of a syringe (Section 2.4) transfer 1 mL of a liquid culture or enrichment of a methanogen to a tube of melted agar medium in the water bath and invert the tube three times to mix the inoculum. With a warm sterile, anoxic syringe quickly transfer 1 mL of agar medium from the first tube to the second tube. Invert this tube and

likewise carry out a serial dilution through six tubes of agar medium. Rotate each tube under cold, flowing water to form a roll tube. Evacuate the atmosphere in each tube and add H_2:CO_2 to 173 kPa (25 psi). Incubate the tubes in an upright position.

3. After incubation, release some of the gas pressure in the tube that contains isolated colonies by penetrating the stopper with a sterile 21G needle. Place the tube in a clamp on a ring stand and remove the Balch stopper as a sterile gassing probe with N_2:CO_2 flowing is inserted into the tube. Place the probe in a clamp to hold it firmly in position. In a similar manner, position a tube that contains a few milliliter of sterile broth medium and insert a sterile gassing probe into the tube.
4. Pick an isolated colony and inoculate the broth. Hungate seal this tube and suspend the cells well; then use 1 mL of this suspension for a serial dilution through melted agar medium. If, after incubation, all "second plate" colonies in a tube of high dilution appear to be of the same type, it may be assumed that cells in an isolated colony represent a pure culture.

2.7. Preparation of a 3-L amount of anoxic medium

The apparatus described in this section for the preparation of large amounts of medium is designed to be fabricated by laboratory personnel from readily available materials for the convenient delivery of media (up to 3 L) from an anoxic reservoir where the pressure is 35 kPa (5 psi), not higher!

1. Obtain a standard, 5 L round-bottom Pyrex flask and a #11 solid black-rubber stopper, which fits snuggly into the neck. Drill two holes through the stopper that will accommodate 5 mm (dia) ss-tubing and one hole for 3 mm(dia) ss-tubing. Space the holes evenly on the stopper (Fig. 1.7A–C). To construct a gas-exit port, insert a piece of ss-tubing about 5 cm long and 5 mm(dia) so that one end is flush with the bottom of the stopper. To the upper end of this tube attach a piece of rubber tubing and a standard plastic stopcock (Fig. 1.7A(1)). The hole through the rotator plastic cylinder of the stopcock should have a diameter of about 3 mm.
2. To prepare a dispensing tube, insert an ss-tube 10 cm long and 5 mm (dia) so that about 5 cm projects below the bottom of the stopper. To this lower end, attach a piece of Nalgene 180 pvc nontoxic-grade plastic tubing about 15 cm long and add a piece of ss-tubing about 10 cm long to serve as a weight, the total length of tubing inside the flask being about 27 cm long, extending to just above the bottom of the flask.
3. To the top end of the ss-tube, which projects above the stopper (Fig. 1.7A(4)) attach a piece of Nalgene tubing about 65 cm long (Fig. 1.7B). To the end of this tubing, add a standard plastic stopcock which has attached a piece of Nalgene tubing about 6 cm long (Fig. 1.7B(8)).

Figure 1.7 Method for preparing 3 L of anoxic medium in a 5-L flask. (A) stopper and attachments: (1) plastic stopcock; (2) stainless-steel connector, upper ferrules replaced with a rubber septum; (3, 4) 65 cm long Nalgene tubing; (5) 5 cm long strong stainless-steel spring, also see (c(9)); (6) #11 black-rubber stopper; (7) stainless-steel hose clamp with cut and bent threads; (B) view of complete apparatus with dispensing stopcock (8); (C) anoxic medium being cooled and gased in an ice bath: (9) 5-cm stainless-steel spring with iron wire attachments; (D) anoxic medium being dispensed into a small bottle; (E) into a 500-mL bottle; and (F) anoxic medium sealed in a (1) 158-mL bottle, (2) 500-mL bottle, and (3), (4) in 1 L bottles.

4. Prepare a port for gas injection by inserting a 3-mm ss-tube about 20 cm long through the rubber stopper so that about 8 cm projects above the stopper. To this end, affix a Swagelok-type ss-union (Fig. 1.7A(2)). Unscrew the top nut of the union, and remove the

front and back ferrules. Insert into the nut a rubber plug 5 mm long prepared from a Balch stopper by use of a No. 4 cork borer. Screw the nut firmly in place. In operation of the system, the rubber plug serves as a septum through which anoxic gas is provided from a 21G needle on the gassing attachment (Fig. 1.7B).

5. To hold the stopper in place, obtain an ss-hose clamp that fits around the neck of the flask, which is about 6 cm (dia). To provide attachments for the spring (Fig. 1.7A(5) and C(9)), cut and bend (Fig. 1.7A(7)), three of the threads on the hose clamp band which are about 4 cm from the screw housing. On the opposite side of the neck, cut and bend a prong about 4 mm wide from the solid ss-band of the clamp. These cuts of the ss-band are easily performed with a small motorized hand tool (Dremel, Racine, WI) equipped with a reinforced, metal cutting, 426 fiber disc.
6. To each end of a strong ss-spring (Fig. 1.7C(9)), which is about 5 cm long, add a loop of iron wire to serve as attachments to the prongs of the hose clamp, when the spring is stretched over the rubber stopper (Fig. 1.7A–C).
7. Clamp the 5 L flask on a ring stand so that the bottom of the flask is supported by a wire gauze on a ring above a Bunsen burner. Add 3 L of medium (without reducing agents) to the flask. The advantage of a 5-L flask is that 1–3 L amounts of medium are easily accommodated.
8. Open gas-exit stopcock (Fig. 1.7A(1)), close medium dispensing stopcock (Fig. 1.7B(8)), seat the rubber stopper firmly in the mouth of the flask, and stretch the ss-spring in place (Fig. 1.7A). Ignite the Bunsen burner gas.
9. Set the N_2:CO_2 tank line pressure so that the pressure gauge at the gassing station reads 35 kPa (5 psi), not higher. This low pressure provides sufficient anoxic gas flow over the contents of the 5 L flask during medium preparation and later provides a sufficient pressure to force the flow of medium, when it is being dispensed.
10. When the medium approaches boiling, insert the 21G needle on the gassing attachment, through the rubber septum in the ss-union (Fig. 1.7A(2)) and start the flow of N_2:CO_2.
11. Previously, prepare a solution of cysteine•HCl by dissolving 1.5 g in 20 mL of hot distilled water in a beaker. Bring the solution into a 20-mL plastic syringe equipped with a 15-cm long, blunt 18G needle. Place the needle through the gas-exit stopcock (Fig. 1.7A(1)), and inject the solution into the 3 L of medium.
12. Bring the medium to a low boil for about 1 min as the resazurin in the medium is reduced to its colorless form. Turn off the Bunsen burner and continue the N_2:CO_2 gassing for about 20 min as the medium cools somewhat; then lower the flask carefully into an ice bath (Fig. 1.7C) as gassing continues.

13. Swirl the medium in the flask and, when it has cooled to the point where it is no longer too-hot-to-touch, place the flask back on the ring stand, and inject 6 mL of 0.2 *M* sodium sulfide through the gas-exit stopcock (Fig. 1.7A(1)). Mix the medium by swirling the flask. (The basal medium is now anoxic without substrate. If H_2 is to be the substrate, the N_2:CO_2 atmosphere can be replaced with H_2:CO_2 after the medium has been dispensed. (If the substrate is to be methanol, acetate, formate, or methylamines, the desired substrate can be injected at this time or can be injected later to the desired bottles of sterile medium from a sterile, anoxic stock solution of substrate.)
14. Prepare to dispense the anoxic medium and Hungate seal each container by assembling the desired bottles, stoppers, and caps. (Tubing of sterile medium into culture tubes is described in Section 2.9) For 50 mL amounts of medium the "125" mL size serum bottles (Wheaton) are useful (Fig. 1.7D). Each bottle has a volume of 158 mL with a Balch stopper in place. For a 250-mL amount of medium, a 500-mL heavy-walled, graduated serum bottle (Wheaton) with a #1 black-rubber stopper that can be crimped in place with a large crimper, is ideal (Fig. 1.7E). For a 500 mL amount of medium, a 1-L graduated media lab bottle (Wheaton) can be modified for use by drilling a 2-cm (dia) hole in the plastic screw cap. Cut off an aluminum-crimp culture tube about 6 cm from the top and insert the cut end through a #9 bored hole in a black-rubber #4 stopper (Fig. 1.7F(3)). These modified bottles have been used successfully for decades (Balch and Wolfe, 1976; Balch *et al.*, 1979). An alternative for a 500-mL amount of medium is a 1-L bottle with a wide mouth, rubber septum, and aluminum screw cap (#191000 Transfusions flaschen, 1L, Glasgerätebau OCHS, 37120 Bovenden, Germany).
15. For anoxic dispensing of the anoxic medium, place a gassing probe (Fig. 1.7D), in each of two receiving bottles with a flow of N_2:CO_2 from each probe. Close stopcock 1 (Fig. 1.7A), open stopcock 8 (Fig.1.7B) and allow some medium to flow into a beaker, replacing air in the dispensing tube. Discard this medium.
16. Insert the plastic tubing at the end of the dispensing tube into the neck of a receiving bottle (Fig. 1.7D), which now has an anoxic atmosphere from the gassing probe, and open the stopcock (Fig. 1.7D) and allow the desired amount of medium to flow into the bottle. (From 3 L of medium, 300 test tubes may be filled, sealed, and crimped. By use of an eight-port gas-dispensing station, the atmosphere in each of eight tubes can be simultaneously exchanged with H_2:CO_2 and pressurized prior to sterilization.)
17. Transfer the dispensing tube to an empty bottle that is being gassed by a probe, and Hungate seal a stopper in the bottle to which the medium has just been added.

Move the gassing probe into an empty bottle. Likewise, dispense the anoxic medium into the desired number of bottles.

18. Prior to sterilization increase the pressure of N_2:CO_2 in each bottle to 70 kPa (10 psi) by use of the gassing attachment (Fig. 1.5). For the 3 L of medium to be steam-sterilized in an autoclave, the time exposure at 121 °C should be increased to 1 h. For safety, the 1-L bottles are isolated in metal containers, should breakage occur. An ss-Sterilizer Box (Fisher) for glass Petri dishes is ideal for use with a 1-L bottle.
19. After sterilization, the gas atmosphere in each bottle can be aseptically adjusted to meet the needs of the experiment.

2.8. Cultivation of marine methanogens

To cultivate methanogens from a marine environment, the medium (Section 2.2 (1–6)) is modified by increasing the amount of NaCl to 400 m*M*, $MgCl_2 \cdot 6H_2O$ to 54 m*M*, and $CaCl_2 \cdot 2H_2O$ to 2 m*M*. Use of the vacuum–vortex system (Section 2.3) is the most simple way to prepare a marine medium. For methods that require boiling of the medium, it is essential to prepare a separate solution that contains the components, $MgCl_2$ and $CaCl_2$, to avoid formation of an insoluble precipitate during the initial boiling step; this separate solution is treated exactly in the manner for the preparation of anoxic media as previously outlined. After cooling the anoxic, divalent-metal solution, it is combined with the cool, anoxic, marine-medium solution and swirled to mix. Add an anoxic substrate such as methanol at this time. Dispense the complete anoxic medium into tubes or bottles. A precipitate will form during heat sterilization, but it will dissolve on standing at room temperature.

2.9. Aseptic transfer of sterile medium

The storage of sterile medium in bottles (Section 2.7) has the advantage that the investigator has flexibility in designing experiments. A bottle of sterile medium can be inoculated directly for culture of a methanogen in larger quantities, or the sterile medium can be modified as desired and aseptically dispensed into tubes.

1. To dispense sterile medium from a bottle use of a male–male Leur Lok, in line, ss-stopcock (Sigma-Aldrich) (Fig. 1.8A) is most efficient. Such a stopcock is expensive but lasts forever. A less expensive alternative consists of a chrome-plated brass, male to male Leur Lok needle connector attached to a chrome-plated, one way, spring-clip stopcock, female Leur to male needle Leur Lok PS-6021 (Sigma-Aldrich), but chrome-plated brass stopcocks must be disassembled and washed thoroughly after

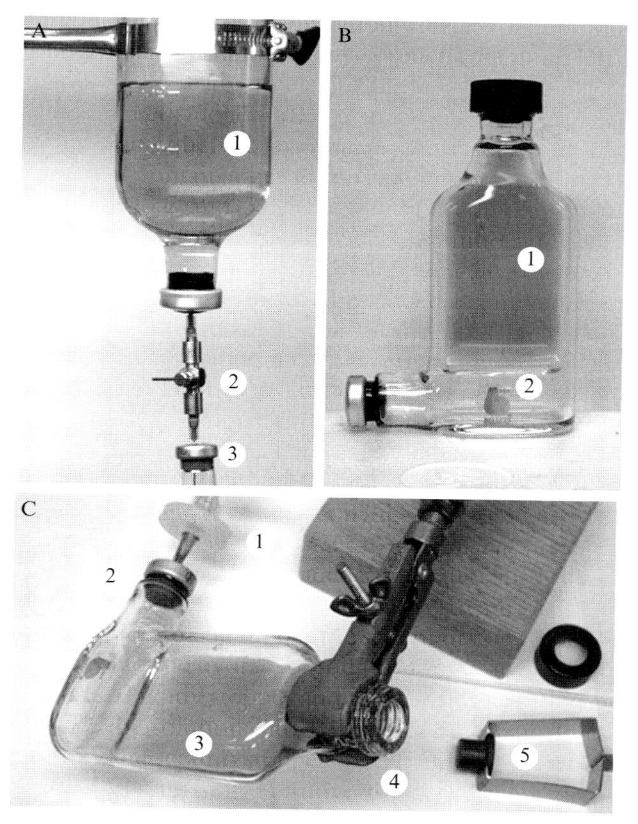

Figure 1.8 Aseptic techniques. (A) Sterile medium being dispensed into a culture tube: (1) bottle with sterile medium under a positive gas pressure, (2) sterile male–male stopcock with attached needles, (3) sterile, empty culture tube with negative gas pressure; (B) bottle plate: (1) agar medium, (2) reservoir for collecting moisture; (C) bottle plate held in position for streaking: (1) 0.2 μm filter, (2) bottom gassing port, (3) agar surface, (4) threads for cap, (5) stopper held in removal tool.

each use to prevent corrosion of brass by sulfide in the medium. Wrap the metal ss-stopcock or the chrome-plated stopcock and connector in aluminum foil and sterilize in an autoclave; fast-exhaust the autoclave.

2. To each test tube, add a drop of distilled water from a 21G needle on a syringe and crimp a Balch stopper in place. This drop of water will provide enough moisture so that the interior of the tube is subjected to moist heat and can be sterilized at 121 °C for 20 min.
3. By use of the gassing attachment (Fig. 1.5), evacuate the air from a tube for 10 s, add anoxic gas for 10 s, and then evacuate the gas for 10 s. Likewise, treat each tube. Sterilize the tubes in an autoclave; then

fast-exhaust the autoclave. Do not store tubes with a negative pressure, but dispense medium into them fairly soon.

4. The atmosphere in the bottle of medium should have a pressure of at least 70 kPa (10 psi), so that the medium flows readily from the bottle. Invert the bottle and flame the rubber stopper. Place the inverted bottle in a clamp on a ring stand (Fig. 1.8A). Remove the aluminum foil from a sterile ss-male–male Leur lok stopcock and add a sterile 21G needle to each end. With the stopcock open, insert one needle through the rubber stopper of the bottle, and immediately close the stopcock after a small amount of medium has passed into a beaker.
5. Insert the other needle of the stopcock through the flamed Balch stopper of a tube which has an internal negative pressure. Open the stopcock and allow 10 mL of medium to flow into the tube. Likewise add sterile medium to each tube.
6. For culture of a methanogen on H_2:CO_2, flame the Balch stopper of the tube. Add a sterile needle to the gassing probe (Fig. 1.5) and aseptically pressurize the atmosphere in the tube to 173 kPa (25 psi). Likewise prepare each tube prior to inoculation.

3. Use of an Anoxic Chamber

The laboratory of W. W. Metcalf at the University of Illinois has pioneered investigations into the genetics of *Methanosarcina*. These studies have reached a level of sophistication such that almost any genetic or molecular procedure that can be performed with *Escherichia coli* can now be performed with *Methanosarcina*. An essential component of such investigations is the Rolf Frêter anoxic chamber (Fig. 1.9A), manufactured by Coy Laboratory Products. Although this double chamber model is a significant financial investment, it is an indispensable tool for serious genetic and molecular studies. A floor space of about 3 × 7 m is required for the chambers, compressed gases and ceramic O_2 scrubbers (Fig. 1.9A). One of the chambers is used for dispensing anoxic liquid or agar medium, and for genetic or molecular studies. The adjacent chamber is used to house a Coy intrachamber incubator (Metcalf *et al.*, 1998) and plastic, anoxic jars (such as GASPAK) for incubation of inoculated Petri plates. Erratic growth of methanogens in such jars occurs unless the jars are continuously stored and used in the chamber where O_2 has had time to diffuse out of the plastic. Use of individual jars provides flexibility for incubation of plates under various conditions, and the Coy intrachamber incubator has space for more than 100 plates.

1. Gases for the chambers are passed through ceramic heated scrubbers (Fig. 1.2D and E). The gas atmosphere in the chambers consists of N_2,

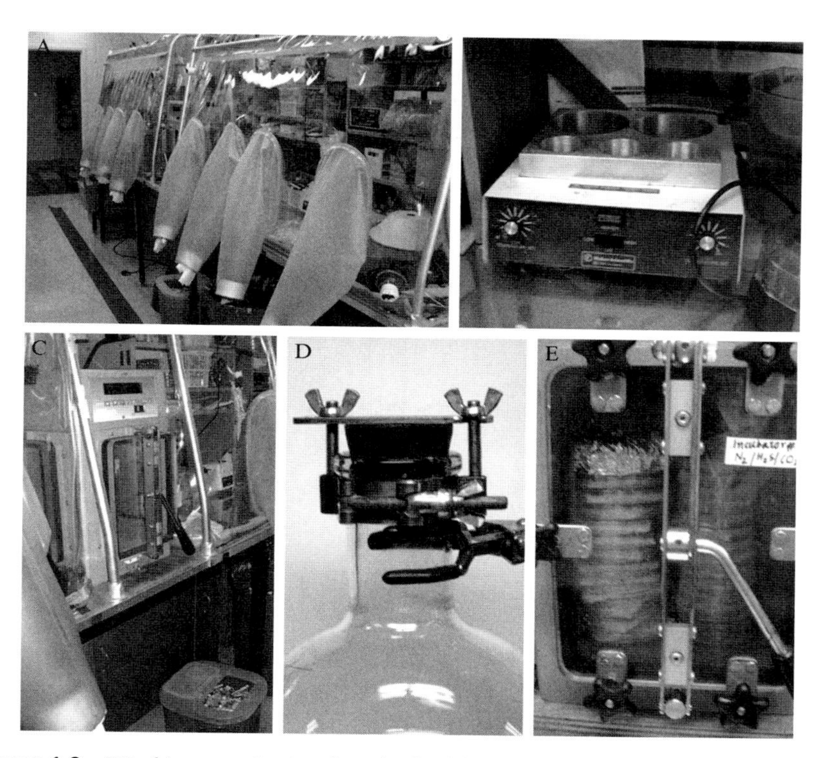

Figure 1.9 Working anoxic chambers in the laboratory of W. W. Metcalf. (A) dual chambers made by Coy, (B) intrachamber dry-heated bath for storing melted agar medium prior to pouring of Petri plates, (C) automatic air lock located between chambers, (D) mechanism for holding a #10 stopper in a 3-L flask that contains anoxic medium as it is passed through the negative pressure of the air lock, (E) intrachamber incubator with inoculated Petri plates under an N_2:CO_2:H_2S atmosphere.

and the presence of 20% CO_2 is essential for handling media with a CO_2–carbonate buffering system. Because O_2 slowly diffuses through the plastic walls, an H_2 component of the atmosphere is maintained below 5% so that two catalytic O_2 scrubbers efficiently remove O_2 from the atmosphere. The pelleted catalyst may be removed and regenerated in an oven at 250 °C. (When using H_2 in an anoxic chamber, there is no substitute for eternal safety vigilance! To ensure that the H_2 concentration in a chamber does not exceed 5%, the Metcalf laboratory uses only a tank mixture of N_2:CO_2 with 5% H_2 to fill the chambers. The most serious potential danger lies in careless operation of the air lock. For example, if a high level of H_2 accumulates in a chamber, and the chamber door of the air lock is opened to an air lock full of air, the catalyst may turn red-hot, igniting an explosion.) A separate electric fan adds to the circulation of the atmosphere. To avoid poisoning the catalyst by fumes of H_2S from the culture media, a bed of charcoal is placed on a tray above the catalyst to scrub H_2S form the atmosphere. It is important to replace the charcoal often, depending on the use of the chamber.

2. An essential component of the internal equipment is a particle scrubber designed to supply germ-free air to a germ-free mouse cage. This scrubber runs continuously so that aseptic procedures can be performed in the chamber.
3. Operation of the air lock involves programmed cycles of negative pressure followed by addition of anoxic gas. Any vessel that contains liquid must have a solid rubber stopper that is securely held in place, when placed in the air lock. Figure 1.9D shows one type of mechanism used in the Metcalf laboratory for clamping a stopper in a 3-L flask.
4. To transfer plastic Petri dishes through the air lock, prick a hole in the plastic cylindrical cover so that gas exchange may take place. (One person, MEM, prefers to place a Band-Aid to serve as a gas filter over the opening.) Prior to use, it is essential to store the dishes in the chamber for days prior to use, so that O_2 may diffuse out of the plastic.
5. One method for the preparation of anoxic agar for Petri plates is to store preweighed amounts of agar folded in aluminum foil in the chamber so that O_2 has time to diffuse out of the dried agar. Then, for example, most of the gas pressure in a stored, 1-L bottle that contains 250 mL of sterile medium is released by inserting a 21G needle through the stopper prior to transfer into the chamber. Inside the chamber remove the stopper and pour the anoxic agar through a dry plastic funnel into the liquid medium. Recrimp the stopper and transfer the bottle out through the air lock. Pressurize the N_2:CO_2 atmosphere, place the bottle into a metal container, and sterilize it; then fast exhaust the autoclave. Carefully use gloves and swirl the hot medium in the container to very gently mix the agar; allow it to cool to about 60 °C.
6. Aseptically release most of the gas pressure in the bottle with a sterile 21G needle. Place the bottle in a metal container and transfer it into the chamber. Aseptically inject any additional components of the medium, such as an antibiotic, at this time. To avoid premature cooling of the agar, the bottle is placed in a heated cavity of a modified dry bath (Fisher) (Fig. 1.9B), until the medium is poured into Petri dishes.
7. For growth of *Methanosarcina*, the Petri plate must be incubated in the presence of 0.1% H_2S in the atmosphere, which if present in the chamber will poison the catalytic O_2 scrubber. Separate gas lines of N_2:CO_2:0.1% H_2S or H_2:CO_2:0.1% H_2S as well as a vacuum line are connected to the intrachamber incubator and to a port with rubber tubing for use with the GASPAK jars. To incubate inoculated Petri plates, the atmosphere in the intrachamber incubator or GASPAK jar is evacuated and replaced with a pressurized, appropriate gas mixture that

also contains 0.1% H_2S. For culture of methanogens that are using substrates other than hydrogen, the N_2:CO_2 atmosphere in the incubation vessel is not pressurized; when H_2:CO_2 is used, the atmosphere is pressurized. (For extra safety precautions in the laboratory of John Leigh, to each inlet and outlet port on the top of an incubation jar for Petri plates is attached a quick-connect flexible metal hose (Swagelok ss-FL4-TA4-24); the male parts (ss-QC4-D-400) allow gas to flow only when the connection is made and shut off when uncoupled. Each metal hose is attached to a four-way cross-over valve with Swagelok fittings, allowing the lines and the jar to be flushed and the jar filled without the possibility of a line disconnection.) Seals on the incubator and jars must be perfect! To examine the plates after incubation, evacuate the atmosphere above the Petri plates and allow the chamber atmosphere to enter the incubator vessel. There is no substitute for visiting a working laboratory and obtaining first-hand training and experience with the operation of an anoxic chamber.

ACKNOWLEDGMENT

I thank W. W. Metcalf for helpful suggestions and for use of his laboratory equipment.

REFERENCES

Balch, W. E., and Wolfe, R. S. (1976). New approach to the cultivation of methanogenic bacteria: 2-Mercaptoethanesulfonic acid (HS-CoM)-dependent growth of *Methanobacterium ruminantium* in a pressurized atmosphere. *Appl. Environ. Microbiol.* **32,** 781–791.

Balch, W. E., Fox, G. E., Magrum, L. J., Woese, C. R., and Wolfe, R. S. (1979). Methanogens: Reevaluation of a unique biological group. *Microbiol. Rev.* **43,** 260–296.

Bryant, M. P. (1972). Commentary on the Hungate technique for culture of anaerobic bacteria. *Am. J. Clin. Nutr.* **25,** 1856–1859.

Herman, M., Noll, K. M., and Wolfe, R. S. (1986). Improved agar bottle plate for isolation of methanogens or other anaerobes in a defined gas atmosphere. *Appl. Environ. Micrbiol.* **51,** 1124–1126.

Holdeman, L. V., Cato, E. P., and Moore, W. E. C. (1977). Anaerobe Laboratory Manual. 4th edn. V. P. I. Anaerobe Laboratory, Blacksburg, VA.

Hungate, R. E. (1950). The anerobic mesophilic cellulolytic bacteria. *Bacteriol. Rev.* **14,** 1–49.

Hungate, R. E. (1969). A roll-tube method for cultivation of strict anaerobes. *Methods Microbiol.* **3B,** 117–132.

Metcalf, W. W., Zhang, J. K., and Wolfe, R. S. (1998). An anerobic, intrachamber incubator for growth of *Methanosarcina* spp. on methanol-containing solid media. *Appl. Environ. Microbiol.* **64,** 768–770.

Olson, K. D. (1992). Modified bottle plate for the cultivation of strict anaerobes. *J. Microbiol. Methods* **14,** 267–269.

Sowers, K. R., and Noll, K. M. (1995). Techniques for anaerobic growth, methanogenic archeae. *In* "Archaea, a Laboratory Manual," (F. T. Robb, A. R. Place, K. R. Soweres, H. J. Schreier, S. DarSarma, and E. M. Fleischmann, eds.), pp. 17–47. Cold Spring Harbor Laboratory Press, NY.

Wolfe, R. S., and Metcalf, W. W. (2010). A vacuum-vortex method for preparation of anoxic solutions or liquid culture media in small amounts for cultivation of methanogens or other strict anaerobes. *Anaerobe* **16,** 216–219.

CHAPTER TWO

Genetic Methods for *Methanosarcina* Species

Nicole Buan,*[,1] Gargi Kulkarni,†[,1,2] *and* William Metcalf†

Contents

Abstract

Unlike most methanogenic microorganisms, *Methanosarcina* species are capable of utilizing a variety of growth substrates, a trait that greatly simplifies genetic analysis of the methanogenic process. The genetic tools and techniques discussed in this chapter form the basis for all genetic experiments in *Methanosarcina acetivorans* C2A and *Methanosarcina barkeri* Fusaro, two methanogens that are routinely used as model organisms for genetic experiments. Based on a number of reports, it is likely that they are portable to other

* Department of Biochemistry and the Redox Biology Center, University of Nebraska-Lincoln, N200 Beadle Center, Lincoln, Nebraska, USA
† Department of Microbiology, University of Illinois, Urbana-Champaign, B103 Chemical and Life Sciences Laboratory, Urbana, Illinois, USA
1 These authors contributed equally
2 Current address: Division of Biology, California Institute of Technology, Pasadena, California, USA

Methods in Enzymology, Volume 494
ISSN 0076-6879, DOI: 10.1016/B978-0-12-385112-3.00002-0

Methanosarcina species, and perhaps to other methanogens as well. Here, we outline the procedures for high-efficiency transformation using liposomes, gene expression from a plasmid, and exploitation of homologous and site-specific recombination to add and delete genes from the chromosome. Finally, we outline the method for testing whether a gene is essential. These methods can be adapted and combined in any number of ways to design genetic experiments in *Methanosarcina*.

1. Introduction

1.1. Genetic tools in methanogens

A variety of genetic tools have been developed for use in diverse methanogenic microorganisms, including ones similar to those used in yeast or *Escherichia coli*. Among these are methods for efficient transformation, selectable and counterselectable markers, autonomously replicating shuttle plasmids, transposons, targeted gene deletion, sequenced genomes, etc. Nevertheless, genetic analysis of methanogens is a relatively recent development, practiced by only a small number of laboratories. This leaves ample opportunities for method refinement and for expanding the scope of genetic research in these unusual and interesting organisms. Further, there are no serious factors limiting the use of methanogens as genetic models for the entire Archaeal domain.

1.2. Advantages and disadvantages of *Methanosarcina* as model Archaea

Members of the genus *Methanosarcina* are particularly attractive model Archaea. *Methanosarcina* species are capable of growth on several simple carbon compounds using any of the known methanogenic pathways: acetoclastic (acetate), carboxydotrophic (CO), hydrogenotrophic ($CO_2 + H_2$), methylotrophic (methanol, methylamines, methylsulfides), and methyl respiration (methylated compounds + H_2). In contrast, most other known methanogen species are restricted to using $CO_2 + H_2$ for hydrogenotrophic growth. A few of these other species are also capable of growth on formate or primary alcohols. Because of their metabolic flexibility, it is relatively easy to delete *Methanosarcina* genes required for growth on one substrate without necessarily affecting the growth on another substrate(s). Therefore, genetic analysis can be used in *Methanosarcina* to test hypotheses about how these organisms grow and generate methane using each of the known methanogenic pathways.

A potential disadvantage to using *Methanosarcina* species derives from their propensity to grow as irregular clumps of cells, sometimes encased in an outer sheath. This trait makes it difficult to isolate clonal populations and presents a potential barrier against exogenous DNA used for genetic manipulations. However, for some species, modulating the osmotic strength of the medium can induce a single-celled morphology (Sowers *et al.*, 1993). This observation has been exploited to allow efficient transformation of DNA and subsequent selection of clonal colonies on solid agar plates (Metcalf *et al.*, 1997).

Methanosarcina acetivorans C2A and *Methanosarcina barkeri* Fusaro are commonly used for genetic experiments. These mesophilic methanogens are readily maintained in high-salt (HS) media that promote single-cell morphology. Despite their close evolutionary relationship, these two methanogens have significant differences in their ability to utilize different growth substrates. *M. acetivorans* C2A can grow on acetate, CO, methanol, methylamines, and methylsulfides. It cannot grow on $CO_2 + H_2$, apparently because the hydrogenases required are not expressed (although the genes are present on the chromosome; Galagan *et al.*, 2002; Guss *et al.*, 2005, 2009). *M. barkeri* Fusaro can grow on all these substrates via four different methanogenesis pathways: acetoclastic, hydrogenotrophic, methylotrophic, and methyl respiration. Genetic analysis of these two methanogens thus affords a complementary and informative view of the different methanogenic pathways and has revealed substantial differences in metabolic pathways and energy conservation mechanisms (Kulkarni *et al.*, 2009).

The parental strains used for most genetic experiments in our laboratories contain deletions of the hypoxanthine phosphoribosyltransferase *hpt* gene (Table 2.1), which is used as a counterselectable marker (Pritchett *et al.*, 2004). Many parental strains also contain the ΦC31 site-specific recombinase gene (*int*) and the ΦC31 *attP* site inserted at the *hpt* locus (Guss *et al.*, 2008). This feature allows for insertion of recombinant plasmids carrying the complementary *attB* sequence into the host chromosome via site-specific recombination, allowing facile complementation of mutations created in the ΦC31 *attB* strain by transformation with *attP* plasmids carrying the deleted gene.

The efficiency of liposome-mediated transformation can be as high as 20% in *M. acetivorans*, but is significantly lower in other *Methanosarcina* species (Metcalf *et al.*, 1997). Deletion of the *hsdR* restriction endonuclease gene from the *M. acetivorans* chromosome results in only a twofold increase in transformation with an autonomously replicating plasmid, and a 10-fold decrease in recovery of puromycin-resistant colonies transformed with a linearized plasmid (Buan, unpublished results). Therefore, endogenous endonuclease activity is not a barrier for transformation and restriction-deficient hosts are not required for routine genetic manipulations.

The methods for genetic manipulation of *Methanosarcina* presented below represent the culmination of many years of refinement. These protocols

Table 2.1 Useful plasmids and strains for genetic analysis of *Methanosarcina* species

E. coli stock #	Name	Genotype	Purpose	Reference
Plasmids				
WM3148	pWM321	*oriS*, *oriR6K*, *bla*, *pac*	Multicopy expression in *Methanosarcina* species	Metcalf *et al.* (1997)
WM1533	pMP44	*oriR6K*, *bla*, *pac*, *hpt*	Creation of markerless deletion mutants	Pritchett *et al.* (2004)
WM3767	pJK301	*oriR6K*, *bla*, *FRT pac-hpt* cassette (*FRT*, *pac*, *hpt*, *FRT*)	Creation of marked or unmarked deletion mutants (in conjunction with pMR55)	Welander and Metcalf (2008)
WM4363	pMR55	*oriR6K*, *bla*, *flp*	Expresses the Flp recombinase	Rother and Metcalf (2005)
WM5002	pAMG82	*oriV*, *oriS*, *cat*, ΦC31 *attB*, *pac*, *hpt*, *uidA*	Creation of transcriptional and translational *uidA* fusion strains	Guss *et al.* (2008)
WM6616	pAB79	*oriV*, *oriS*, *cat*, ΦC31 *attB*, *pac*, *hpt*, *uidA*	Creation of transcriptional and translational *uidA* fusion strains	Guss *et al.* (2008)
WM5262	pJK026A	*oriV*, *oriS*, ΦC31 *attB*, *cat*, *FRT5 pac-hpt* cassette (*FRT5*, *pac*, *hpt*, *FRT5*) P_{mcrB}*uidA*	Expression of genes from the P_{mcrB} promoter	Guss *et al.* (2008)
WM5263	pJK027A	*oriV*, *oriS*, ΦC31 *attB*, *cat*, *FRT5 pac-hpt* cassette (*FRT5*, *pac*, *hpt*, *FRT5*), $P_{mcrB(tetO1)}$*uidA*	Expression of genes from the $P_{mcrB(tetO1)}$promoter	Guss *et al.* (2008)
WM5265	pJK028A	*oriV*, *oriS*, ΦC31 *attB*, *cat*, *FRT5 pac-hpt* cassette (*FRT5*, *pac*, *hpt*, *FRT5*), $P_{mcrB(tetO3)}$*uidA*	Expression of genes from the $P_{mcrB(tetO3)}$ promoter	Guss *et al.* (2008)
WM5267	pJK029A	*oriV*, *oriS*, ΦC31 *attB*, *cat*, *FRT5 pac-hpt* cassette (*FRT5*, *pac*, *hpt*, *FRT5*), $P_{mcrB(tetO4)}$*uidA*	Expression of genes from the $P_{mcrB(tetO4)}$ promoter	Guss *et al.* (2008)
WM5775	pGK050A	*oriR6K*, *bla*, *tetR*, *FRT5 pac-hpt* cassette (*FRT5 pac-hpt FRT5*), $P_{mcrB(tetO1)}$	Replacement of promoters with the $P_{mcrB(tetO1)}$ promoter	Guss *et al.* (2008)

WM5776	pGK051A	*oriR6K*, *bla*, *tetR*, *FRT5 pac-hpt* cassette (*FRT5 pac-hpt FRT5*), $P_{mcrB(tetO3)}$	Replacement of promoters with the $P_{mcrB(tetO3)}$ promoter	Guss *et al.* (2008)
WM5779	pGK052A	*oriR6K*, *bla*, *tetR*, *FRT5 pac-hpt* cassette (*FRT5 pac-hpt FRT5*), $P_{mcrB(tetO4)}$	Replacement of promoters with the $P_{mcrB(tetO4)}$ promoter	Guss *et al.* (2008)

Stock #	Organism	Purpose
Methanosarcina strains		
DSM2834	*M. acetivorans* C2A	Wild type
DSM804	*M. barkeri* Fusaro	Wild type

Methanogen stock #	Genotype	Purpose	Reference
M. acetivorans derivatives			
WWM73	*Δhpt::(*P_{mcrB}*tetR* ΦC31 *int attP)*	8-ADP resistant parent, integration of ΦC31 *attB* plasmids, expresses TetR repressor	Guss *et al.* (2008)
WWM75	*Δhpt::(*P_{mcrB}*tetR* ΦC31 *int attB)*	8-ADP resistant parent, integration of ΦC31 *attP* plasmids, expresses TetR repressor	Guss *et al.* (2008)
WWM82	*Δhpt::(*P_{mcrB}ΦC31 *int attP)*	8-ADP resistant parent, integration of ΦC31 *attB* plasmids	Guss *et al.* (2008)
WWM83	*Δhpt::(*P_{mcrB}ΦC31 *int attB)*	8-ADP resistant parent, integration of ΦC31 *attP* plasmids	Guss *et al.* (2008)
M. barkeri derivatives			
WWM85	*Δhpt::(*P_{mcrB} ΦC31 *int attP)*	8-ADP resistant parent, integration of ΦC31 *attB* plasmids	Guss *et al.* (2008)
WWM86	*Δhpt::(*P_{mcrB} ΦC31 *int attB)*	8-ADP resistant parent, integration of ΦC31 *attP* plasmids	Guss *et al.* (2008)
WWM154	*Δhpt::(*P_{mcrB}*tetR* ΦC31 *int attB)*	8-ADP resistant parent, integration of ΦC31 *attB* plasmids, expresses TetR repressor	Guss *et al.* (2008)
WWM155	*Δhpt::(*P_{mcrB}*tetR* ΦC31 *int attP)*	8-ADP resistant parent, integration of ΦC31 *attP* plasmids, expresses TetR repressor	Guss *et al.* (2008)

provide a starting point for nearly any kind of genetic experiment that one might want to perform in these methanogens. It is essential to note that *Methanosarcina* species, like all methanogens, are exquisitely sensitive to oxygen. Accordingly, all procedures are designed to be performed in an anaerobic chamber (Coy Laboratory products, MI, USA) under an atmosphere of $N_2/CO_2/H_2$ (75%/20%/5%). Methods for cultivation in prereduced, anaerobic liquid media, and for anaerobic plating on solid media, have been previously described (Balch *et al.*, 1979; Metcalf *et al.*, 1998; Sowers *et al.*, 1993). Glassware should be placed in the anaerobic chamber at least 1 day before use to remove adsorbed oxygen. Plasticware should be placed in the anaerobic chamber for at least 1 week to render it sufficiently anaerobic.

2. Liposome-Mediated Transformation

The protocol outlined below (adapted from Metcalf *et al.* (1997)) is a reliable method for transforming *Methanosarcina* species including *M. barkeri* and *M. acetivorans*.

1. Harvest *Methanosarcina* cells from a late exponential-phase culture grown in 20 ml HS medium (Metcalf *et al.*, 1997) containing a suitable growth substrate by centrifugation in a clinical centrifuge for 15 min. (We maintain a small clinical centrifuge within the anaerobic chamber for this purpose.) Decant the supernatant and remove any residual growth HS medium by aspiration with a sterile pipette. Carefully resuspend the cell pellet in 1 ml bicarbonate-buffered 0.85 *M* sucrose solution (pH 6.8).
2. For each transformation to be performed, combine 110 μl bicarbonate-buffered sucrose solution, 30 μl DOTAP liposomal transfection reagent (Roche Applied Science, IN, USA), and 2 μg plasmid DNA (dissolved in ~10 ml sterile distilled water) in a 13 × 100 mm disposable borosilicate glass tube (Fisher Scientific, PA, USA). Incubate for 15–30 min at room temperature. Two controls are recommended: the shuttle vector pWM321 (Metcalf *et al.*, 1997) is a useful positive control and sterile H_2O can be used in place of DNA as the negative control.
3. Add 1 ml of the cell suspension from step #1 to each tube created in step #2. Mix gently and incubate for 2–4 h at room temperature.
4. Transfer the entire transformation mixture to 10 ml HS broth medium with an appropriate growth substrate and incubate for 12–16 h at 37 °C.
5. Centrifuge 1 and 9 ml aliquots of each transformation for 15 min, then:

   For M. acetivorans transformations: Decant the supernatant and resuspend cells in the residual growth medium. Spread cells on 1.4% HS agar supplemented with puromycin (2 μg ml^{-1}) and an appropriate growth substrate.

   For M. barkeri transformations: Decant the supernatant and remove any residual growth medium. Resuspend cells in 2 ml of molten 0.5% HS agar (no substrate) maintained at ~45 °C in a heating block, then pour onto 1.4% HS-agar plates supplemented with puromycin (2 μg ml^{-1}) and an appropriate growth substrate.

Note: We have observed greater transformation efficiencies with trimethylamine as the growth substrate. If methanol is desired as the growth substrate, we have found that reduction of the concentration from 125 to 50 m*M* results in higher transformation efficiencies (but with smaller colonies).

6. Incubate the plates in an anaerobic jar or an anaerobic incubator under an atmosphere of $N_2/CO_2/H_2S$ (80%/20%/0.1%) or $H_2/CO_2/H_2S$ (80%/20%/0.1%). Check for colonies after 10–14 days of incubation and then streak them for single-colony isolation on HS-agar plates using sterile, anaerobic sticks. (*Note*: H_2S is highly toxic. Appropriate measures should be in place to prevent exposure to the gas phase. Also, H_2S inactivates the expensive palladium catalyst used to maintain oxygen-free conditions in many anaerobic chambers and measures should be in place to allow replacement of the gas phase prior to opening the incubation vessels inside the anaerobic chamber (see Metcalf *et al.* (1998) for one approach).
7. Colonies can be transferred from agar plates directly into liquid media in sealed Balch tubes (Balch and Wolfe, 1976), without opening the tubes, using sterile, anaerobic syringes. To do this, remove the needle from the syringe (keeping the protective sheath on the needle) and pull the plunger out slightly. Place the opening of the syringe over the colony and push into the agar while pulling up on the syringe plunger. The colony with the surrounding agar will be aspirated into the syringe. Replace the sterile needle. Sterilize the Balch tube stopper using a heating coil (Coy Labs, MI, USA). Pierce the stopper with the syringe needle and wash the colony and agar into the medium by pushing and pulling the plunger of the syringe three to four times. The liquid medium is now inoculated.

3. Deleting Genes

We routinely use two methods to create targeted gene deletions in methanogens (Pritchett *et al.*, 2004; Welander and Metcalf, 2008). The first method uses pMP44, which relies on homologous recombination to create a merodiploid mutant (Fig. 2.1). Resolution of the unstable merodiploid by a second recombination event produces two types of progeny. Roughly, half the offspring will retain the parental locus, while the other half will retain the mutation of interest. It should be noted, however, that the colonies obtained from this method will only have a 50/50 ratio of

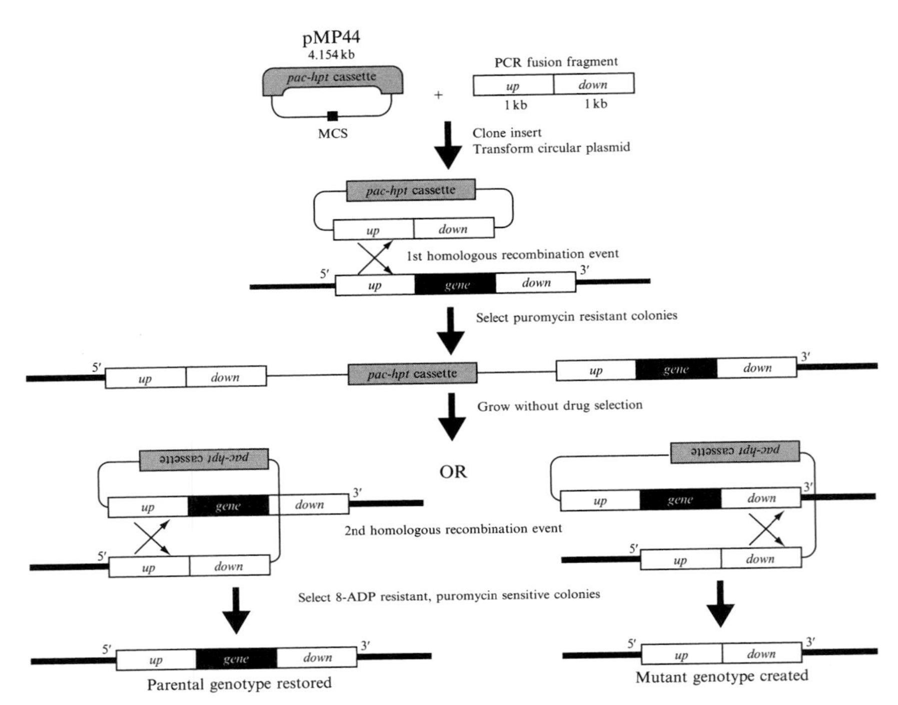

Figure 2.1 Creation of deletion mutant strains using pMP44. Markerless deletion mutant strains can be created using the plasmid pMP44. A deletion construct PCR fragment (~1 kb each upstream and downstream genomic DNA sequence) is cloned into the pMP44 multiple-cloning site. The circular plasmid is transformed into the strain of interest and after successive selection for puromycin resistance and loss of 8-ADP sensitivity, isolated colonies are inoculated into liquid medium. Cultures are then screened by PCR and DNA hybridization to determine if 8-ADP resistant, puromycin-sensitive cells contain the desired deletion mutation or if they have restored the parental genotype.

parent/mutant if there is no difference in growth rate between the parent strain and the mutant. If there is a growth defect in the mutant, the fraction with the parental genotype will be much greater. If the gene to be deleted is essential, only the parental genotype will be recovered.

In the second method, the targeted gene is replaced by the *pac-hpt* cassette of plasmid pJK301 (Fig. 2.2). To do this, regions of homology upstream and downstream of the desired deletion are cloned on either side of the *pac* cassette. The resulting plasmid is then linearized before transformation. Thus, only colonies that have incorporated the mutant allele by double-recombination event will be recovered. Due to the paucity of selectable markers that have been developed for use in *Methanosarcina*, the *pac-hpt* cassette must be removed if additional genetic manipulations are to

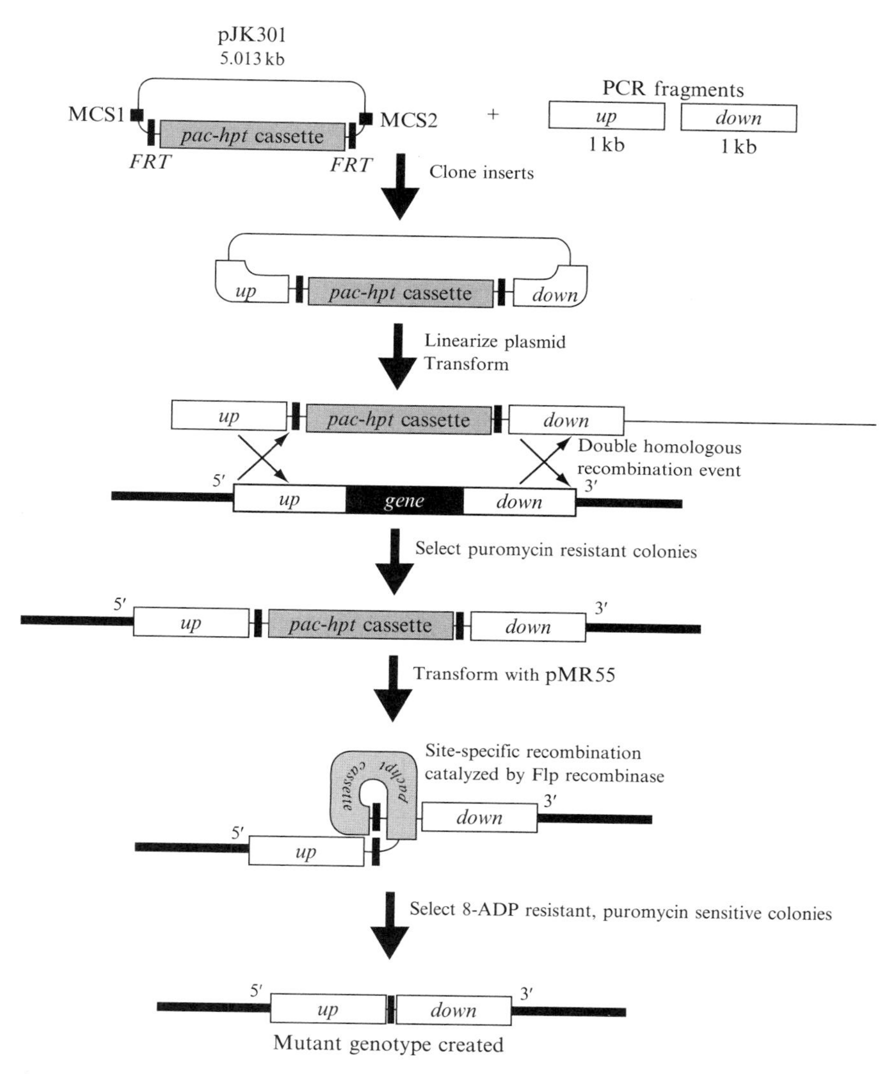

Figure 2.2 Creation of deletion mutant strains using pJK301. Marked deletion mutations in *Methanosarcina* can be created using pJK301 via double-homologous recombination. Approximately 1 kb upstream genomic DNA sequence and 1 kb downstream genomic DNA sequence are cloned into the multiple-cloning sites (at MCS1 and MCS2). The linearized plasmid is transformed into the parent strain and puromycin-resistant cells are selected. These colonies should be screened by PCR and DNA hybridization to ensure that the *pac-hpt* cassette has replaced the gene of interest. This marked deletion strain can either be used for experiments, or it can be subsequently transformed with pMR55 to remove the *pac-hpt* cassette. After removal of the marker, a FRT "scar" will remain on the deletion mutant chromosome.

be performed in the mutant strain using the puromycin selection (e.g., introduction of additional mutations or complementation of the mutation). This can be done by subsequent transformation with a plasmid encoding the Flp recombinase (pMR55), which will excise the cassette by site-specific recombination between the flanking FRT sites. Due to the loss of the *hpt* gene, the desired recombinants can be selected as 8-ADP (8-aza-2,6-diaminopurine) resistant clones (Rother and Metcalf, 2005).

PCR-based screens are a quick and easy method to verify that the desired mutations are present, but it is recommended that strains be confirmed by DNA hybridization experiments (Southern blotting) to rule out the presence of large duplications in the genome. Rarely, incomplete digestion of the pJK301-derivative deletion plasmid results in creation of a merodiploid by single recombination of a circular plasmid onto the chromosome. In a few cases, the deletion plasmid has integrated in a nontarget site. We have also observed that deletion strains carrying the *hpt* gene can develop 8-ADP resistance without deletion of the *pac* gene. These problems are rare, but nevertheless should be ruled out before continuing experiments with any mutant strains.

3.1. Using pMP44

pMP44 is an ampicillin-resistance plasmid that replicates in *E. coli* pir^+ strains but does not replicate in methanogens (Table 2.1). It carries a puromycin-resistance cassette to allow selection of methanogen clones that have incorporated the plasmid onto the chromosome (Fig. 2.1). It also contains a copy of the *hpt* gene to allow for subsequent counter-selection to screen for unmarked deletion mutants (Pritchett *et al.*, 2004).

1. To delete a gene of interest, ~1 kb of upstream sequence and 1 kb of downstream sequence are cloned adjacent to each other into the pMP44 multiple-cloning site (MCS). We regularly use overlap extension PCR (Horton *et al.*, 1989) to create an up–down fusion construct to simplify the cloning procedure.
2. Once the correct plasmid has been obtained and verified, it is purified from *E. coli* and resuspended in sterile H_2O at a concentration of 0.2 μg plasmid DNA per microliter.
3. The circular plasmid is transformed using the liposome transformation method described above. Merodiploid recombinants are selected by plating on HS agar with puromycin (2 μg ml^{-1}) and an appropriate growth substrate.
4. To allow resolution of the unstable merodiploid locus, colonies are picked to liquid medium without puromycin and grown to stationary phase. Aliquots of this culture are then plated onto HS agar containing an appropriate growth substrate and 8-ADP (20 μg ml^{-1}). Resolution

occurs by homologous recombination between the repeated sequences on either side of the integrated pMP44 vector, which results in loss of the plasmid. Although these recombinants comprise a small fraction of the population, they can be easily selected due to loss of the *hpt* gene, which confers sensitivity to the toxic base analog 8-ADP.

5. Recombinants should be purified multiple times by streaking for isolated colonies on HS agar containing 8-ADP, then verified to have lost the plasmid backbone by streaking onto HS puromycin agar (the desired recombinants will also be puromycin-sensitive).
6. Single colonies from 8-ADP resistant, puromycin-sensitive clones can be screened for the deletion of interest by PCR and confirmed by DNA hybridization experiments.

3.2. Using pJK301 and pMR55

pJK301 is also a plasmid that can replicate in *E. coli* DH5α but cannot replicate in methanogens (Table 2.1). It carries a *pac-hpt* cassette encoding puromycin resistance and 8-ADP sensitivity. pJK301 can be used to create both marked and unmarked deletions (Fig. 2.2).

1. Approximately 1 kb of upstream sequence is cloned into the first MCS. In a second cloning event, ~1 kb of downstream sequence is cloned into the second MCS. The plasmid is sequenced to verify that it is free from error and linearized by restriction enzyme digestion using an enzyme that cuts within the vector backbone.
2. The digested plasmid is purified via ethanol precipitation, or a commercially available kit and then used to transform the recipient *Methanosarcina* strain to puromycin resistance as described above.
3. Puromycin-resistant colonies are purified by streaking on HS agar with puromycin and then verified by PCR and DNA hybridization.

These marked deletions are suitable for most purposes; however, as described above, it is often necessary to delete the *pac-hpt* cassette to allow subsequent genetic manipulations using puromycin resistance as a selectable marker. The *pac-hpt* cassette on pJK301 is flanked by FRT recombination sites (3). These FRT sites are recognized by the Flp recombinase, which will excise the *pac* cassette and create a puromycin-sensitive strain carrying the unmarked deletion mutation of interest. An FRT "scar" will remain at the site of deletion. pMR55 is a plasmid that cannot replicate in *Methanosarcina* and carries the Flp recombinase cassette. To remove the FRT-flanked *pac-hpt* markers created using pJK301:

1. Transform the strain of interest with pMR55 as described above, and plate 10 μl, 100 μl, and 1 ml aliquots onto HS agar containing an appropriate growth substrate and 8-ADP (20 μg ml^{-1}).

2. Putative recombinants should be purified multiple times by streaking for isolated colonies on HS agar containing 8-ADP and then verified to have lost the plasmid backbone by streaking onto HS puromycin agar.
3. Single colonies from 8-ADP resistant, puromycin-sensitive clones can be screened for the deletion of interest by PCR and confirmed by DNA hybridization experiments.

4. Test for Gene Essentiality

The methods described in the preceding sections will fail in cases where the gene of interest is essential for viability. To provide a positive test for gene essentiality in *Methanosarcina*, we developed a series of tetracycline (Tc)-dependent promoters ($P_{mcrB(tetO)}$ promoters; Fig. 2.3) (Guss *et al.*, 2008). We also created a series of strains that constitutively express the *E. coli* TetR protein, which is required for regulated expression of these promoters (Table 2.1). In the absence of Tc, the TetR repressor protein binds to the operator sequence (*tet*O) within these promoters, thus blocking gene expression. In the presence of Tc, TetR binds to Tc and dissociates from the promoter, thus allowing the downstream gene to be expressed. Therefore, if the strain harboring a $P_{mcrB(tetO)}$-gene fusion (hereafter referred to as P_{tet}) is unable to grow in the absence of Tc, the gene of interest is required for growth of the organism.

4.1. Construction of *Methanosarcina* P_{tet} strains

To approximate the native expression level of the gene to be tested, three $P_{mcrB(tetO)}$ promoters ($P_{mcrB(tetO1)}$, $P_{mcrB(tetO3)}$, and $P_{mcrB(tetO4)}$) have been incorporated into a series of vectors (pGK050A, pGK051A, and pGK052A). Each of these $P_{mcrB(tetO)}$ promoters has a different expression level in *Methanosarcina*. The vectors replicate in *E. coli* DH5α strains and carry a selectable *pac* gene, a counterselectable *hpt* gene and a *tetR* gene encoding the TetR repressor protein (Guss *et al.*, 2008). Follow the steps outlined below to construct P_{tet} strains using these vectors.

1. Clone the gene of interest downstream of the $P_{mcrB(tetO)}$ promoter using NdeI and another restriction enzyme at the MCS. Use of the NdeI site (CATATG) allows construction of in-frame translational fusions to the promoter.
2. Clone ~1.5 kb region upstream of the native promoter of the gene of interest adjacent to *tetR* using NcoI and/or ApaI sites.
3. Linearize the resulting plasmid and transform into a *Methanosarcina* strain carrying a second copy of *tetR* in its chromosome; *M. acetivorans* C2A, WWM74 and WWM75; *M. barkeri* Fusaro, WWM154 and WWM155

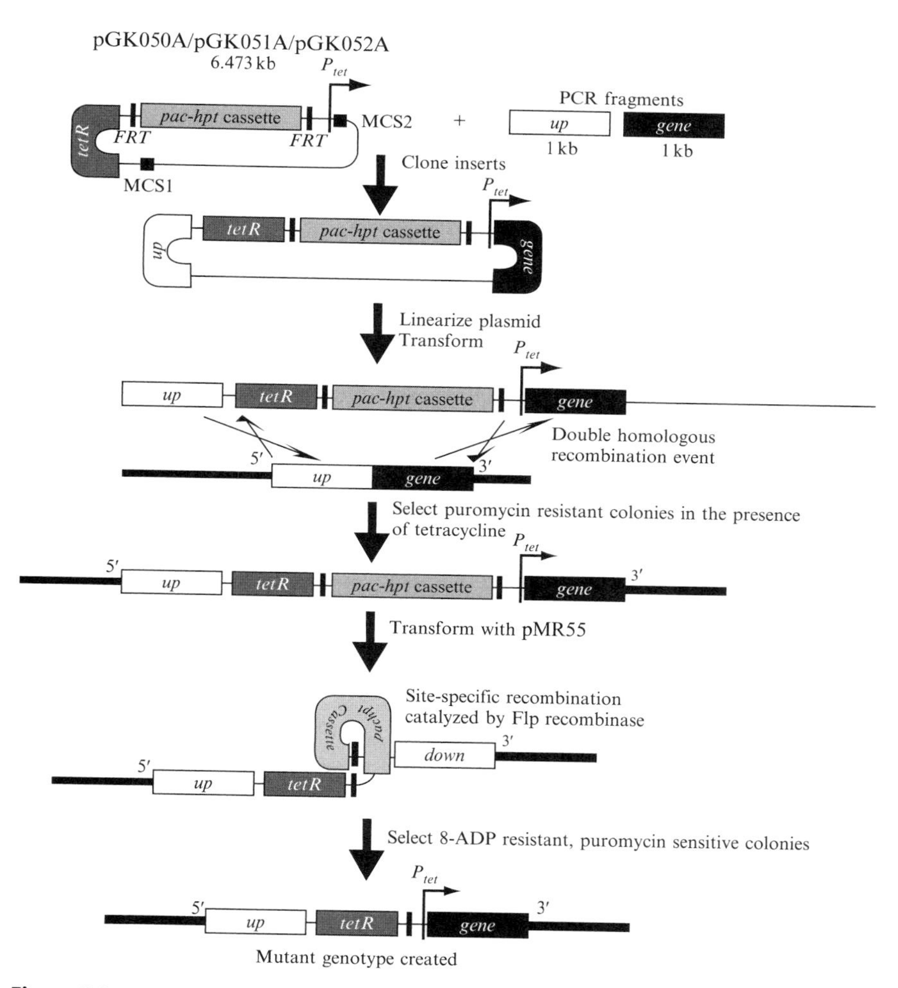

Figure 2.3 Creating promoter fusion strains using pGK050A, pGK051A, or pGK052A plasmids. Each of the pGK050A, pGK051A, and pGK052A plasmids carries a P_{tet} promoter of varying expression level. These plasmids are used for putting a gene(s) under control of the P_{tet} promoter in place of the native promoter. pGK050A, pGK051A, and pGK052A all contain a *pac-hpt* cassette for puromycin resistance and 8-ADP sensitivity. Also on the plasmids is a *tetR* repressor cassette. Approximately 1 kb upstream genomic DNA sequence is cloned into the MCS1 multiple-cloning site, and 1 kb of coding sequence is cloned into MCS2. The linearized plasmid is transformed into the parent strain, and transformed cells are screened by PCR and/or Southern blot to verify the expected genotype. Strains carrying the P_{tet} promoter fusions and the *pac-hpt* cassette can be used in experiments as-is. Alternatively, the *pac-hpt* cassette can be removed to create a markerless P_{tet} promoter fusion strain. The *pac-hpt* cassette is flanked by FRT5 sites, which serve as recombination sites for the Flp recombinase encoded on pMR55.

(Guss *et al.*, 2008). The presence of two *tetR* copies reduces the probability of obtaining constitutive *tetR*$^{-}$ mutants. Select transformants on HS-agar plates with an appropriate substrate, puromycin, and 100 μg ml^{-1} Tc. Tc must be included in the recovery and plating medium to allow expression of the putative essential gene. Verify the resulting P_{tet} strain using PCR and/or DNA hybridization.

4. The *pac-hpt* cassette in the P_{tet} strain is flanked by recognition sites for Flp site-specific recombination system (FRT). Therefore, this cassette can be removed by transforming the strain with pMR55 as described above (Rother and Metcalf, 2005). The markerless P_{tet} strain can then be used for future experiments requiring puromycin selection.

4.2. Testing gene essentiality on solid medium

Streak the P_{tet} strain and the Tc-independent parental strain (control) on HS-agar plates containing the substrate of interest with and without Tc. If the P_{tet} strain can only grow in the presence of Tc, then the gene is essential for growth of *Methanosarcina* on the substrate tested. For instance, as shown in Fig. 2.4, the gene encoding the methyl-CoM reductase, *mcr*, is essential for growth of *M. barkeri* on methanol.

We have observed that *Methanosarcina* often grow on endogenous substrates found in agar. This complication can be avoided using the spot-plate method:

1. Grow the parental strain and the P_{tet} strain(s) to be tested on the substrate of interest (with Tc) for at least 15 generations. Pellet the cells from a 10-ml saturated culture in a clinical centrifuge and resuspend in 5 ml plain HS medium. Repeat the centrifugation and resuspension two more times, for a total of three washes. Serially, dilute the culture 10-fold from 10^{-1} to 10^{-7} in a 96-well microtiter plate using plain HS medium (no substrate or Tc).

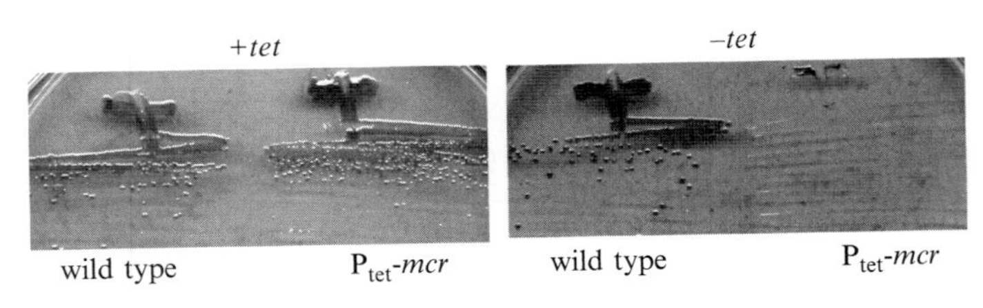

Figure 2.4 Essentiality of the *mcr* operon in *Methanosarcina barkeri*. $P_{mcrB(tetO1)}$::*mcrBCDGA* strain (WWM235) and its parental strain (WWM155) were streaked on HS-methanol agar in the presence or absence of tetracycline. Growth of the $P_{mcrB(tetO1)}$::*mcrBCDGA* only in the presence of tetracycline indicates that the *mcr* operon is essential.

2. Set up the spotting plate by placing three layers of GB004 paper (Whatman, NJ, USA), two layers of GB002 paper (Schleicher & Schuell BioScience, NH, USA) and one layer of 3MM paper (Whatman, NJ, USA) in a 70 × 10 mm petri dish (Fisher Scientific, PA, USA).
3. Soak the stack of papers with 30 ml HS medium containing the substrate of interest with and without Tc and immediately overlay a 0.22-μm nylon membrane (GE Water and Process Technologies, PA, USA). It is important that there are no air bubbles under the nylon membrane to allow for efficient wicking of medium through the paper to the growing cells.
4. Spot 10 μl of each serial dilution onto the membrane using a multichannel pipettor. Let the spots soak into the membrane for ~10 min. Carefully add 10–15 ml more medium (plus appropriate substrate and/or Tc) to saturate the papers.
5. Seal the plate with parafilm and incubate upright (do NOT invert) for at least 2 weeks to check growth. Lack of growth of the P_{tet} strain in the absence of Tc indicates essentiality of the gene for growth of *Methanosarcina* (Fig. 2.5).

4.3. Testing gene essentiality in broth medium

Gene essentiality can also be tested in broth medium. For this, adapt the P_{tet} strain to the substrate of interest for at least 15 generations in the presence of Tc. Grow the adapted strain until midexponential phase (OD_{600} ~ 0.5 for methanol and methanol plus H_2/CO_2 and OD_{600} ~ 0.25 for H_2/CO_2 and acetate) and serially dilute to 10^{-4} with and without Tc in four replicates. At this dilution, the concentration of residual Tc contributed by the

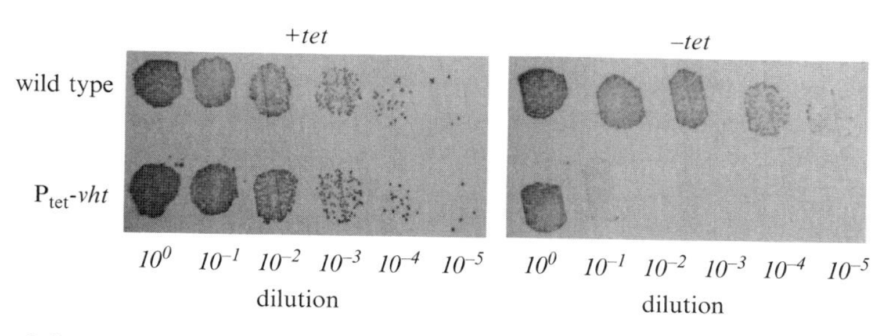

Figure 2.5 Essentiality of the *vht* operon in *Methanosarcina barkeri*. Serial dilutions of wild type and P_{tet}-*vhtGACD* were spotted onto nylon membranes supported by filter pads soaked in HS-methanol medium with and without tetracycline. Cultures were then incubated for 2 weeks. Lack of growth in the absence of tetracycline indicates that the *vht* operon is essential under these growth conditions.

inoculum is less than that required for induction of the P_{tet} promoter (1 μg ml^{-1}; Guss *et al.*, 2008). Monitor growth at OD_{600}.

The spot-plate method is preferred for doing gene essentiality tests over the broth method, because strains that constitutively express the gene of interest due to suppressor mutations in the P_{tet} promoter or in the *tetR* gene outgrow the unmutated strain in broth, thereby contributing to OD_{600} and resulting in a false-positive result (Buan and Metcalf, 2009). In contrast, on spot plates, suppressor mutants can be seen as distinct colonies.

5. Adding Genes

There are several genetic tools available for expressing a gene of interest in *Methanosarcina*. It is possible to express gene(s) from autonomous multicopy plasmids or from single-copy insertions into the chromosome. Numerous variations of copy number, reporter genes, promoter, recombinase, etc. can be explored using the few tools discussed here.

5.1. Multicopy expression from a plasmid

pWM321 is an *E. coli/Methanosarcina* shuttle vector created from the naturally occurring plasmid pC2A from *M. acetivorans* C2A (Galagan *et al.*, 2002; Metcalf *et al.*, 1997). pC2A is present in ~6 copies/*Methanosarcina* cell (Guss *et al.*, 2008; Sowers and Gunsalus, 1988). It has been engineered to carry a puromycin-resistance cassette for selection in *Methanosarcina*, as well as an R6K origin of replication and an ampicillin-resistance cassette for replication and selection in *E. coli pir*$^{+}$ DH5α. A large MCS is available for cloning in promoters and genes of interest.

5.2. Integration at a single site on the chromosome

We have engineered a variety of *M. acetivorans* C2A and *M. barkeri* Fusaro strains that carry a constitutively expressed ΦC31 site-specific recombinase gene (*int*) and a ΦC31 *attB* site integrated into the chromosomal *hpt* locus (Table 2.1). Plasmids carrying the corresponding ΦC31 *attP* site can recombine with the *attB* site, resulting in integration of the entire plasmid at the *hpt* locus (Guss *et al.*, 2008) (Fig. 2.6).

5.2.1. Reporter gene fusions using pAB79 and derivatives

For transcriptional or translational reporter-gene fusions, pAB79 is commonly used (Guss *et al.*, 2008). pAB79 carries a chloramphenicol-resistance marker and replicates as a single-copy plasmid in most *E. coli* strains. However, it can be replicated as a high-copy plasmid in strains carrying

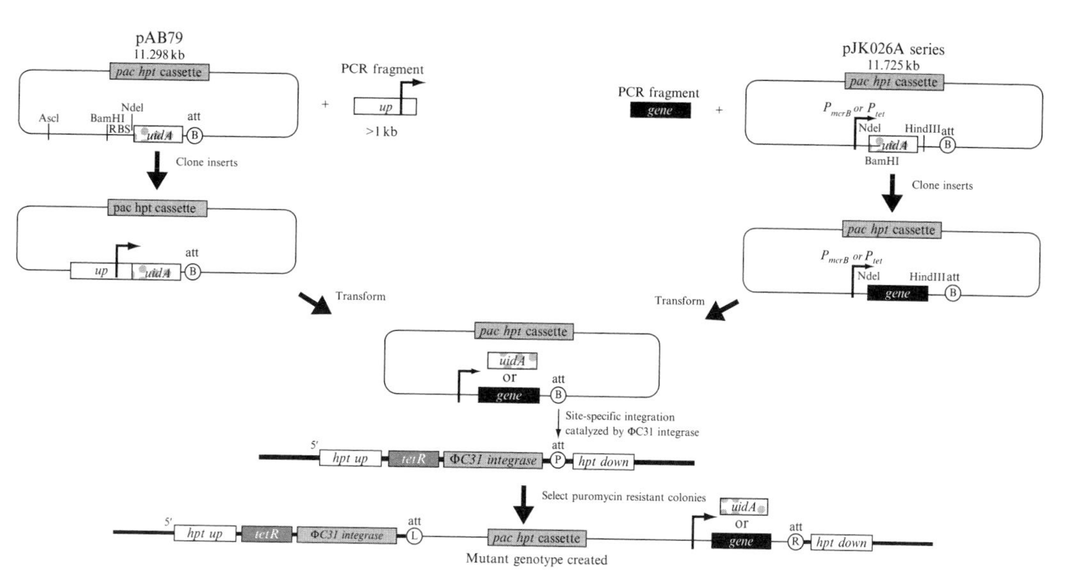

Figure 2.6 Site-specific integration using ΦC31 integrase and ΦC31 att sequences. *Methanosarcina* strains have been created that constitutively express the ΦC31 integrase and which carry *att* sequences on the chromosome (*attB*, in this example). Using these strains, it is possible to integrate plasmids that carry the complementary ΦC31 *att* site onto the chromosome (*attP*, in this figure). pAB79 is commonly used to create transcriptional or translational reporter fusions to the β-glucuronidase gene (*uidA*). pJK026A series plasmids are used for expression of any gene of interest. Genes put into pJK026A will be expressed constitutively in *Methanosarcina* from the P_{mcrB} promoter. Genes put into pJK027A, pJK028A, and PJK029A will be expressed from one of the P_{tet} promoters.

an inducible copy of the *trfA* gene (Table 2.1). Plasmid pAB79 can integrate into the *Methanosarcina* chromosome via ΦC31 site-specific recombination and can be used to create either transcriptional or translational fusions to *uidA*. By cloning promoters of interest into the BamHI site, the *mcrB* ribosome-binding site (RBS) is maintained to allow efficient translation initiation of *uidA*; thus, expression of the reporter-gene fusion is dependent only on transcription initiating within the cloned segment. Tandem translation stop codons are maintained in this case to prevent translational read through into the reporter gene. Alternatively, one can maintain the RBS from the gene of interest by cloning into the NdeI site, thus creating a translational fusion that requires both transcriptional and translational signals to be present in the cloned fragment. pAB79 is used as follows:

1. A DNA fragment containing a putative promoter (usually 1 kb of sequence from upstream of a gene of interest) is cloned upstream of the *uidA* gene. For translational fusions an AscI site is added to the upstream primer. The native RBS is maintained by converting the start codon of the gene of interest into an NdeI site. The PCR fragment is then cloned into AscI-, NdeI-cut pAB79. Transcriptional fusions are constructed by addition of an AscI site to the upstream primer and a BamHI site to the downstream primer. This fragment is then cloned into AscI-, BamHI-cut pAB79.
2. The circular pAB79-derived plasmid is transformed into the appropriate host strain and plated onto HS puromycin agar.
3. Puromycin-resistant clones are purified by streaking for isolation on HS puromycin agar. Because it is possible to obtain recombinants with more than one copy of the inserted plasmid, it is advisable to verify the resulting strains by PCR (Guss *et al.*, 2008). *M. acetivorans* integrants (WWM82-derived strains) can be screened with the primers "C31 screen-all#1" (GAAGCTTCCCCTTGACCAAT), "C31 screen-C2A#1" (TTGATTCGGATACCCTGAGC), "C31 screen-pJK200#1" (GCAAAGAAAAGCCAGTATGGA), and "C31 screen-pJK200#2" (TTTTTCGTCTCAGCCAATCC). *M. barkeri* integrants (WWM85-derived strains) can be screened with the primers "C31 screen-all#1", "C31 screen-Fus#1" (CGAACTGTGGTGCAAAAGAC), "C31 screen-pJK200#1", and "C31 screen-pJK200#2". The PCRs can be performed using Taq polymerase in Failsafe buffer J (Epicentre, Madison WI). Template DNA can be obtained by resuspending cells from a colony grown on agar-solidified media in sterile H_2O, which causes immediate cell lysis. After a 4-min preincubation at 94 °C, 35 cycles of 94 °C for 30 s, 53 °C for 30 s, and 72 °C for 90 s are performed, followed by a final incubation at 72 °C for 2 min. The expected bands are as follows: parental strain control, 910 bp; plasmid control, 510 bp; single plasmid integrations, 679 and 740 bp; integration of plasmid multimers, 680, 741, and 511 bp.

4. Promoter activity can then be assessed by assay of β-glucuronidase activity as described in Rother *et al.* (2005) in extracts of cells grown under the desired conditions (see for example Bose and Metcalf, 2008; Bose *et al.*, 2008).

5.2.2. Expression of genes using pJK026A and derivatives

Plasmids pJK026A through pJK029A are used to express a gene of interest from a series of constitutive or tightly regulated promoters. These plasmids are similar to pAB79 in that they encode chloramphenicol resistance and can replicate in single or multicopy in *E. coli*, and that they do not replicate in *Methanosarcina*, but can integrate into the chromosome via ΦC31-mediated site-specific recombination (Table 2.1). They also carry a puromycin-resistance cassette for selection in *Methanosarcina*. In these plasmids, the gene of interest can be cloned in place of the *uidA* gene on the plasmid. High level constitutive expression is achieved using the P_{mcrB} promoter (pJK026A), sequentially lower levels of constitutive expression can be obtained using the $P_{mcrB(tetO1)}$ (pJK027A), $P_{mcrB(tetO3)}$ (pJK028A), and $P_{mcrB(tetO4)}$(pJK029A) promoters. If the latter three plasmids are used in hosts that express the TetR protein, expression of the cloned genes is rendered tetracycline-dependent. Similar to pAB79-derived plasmids, circular derivative plasmids can be transformed into the appropriate parent and screened by PCR and/or Southern blot to confirm the genotype.

ACKNOWLEDGMENTS

This work was supported in part by a Department of Energy grant DE-FG02-02ER15296, National Science Foundation Grant MCB0517419 and by NIH NIGMS Kirschstein NRSA Postdoctoral Fellowship F32-GM078796. Any opinions, findings, and conclusions or recommendations expressed in this material are those of the authors and do not necessarily reflect the views of the National Science Foundation or the Department of Energy.

REFERENCES

Balch, W. E., and Wolfe, R. S. (1976). New approach to the cultivation of methanogenic bacteria: 2-mercaptoethanesulfonic acid (HS-CoM)-dependent growth of *Methanobacterium ruminantium* in a pressurized atmosphere. *Appl. Environ. Microbiol.* **32,** 781–791.

Balch, W. E., *et al.* (1979). Methanogens: Reevaluation of a unique biological group. *Microbiol. Rev.* **43,** 260–296.

Bose, A., and Metcalf, W. W. (2008). Distinct regulators control the expression of methanol methyltransferase isozymes in *Methanosarcina acetivorans* C2A. *Mol. Microbiol.* **67,** 649–661.

Bose, A., *et al.* (2008). Genetic analysis of the methanol- and methylamine-specific methyltransferase 2 genes of *Methanosarcina acetivorans* C2A. *J. Bacteriol.* **190,** 4017–4026.

Buan, N. R., and Metcalf, W. W. (2009). Methanogenesis by *Methanosarcina acetivorans* involves two structurally and functionally distinct classes of heterodisulfide reductase. *Mol. Microbiol.* **75,** 843–853.
Galagan, J. E., *et al.* (2002). The genome of *M. acetivorans* reveals extensive metabolic and physiological diversity. *Genome Res.* **12,** 532–542.
Guss, A. M., *et al.* (2005). Genetic analysis of mch mutants in two *Methanosarcina* species demonstrates multiple roles for the methanopterin-dependent C-1 oxidation/reduction pathway and differences in H_2 metabolism between closely related species. *Mol. Microbiol.* **55,** 1671–1680.
Guss, A. M., *et al.* (2008). New methods for tightly regulated gene expression and highly efficient chromosomal integration of cloned genes for *Methanosarcina* species. *Archaea* **2,** 193–203.
Guss, A. M., *et al.* (2009). Differences in hydrogenase gene expression between *Methanosarcina acetivorans* and *Methanosarcina barkeri*. *J. Bacteriol.* **191,** 2826–2833.
Horton, R. M., *et al.* (1989). Engineering hybrid genes without the use of restriction enzymes: Gene splicing by overlap extension. *Gene* **77,** 61–68.
Kulkarni, G., *et al.* (2009). Hydrogen is a preferred intermediate in the energy-conserving electron transport chain of *Methanosarcina barkeri*. *Proc. Natl. Acad. Sci. USA* **106,** 15915–15920.
Metcalf, W. W., *et al.* (1997). A genetic system for Archaea of the genus *Methanosarcina*: Liposome-mediated transformation and construction of shuttle vectors. *Proc. Natl. Acad. Sci. USA* **94,** 2626–2631.
Metcalf, W. W., *et al.* (1998). An anaerobic, intrachamber incubator for growth of *Methanosarcina* spp. on methanol-containing solid media. *Appl. Environ. Microbiol.* **64,** 768–770.
Pritchett, M. A., *et al.* (2004). Development of a markerless genetic exchange method for *Methanosarcina acetivorans* C2A and its use in construction of new genetic tools for methanogenic archaea. *Appl. Environ. Microbiol.* **70,** 1425–1433.
Rother, M., and Metcalf, W. W. (2005). Genetic technologies for Archaea. *Curr. Opin. Microbiol.* **8,** 745–751.
Rother, M., *et al.* (2005). Methanol-dependent gene expression demonstrates that methyl-coenzyme M reductase is essential in *Methanosarcina acetivorans* C2A and allows isolation of mutants with defects in regulation of the methanol utilization pathway. *J. Bacteriol.* **187,** 5552–5559.
Sowers, K. R., and Gunsalus, R. P. (1988). Plasmid DNA from the acetotrophic methanogen *Methanosarcina acetivorans*. *J. Bacteriol.* **170,** 4979–4982.
Sowers, K. R., *et al.* (1993). Disaggregation of *Methanosarcina* spp. and growth as single cells at elevated osmolarity. *Appl. Environ. Microbiol.* **59,** 3832–3839.
Welander, P. V., and Metcalf, W. W. (2008). Mutagenesis of the C1 oxidation pathway in *Methanosarcina barkeri*: New insights into the Mtr/Mer bypass pathway. *J. Bacteriol.* **190,** 1928–1936.

CHAPTER THREE

Genetic Systems for Hydrogenotrophic Methanogens

Felipe Sarmiento B.,* John A. Leigh,† *and* William B. Whitman*

Contents

Abstract

Methanogens are obligate anaerobic Archaea that produce energy from the biosynthesis of methane. These lithotrophic microorganisms are widely distributed in oxygen-free environments and participate actively in the carbon cycle. Indeed, methanogenesis plays a major role in the last step of the anoxic degradation of organic substances, transforming acetate, CO_2, and H_2 to

* Department of Microbiology, University of Georgia, Athens, Georgia, USA
† Department of Microbiology, University of Washington, Seattle, Washington, USA

Methods in Enzymology, Volume 494
ISSN 0076-6879, DOI: 10.1016/B978-0-12-385112-3.00003-2

methane. The vast majority of the known methanogens are classified as hydrogenotrophic because they use principally H_2 as the electron donor to drive the reduction of CO_2. Unlike many other cultured Archaea, many methanogens thrive in neutral pH, low salinity, and temperate environments. This has been a great advantage in cultivating these organisms in laboratory conditions and in the development of genetic tools. Moreover, the hydrogenotroph *Methanococcus maripaludis* is currently a model organism among Archaea, not only for its utility in genetic but also for biochemical and physiological studies. Over time, a broad spectrum of genetic tools and techniques has been developed for methanococci, such as site-directed mutagenesis, selectable markers, transformation methods, and reporter genes. These tools have contributed greatly to the overall understanding of this group of microorganisms and the processes that govern its life style. In this chapter, we describe in detail the available genetic tools for the hydrogenotrophic methanogens.

1. Introduction

Methanogens are a diverse group of microorganisms that are distinguished by their ability to obtain most of their metabolic energy from the biosynthesis of methane or methanogenesis. These microorganisms share specific characteristics that make them unique. First, they are Archaea, belonging to the phylum *Euryarchaeota*. Second, they are obligate anaerobes, being found only in oxygen-free environments. Finally, they are obligate methane producers. Indeed, to date, no methanogens have been identified that can grow under physiological conditions without producing methane (Hedderich and Whitman, 2006).

Methanogens are widely distributed and are found in different anaerobic habitats on Earth. Those habitats range from temperate environments, such as fresh water sediments, to environments with extreme salinity, temperature, and pH, such as hot springs. In these anaerobic habitats, different microorganisms degrade organic matter to produce H_2, CO_2, and acetate. In the absence of sulfate, oxidized metals, and nitrite, methanogens consume these substrates, contributing in two different ways to the anaerobic food chain. First, methanogens catalyze the last step of the anoxic degradation of organic substances, producing methane, which is released to the atmosphere. Second, by maintaining an extremely low partial pressure of H_2, they keep the fermentative pathways energetically favorable (Hedderich and Whitman, 2006).

Taxonomic classification of methanogens places them in the phylum *Euryarchaeota* and divides them into six well-established orders (*Methanobacteriales*, *Methanocellales*, *Methanococcales*, *Methanomicrobiales*, *Methanosarcinales*, and *Methanopyrales*) and 31 genera. This taxonomy is supported by comparative 16S rRNA gene sequence analysis and some distinct phenotypic properties, such as lipid composition, substrate range, and cell envelope structure (Liu, 2010; Sakai *et al.*, 2008).

In spite of this diversity in taxonomy and physiology, methanogens can only utilize a restricted number of substrates. Methanogenesis is limited to three main types of substrates: CO_2, methyl-group containing compounds, and acetate. Most organic substances cannot be utilized by methanogens directly and must be converted by other microorganisms to the substrates for methanogenesis.

Most methanogens that belong to the orders *Methanobacteriales*, *Methanococcales*, *Methanomicrobiales*, *Methanocellales*, and *Methanopyrales* are hydrogenotroph and can reduce CO_2 to CH_4 using H_2 as the electron donor. In addition, a few representatives of the order *Methanosarcinales* are facultative hydrogenotrophs. Because these organisms are primary methylotrophs or acetotrophs, they are not discussed further in this chapter. The reduction of CO_2 is driven in four steps, some of which involve the unusual or unique coenzymes that function as C-1 moiety carriers. The process begins with the binding of CO_2 to the coenzyme methanofuran (MFR) and its subsequent reduction to the formyl level. Then, the formyl group is transferred to the coenzyme H_4MPT and dehydrated to methenyl-H_4MPT. The carbon group is subsequently reduced from methenyl-H_4MPT to methylene-H_4MPT and methyl-H_4MPT. Consequently, the methyl group is transferred to the thiol group of coenzyme M (CoM). In the last step, methyl CoM reductase (MCR) catalyzes the reduction of methyl-CoM to methane using coenzyme B (CoB) as an electron donor. The oxidation of CoB yields a heterodisulfide with CoM, which must then be reduced to regenerate the thiols (Liu and Whitman, 2008).

While H_2 is the main electron donor for methanogenesis, many hydrogenotrophic methanogens can use formate, ethanol, or some secondary alcohols as electron donors. In the first case, four molecules of formate are oxidized to CO_2, and one molecule of CO_2 is reduced to methane. For the alcohols, ethanol is oxidized to acetate, and the secondary alcohols are oxidized to ketones to drive the reduction of CO_2. However, methanogens grow poorly using alcohols as electron donors (Hedderich and Whitman, 2006).

Among the hydrogenotrophic methanogens, *Methanococcus maripaludis* has become one of the most useful model organisms to develop new molecular tools. This microorganism, which belongs to the order *Methanococcales*, presents several advantages for genetic studies. (1) It is a mesophile that grows best at 37 °C. (2) It has a relatively fast growth rate, with a doubling time of 2 h. (3) It is a facultative autotroph that takes up different organic substrates from the medium. (4) Plating techniques have been optimized to reach efficiency near 100%. (5) Cells lyse in low-ionic-strength buffers or in very low concentration of detergents, facilitating the isolation of DNA and other cellular components (Tumbula and Whitman, 1999).

A broad collection of molecular biology tools, including transformation, *in-frame* deletions, transposon mutagenesis, and auxotrophic selection have been developed. These have contributed greatly on our understanding of the genetics of methanogens. In this chapter, we present and describe the principal advances that have been made in genetic systems for hydrogenotrophic methanogens, taking as the main reference point the order *Methanococcales*.

2. Genome Sequences

In 1996, the genome of *Methanocaldococcus jannaschii* was completely sequenced (Bult *et al.*, 1996). This milestone represented the first sequenced genome of a representative of the Archaea as well as a methanogen. Since then, 24 genomes of hydrogenotrophic methanogens have been sequenced, belonging to five of the six corresponding taxonomic orders. Based on genome characteristics, the currently available genomes can be clustered in two groups. The first group consists of the genomes of the orders *Methanococcales*, *Methanobacteriales*, and *Methanopyrales*, which generally have small genomes that range between 1.3 and 1.9 Mb and a low G + C content between 27 and 34 mol%. Exceptions include only *Methanobrevibacter ruminantium M1*, which has a genome size of 2.9 Mb, and *Methanothermobacter thermautotrophicus ΔH* and *Methanopyrus kandleri AV19*, which have mol% GC of 49.54% and 61.6%, respectively. The second group consists of the orders *Methanomicrobiales* and *Methanocellales*. These hydrogenotrophic methanogens possess larger genomes that range between 1.8 and 3.5 Mb and higher genome GC contents between 45 and 62 mol%.

A complete list of the hydrogenotrophic methanogens that have fully sequenced genomes with their principal features are presented in Table 3.1.

3. Growth and Storage of Methanogen Cultures

3.1. Liquid media preparation

Methanogens are obligate anaerobic microorganisms and require two main conditions to be successfully grown: low oxygen partial pressure and a low redox potential of −0.33 V. A three-step system to prepare prereduced media that meets these conditions was developed 60 years ago (Hungate, 1950). First, media were boiled to expel dissolved oxygen. Second, to impede oxygenation of the medium, it was kept under an anaerobic gas phase. Finally, a reducing agent and a redox indicator were added to keep a low redox potential and to indicate the oxidative state of the medium, respectively. This Hungate

Table 3.1 Characteristics of the sequenced genomes of hydrogenotrophic methanogens

Hydrogenotrophic methanogen	Order	Genome size (bp)	Gene count	G + C percentage	NCBI Accession number	Reference
Methanosphaera stadtmanae DSM 3091	*Methanobacteriales*	1767403	1592	27.63	NC_007681	Fricke *et al.* (2006)
Methanobrevibacter ruminantum M1	*Methanobacteriales*	2937203	2283	32	NC_013790	np
Methanobrevibacter smithii ATCC 35061	*Methanobacteriales*	1853160	1837	31.03	NC_009515	Samuel *et al.* (2007)
Methanobrevibacter smithii DSM 2375	*Methanobacteriales*	1704865	1747	31.28	NZ_ABYW00000000	np
Methanobrevibacter smithii F1, DSM 2374	*Methanobacteriales*	1727775	1755	31.19	NZ_ABYV00000000	np
Methanothermobacter thermautotrophicus Delta H	*Methanobacteriales*	1751377	1921	49.54	NC_000916	Smith *et al.* (1997)
Methanocella paludicola SANAE	*Methanocellales*	2957635	3064	54.92	NC_013665	np
Methanocella sp. RC-I	*Methanocellales*	3179916	3170	54.60	NC_009464	Erkel *et al.* (2006)
Methanocaldococcus fervens AG86	*Methanococcales*	1485061	1630	32.22	NC_013156	np
Methanocaldococcus infernus ME	*Methanococcales*	1328194	1538	33.53	NC_014122	np
Methanocaldococcus jannaschii DSM 2661	*Methanococcales*	1664970	1765	31.29	NC_000909	Bult *et al.* (1996)
Methanocaldococcus vulcanis M7	*Methanococcales*	1746329	1729	31	NC_013407	np
Methanococcus aeolicus Nankai-3	*Methanococcales*	1569500	1552	30.04	NC_009635	np
Methanococcus maripaludis C5	*Methanococcales*	1780761	1896	32.98	NC_009135	np
Methanococcus maripaludis C6	*Methanococcales*	1744193	1888	33.42	NC_009975	np
Methanococcus maripaludis C7	*Methanococcales*	1772694	1855	33.28	NC_009637	np
Methanococcus maripaludis S2	*Methanococcales*	1661137	1772	33.10	NC_005791	Hendrickson *et al.* (2004)

(continued)

Table 3.1 (*continued*)

Hydrogenotrophic methanogen	Order	Genome size (bp)	Gene count	G + C percentage	NCBI Accession number	Reference
Methanococcus vannielii SB	*Methanococcales*	1720048	1752	31.33	NC_009634	np
Methanococcus voltae A3	*Methanococcales*	1936387	1768	28.56	NC_014222	np
Methanocorpusculum labreanum Z	*Methanomicrobiales*	1804962	1822	50.01	NC_008942	Anderson *et al.* (2009a)
Methanoculleus marisnigri JR1	*Methanomicrobiales*	2478101	2557	62.06	NC_009051	Anderson *et al.* (2009b)
Methanosphaerula palustris E1-9c	*Methanomicrobiales*	2922917	2866	55.35	NC_011832	np
Methanospirillum hungatei JF-1	*Methanomicrobiales*	3544738	3304	45.15	NC_007796	np
Methanopyrus kandleri AV19	*Methanopyrales*	1694969	1729	61.16	NC_003551	Slesarev *et al.* (2002)

np = not published

culturing technique and its later modifications, such as the introduction of the modified Wolin–Miller tube and the development of commercially available anaerobic glove boxes, greatly accelerate the research on anaerobic microorganisms including methanogens (Balch and Wolfe, 1976).

Based on this technique, *M. maripaludis* is routinely cultured in a basal medium (McN) that contains the minimal requirements for this methanogen to grow (Tables 3.2 and 3.3). However, a complex medium can be prepared by adding some nutrients, such as Casamino acids, vitamins, yeast extract, and acetate, which enhance growth. Normally, the preferred source of carbon and energy is H_2/CO_2, which is added as a mixed gas at a pressure of 275 kPa. However, recently *M. maripaludis* has been cultivated in a similar medium but in the presence of formate instead of H_2/CO_2 and with glycylglycine buffer. In this condition, *M. maripaludis* has a lower growth yield but with a similar growth rate (Boguslaw Lupa, personal communication). From a practical and safety perspective, formate presents an improvement to the cultivation method of *M. maripaludis* because it avoids working with high pressure of a flammable gas.

Preparation of 100 ml of McN, McCV, or McF liquid media: In a 500-ml round-bottom flask, mix the components indicated in Table 3.2, stopper loosely, and boil the solution for 5–10 s under a stream of N_2/CO_2 (80:20, v/v). Allow the solution to cool under the stream of N_2/CO_2 and add 0.05 g of L-cysteine hydrochloride hydrate or sodium 2-mercaptoethanosulfonate to reduce the medium. Tightly stopper and transfer the flask into the anaerobic chamber. Dispense the solution into Balch tubes or serum bottles. Seal the tubes with 20 mm blue septum stoppers (Bellco Glass, Inc.) and secure each stopper with an aluminum seal. Remove the tubes from the chamber and exchange gases through a three-cycle procedure of H_2/CO_2 and vacuum. Pressurize each tube to 137 kPa of H_2/CO_2 and sterilize the tubes by autoclaving at 121 °C for 20 min.

For the preparation of McF, a few modifications should be introduced. First, the medium composition is similar to the McN and McCV but NaCl is replaced by sodium formate and glycylglycine is added (Table 3.2). In addition, instead of H_2/CO_2 the gas in the tubes is exchanged with N_2/CO_2 at 35–70 kPa. Since cells growing in the presence of formate do not require high concentrations of H_2/CO_2, the tubes can also be sealed with stoppers that have less resistance to pressure, such as the gray butyl rubber septum stoppers (Bellco Glass, Inc.).

Before inoculation of all media, add anaerobically 0.1 ml of 2.5% $Na_2S \cdot 9H_2O$ to each 5 ml of medium using anaerobic procedures. In the case of McN and McCV, after inoculation, pressurize each tube to 275 kPa. Incubate at 37 °C overnight.

Other representatives of the order *Methanococcales*, such as *Methanococcus voltae* and *Methanococcus vannielii* are cultured following the same procedure using media prepared with the modifications that are indicated in Table 3.2.

Table 3.2 Media components

Composition per 100 ml	Amount
Basal medium for Methanococcus maripaludis *(McN)*	
Glass-distilled water	50 ml
General salt solution	50 ml
K_2HPO_4 (14 g/l)	1 ml
Trace mineral solution	1 ml
Iron stock solution	0.5 ml
Rezasurin (1 g/l)	0.1 ml
NaCl (293 g/l)	7.5 ml
$NaHCO_3$	0.5 g
Complex medium for M. maripaludis *(McCV)*	
McN	100 ml
Sodium acetate·$3H_2O$ (136 g/l)	1 ml
Yeast extract	0.2 g
Casamino acids	0.2 g
Trace vitamin solution	1 ml
Modification of basal medium for M. voltae[a]	
Add sodium acetate·$3H_2O$ (136 g/l)	1 ml
Add isoleucine	0.05 g
Add leucine	0.1 g
Add pantoyl lactone (1 m*M*)	1 ml
Modification of Basal medium for M. vannielii[a]	
Reduce NaCl (293 g/l)	1.8 ml
Modification of basal medium for growth on formate (McF)[a]	
Reduce glass-distilled water	30 ml
Omit NaCl (293 g/l)[b]	0 ml
Add sodium formate (5 *M*)	8 ml
Add sodium glycylglycine (1 *M*, pH 7)	20 ml

[a] All modifications are based in 100 ml volume of the solution.
[b] NaCl is replaced by sodium formate.

Side notes:

- There is always a risk of an explosion when autoclaving media in seal glass tubes and bottles. For this reason, avoid chipped or heavily scratched glassware, autoclave tubes and bottles in protective wire baskets, and wear safety glasses.
- Metabolism of formate yields CO_2 and methane that are accumulated inside the tube. Approximately, 1 mmol of formate generate 25 ml of gas. Therefore, in a 30-ml tube with 5 ml of medium, the gas presence could increase by 100 kPa or 15 psi. Depending upon the pH, some of the CO_2 will be dissolved in the medium. Nevertheless, the glassware used must be able to withstand this increase in pressure without exploding.

Table 3.3 Composition of stock solutions for media

Composition per 1 l	Amount
General salt solution	
KCl	0.67 g
$MgCl_2 \cdot 6H_2O$	5.50 g
$MgSO_4 \cdot 7H_2O$	6.90 g
NH_4Cl	1.00 g
$CaCl_2 \cdot 2H_2O$	0.28 g
Trace mineral solution	
Nitriloacetic acid	1.5 g
$MnSO_4 \cdot H_2O$	0.1 g
$Fe(NH_4)_2(SO_4)_2 \cdot 6H_2O$	0.2 g
$CoCl_2 \cdot 5H_2O$	0.1 g
$ZnSO_4 \cdot 7H_2O$	0.1 g
$CuSO_4 \cdot 5H_2O$	0.01 g
$NiCl_2 \cdot 6H_2O$	0.025 g
Na_2SeO_3	0.2 g
$Na_2MoO_4 \cdot 2H_2O$	0.1 g
$Na_2WO_4 \cdot 2H_2O$	0.1 g
Iron stock solution	
$Fe(NH_4)_2(SO_4)_2 \cdot 6H_2O$	2 g
Concentrated HCl	100 μl
Trace vitamin solution	
Biotin	2 mg
Folic acid	2 mg
Pyridoxine hydrochloride	10 mg
Thiamine hydrochloride	5 mg
Riboflavin	5 mg
Nicotinic acid	5 mg
DL-calcium pantothenate	5 mg
Vitamin B_{12}	0.1 mg
p-Aminobenzoic acid	5 mg
Lipoic acid	5 mg

3.2. Solid media preparation

Different methods have been described to grow methanogens on solid medium. In this chapter, the two most widely used methods will be described: the Petri dish and the serum bottle (Jones *et al.*, 1983; Uffen and Wolfe, 1970).

Preparation of 100 ml of McN or McCV solid media in petri dishes: In a 500-ml round-bottom flask, mix the components indicated in Table 3.2 but reduce the $NaHCO_3$ to 0.2 g and add 1% of agar. Boil the solution under a stream of N_2/CO_2 (80:20, v/v). Heat slowly and mix gently to avoid burning the agar and foaming. Under a stream of N_2/CO_2, allow the solution to cool and add 0.05 g of L-cysteine hydrochloride hydrate or sodium 2-mercaptoethanosulfonate to reduce the medium. Secure the stopper with a suitable clamp, place the bottles in metal safety cages, and autoclave them for 20 min at 121 °C. Allow the medium to cool, and transfer the flask into the anaerobic chamber. Dispense 20 ml of medium in 100 × 15 mm plates, and allow the plates to dry in the chamber atmosphere (20% CO_2, 3–5% H_2, balance N_2) for 2–3 days. One day before use, incubate the plates in a canister or pressure tank under an atmosphere of H_2S (see side notes).

Inside the chamber, transfer 0.5 ml of a cell suspension to each plate, spread it by gently moving the plate in circles or using a sterile bent glass rod. To prevent killing the cells at high dilutions, the glass rods should be immersed in reduced medium before use. Tape the plates together, invert and place them into the pressure tank. Pressurize the tank using H_2/CO_2 (80:20, v/v) to 100 kPa. The H_2/CO_2 gas mixture neutralizes the alkaline sulfide solution, releasing volatile H_2S. Incubate the canister at 37 °C for 5–7 days, maintaining the pressure at 100 kPa by the addition of H_2/CO_2. Finally, release the pressure of the canister in a fume hood and transfer inside the chamber to pick isolated colonies.

Side notes:

- Plastic petri plates are made anaerobic by storage in the anaerobic chamber for at least 1 day before use. To remove air from the sleeves of plates, two corners are cut prior to transferring the plates through the anaerobic air lock.
- To generate an atmosphere of H_2S inside the canister, insert a paper towel inside a Balch tube containing 10 ml of 20% solution of $Na_2S \cdot 9H_2O$ and allow the paper towel absorb the solution. Then place the tube inside the canister in contact with the plates 1 day before inoculation. Allow the tube to remain in the canister for the entire incubation period.
- Before opening the canister inside the anaerobic chamber, the sulfide is removed to avoid poisoning the catalyst. First, release the pressure in the fume hood. Once inside the chamber, flush the canister three cycles with 100 kPa N_2 and partial vacuum of −60 kPa.
- Water vapor in the anaerobic chamber inactivates the catalyst. To remove the water vapor produced while pouring the hot agar medium, a beaker with calcium chloride is kept inside the chamber. It absorbs the water vapor. This desiccant should be replaced regularly and specifically before plating.

Preparation of 100 ml of McN or McCV solid medium by the serum bottle plate method: Prepare McN or McCV medium as described above, but in only

50 ml of distilled water. In a separate flask, prepare a 2% agar solution with 50 ml of distilled water and heat until the agar is completely melted. Combine both solutions, loosely stopper in a round-bottom flask and boil the solution under a stream of N_2/CO_2 (80:20, v/v). Under the stream of N_2/CO_2, allow the solution to cool and add 0.05 g of L-cysteine hydrochloride hydrate or sodium 2-mercaptoethanosulfonate. Before autoclaving, transfer the hot medium to the anaerobic chamber and dispense 20 ml of the solution to 70-ml serum bottles. Seal the serum bottles with 20 mm blue stoppers (Bellco Glass, Inc.) and secure each stopper with an aluminum seal. Remove the bottles from the chamber and exchange gases through a three-cycle procedure of H_2/CO_2 and vacuum. Transfer the bottles to a wire safety basket and sterilize by autoclaving at 121 °C. Agar bottles can be stored inside the anaerobic chamber for later use.

Prior to plating, melt the agar by placing the bottles in boiling water or autoclaving in a 5-min cycle at 121 °C. After they cool to 48–55 °C, add 0.4 ml of 2.5% solution of an $Na_2S \cdot 9H_2O$, mix the components, and allow the bottle to cool on its side to solidify the medium.

Anaerobically transfer 0.5 ml of culture suspension to the bottles, and incubate them on their sides with the agar side up at 37 °C for 3–5 days.

Side notes:

- The addition of 137 kPa of H_2/CO_2 to the medium changes the pH of the solution and prevents the precipitation of some salts, such as phosphate salts.
- The maximum capacity of a 500-ml round-bottom flask is 300 ml of liquid solution. Over this limit can result in the explosion of the bottles during autoclaving.
- For any anaerobic work, all plastic and glass materials (plates, tips, syringes, bottles, and tubes) should be brought into the anaerobic chamber 24 h before use to allow oxygen to diffuse from the plastic.

3.3. Glycerol stock culture preparation

Facile storage of stock cultures is essential for maintaining collections of mutants and other strains useful for genetic studies. The original protocol for long-term storage of methanogens in glycerol stocks was designed for *M. maripaludis* (Whitman *et al.*, 1986) and has been successfully implemented for all the *Methanococcus* species. However, this method has not proven suitable for all other methanogens. Here is described the most updated version of this protocol.

Preparation of 100 ml of 60% glycerol stock solution in McCV media: One day before the preparation of the glycerol stock solutions, take 3-ml serum bottles and 13-mm stoppers inside the anaerobic chamber.

In a 160-ml serum bottle, prepare 60 ml of anaerobic 100% glycerol by sparging it with N_2 for 3–4 h or incubating it in the anaerobic chamber for 2–3 days. In a 500-ml round-bottom flask, prepare 100 ml of McCV medium as indicated above. Add 40 ml of McCV medium to the anaerobic glycerol into the anaerobic chamber and mix them thoroughly in the serum bottle to obtain a final concentration of 60% (vol/vol) glycerol. Aliquot 1 ml of the solution to the 3-ml serum bottles, and seal the bottles using 13-mm stoppers and aluminum seals. Sterilize by autoclaving at 121 °C for 20 min. After cooling, transfer the bottles to the anaerobic chamber until use.

Stock cultures are prepared inside the anaerobic chamber by transferring 1 ml of culture to each 3-ml serum bottle. After thorough mixing, the bottles are stored in the −80 °C freezer. To inoculate a cell culture from a frozen stock, transfer the stock to a cold bath at −20 °C. Scratch the surface of the frozen sample with a syringe needle and then flush the syringe with McCV medium in an anaerobic culture tube.

In the cases of other hydrogenotrophic methanogens, such as representatives of the order *Methanobacteriales*, storage of frozen suspensions requires anoxic and reducing conditions. Hippe (1984) and Miller (1989) describe two different protocols for long-term storage of cultures that have been successfully implemented.

4. Genetic Tools

4.1. Genetic markers

Because of the absence of peptidoglycan in their cell walls and their unique ribosome structure, methanogens are naturally resistant to many of the common antibiotics. Thus, the quest for functional genetic markers for methanogens was a difficult task. Furthermore, some antibiotics that inhibit methanogens base their modus operandi on the toxic effect of their side groups or act as cell wall detergents, such as chloramphenicol and tetracycline, respectively (Beckler *et al.*, 1984; Böck and Kandler, 1985). Thus, common resistance genes are not effective. To solve this problem, Possot *et al.* (1988) tested the resistance of *M. voltae* to 12 different antibiotics and found that puromycin was an ideal candidate for use as a genetic marker. Puromycin transacetylase provides resistance to this antibiotic in many microorganisms, and the structural gene from *Streptomyces alboniger* was cloned under the control of the *M. voltae* methyl CoM reductase promoter, generating the *pac* cassette that confers puromycin resistance to methanogens (Gernhardt *et al.*, 1990). Puromycin is still widely used in different representatives of methanogens because it is efficient, stable, and reliable, usually effective at a final concentration of 2.5 μg/ml in both liquid medium and plates.

The other antibiotic that is widely used in methanogens is neomycin (Böck and Kandler, 1985; Weisberg and Tanner, 1982). Neomycin resistance was first reported in *M. maripaludis* (Argyle *et al.*, 1995) when the aminoglycoside phosphotransferase genes APH3′I and APH3′II were cloned under the control of the *M. voltae* methyl reductase promoter and transformed into *M. maripaludis*. This antibiotic is usually used at a final concentration of 500 μg/ml in plates and 1 mg/ml for liquid medium.

Puromycin and neomycin preparation: Both solutions are prepared following the same protocol at a 200× concentration. Dissolve puromycin or neomycin to a concentration of 0.5 or 100 mg/ml, respectively, with distilled water. Transfer the solution to a serum bottle and sparge with N_2 for 1 h. Inside the anaerobic chamber, the solution is filter-sterilized by passage through a 0.2-μm filter. Aliquots of 10 ml are transferred to sterile Balch tubes, pressurized to 137 kPa using H_2/CO_2, and stored at −20 °C.

4.2. Shuttle vectors

Diverse plasmids of hydrogenotrophic methanogens have been fully described. These plasmids include pME2001 and pME2200 from *Methanobacterium thermoautotrophicum* (Bokranz *et al.*, 1990; Stettler *et al.* 1994), pFVI and pFZI from *Methanobacterium formicicum* (Nolling *et al.*, 1992), and two extrachromosomal elements from *M. jannaschii* (Bult *et al.*, 1996). Tumbula *et al.* (1997) described the complete sequence of a cryptic 8.285-bp low-copy number plasmid from *M. maripaludis* (pURB500) that provided good insights about the minimal replication regions for methanogen plasmids. The plasmid pURB500 contains two large noncoding regions named ORFLESS1 and ORFLESS2, which possess an impressive number of direct and imperfect inverted repeats and potential stem-loop structures.

The first replicating shuttle vector (pDLT44) for methanogens was constructed based on the ligation of the cryptic plasmid pURB500 and an *Escherichia coli* pUC18 plasmid that contained the methanococcal puromycin resistance (*pac*) cassette, named pMEB.2 (Gernhardt *et al.*, 1990). This plasmid replicates in different strains of *M. maripaludis* and is stable for storage (under ampicillin selection) in *E. coli*, but it does not replicate in *M. voltae* (Tumbula *et al.*, 1997). An expression shuttle vector (pWLG30 + lacZ) was then developed for *M. maripaludis* from pDLT44 (Gardner and Whitman, 1999). This plasmid contains *lacZ* under the control of a methanococcal promoter, which can be used for screening in both *E. coli* and methanococci.

The two shuttle expression vectors most frequently used in *M. maripaludis* are plasmids pMEV1 and pMEV2 (Figs. 3.1 and 3.2). They are derivatives of pWLG30 + lacZ plasmid. Both contain most of pURB500, including the origin of replication in *M. maripaludis*, as well as an origin of replication and ampicillin resistance marker for *E. coli*. Furthermore, both

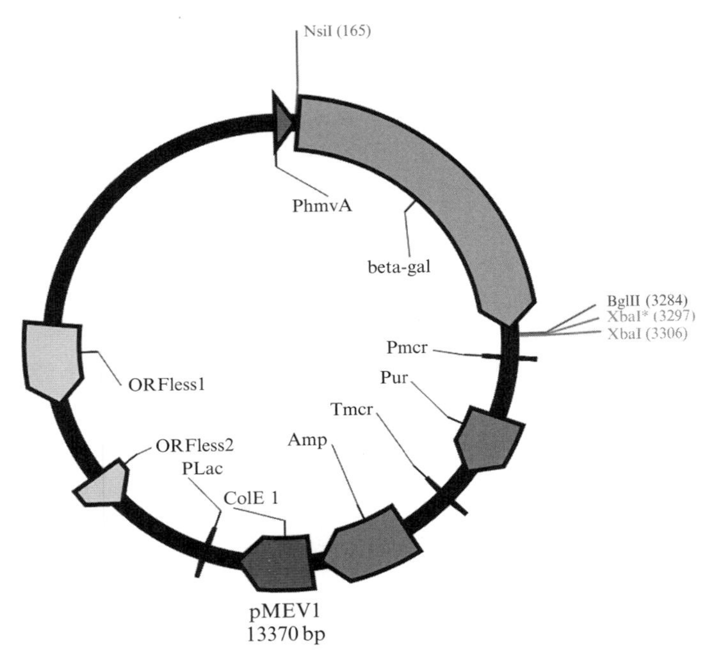

Figure 3.1 *Map of the methanococcal shuttle vector pMEV1.* The vector contains the puromycin resistance marker (Pur) for positive selection. P*hmva*, histone gene promoter from *M. voltae*; beta-gal, *lacZ* gene; P*mcr*, methyl coenzyme M reductase promoter from *M. voltae*; T*mcr*, methyl coenzyme M reductase terminator from *M. voltae*; Amp, ampicillin resistance cassette for *E. coli*; ColE1, origin of replication for *E. coli*; P*lac*, lac promoter; ORFless1 and ORFless2, possible origins of replication for *M. maripaludis*.

plasmids possess the *lacZ* gene under the control of the strong *M. voltae* histone promoter P_{hmvA} (Agha-Amiri and Klein, 1993). To clone and express a gene under this promoter, replace the *lacZ* gene using the upstream restriction site *Nsi*I and the downstream restriction site *Xba*I. Finally, both plasmids only differ on the resistance marker for *M. maripaludis*, pMEV1 carries the *pac* cassette for puromycin resistance, and pMEV2 carries neomycin resistance.

Side notes:

- To clone a gene under the control of the P_{hmvA} promoter, design the 5′ primer, to amplify the target gene, *in-frame* with the *Nsi*I site. This enzyme cuts in the restriction sequence ATGCA′T. The ATG for transcription initiation is the A′T. If the first ATG is used for transcription, the ORF will be too close to the ribosome binding site and will not be expressed.

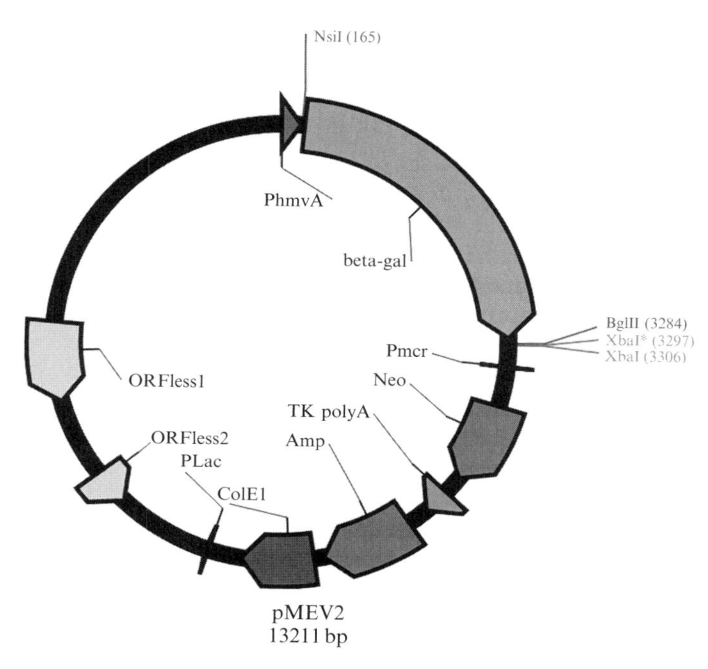

Figure 3.2 *Map of the methanococcal shuttle vector pMEV2.* The vector contains the neomycin resistance marker (Neo) for positive selection. P*hmva*, histone gene promoter from *M. voltae*; beta-gal, *lacZ* gene; P*mcr*, methyl coenzyme M reductase promoter from *M. voltae*; TK polyA, thymidine kinase polyadenylation signal sequence; Amp, ampicillin resistance cassette for *E. coli*; ColE1, origin of replication for *E. coli*; P*lac*, lac promoter; ORFless1 and ORFless2, possible origins of replication for *M. maripaludis*.

- For the 3′ primer, include an *Xba*I site for cloning in the pMEV vectors.
- To insert a PCR product into the pMEV vectors, triple digest the pMEV vector with *Nsi*I, *Bgl*II, and *Xba*I and use the vector without gel purification. The triple digestion with *Bgl*II reduces the cloning background due to the inefficient digestion by *Nsi*I. Thus, *Bgl*II digestion prevents self-ligation of the incomplete digestion product.

4.3. Integration plasmid and gene replacement mutagenesis

The first integration plasmids for methanogens (Mip1 and Mip2) were constructed based on the *E. coli* pUC18 plasmid and pMEB.2, which carries the *pac* cassette. These plasmids replicate in *E. coli* and integrate into the genome of *M. voltae*. In order to allow for homologous recombination into the *M. voltae* genome, the plasmids possessed the *his*A gene from this methanogen (Gernhardt *et al.*, 1990).

The plasmid pIJ03 is used commonly in *M. maripaludis* for single or double recombination into the genome (Stathopoulos *et al.*, 2001). This suicide plasmid lacks an origin of replication for methanococci. The *pac* cassette for selection is flanked by two different multicloning sites to facilitate cloning of PCR products of any DNA sequence (Fig. 3.3). Single homologous recombination of this plasmid leads to the incorporation of the whole plasmid into the genome. On the other hand, double homologous recombination leads to a deletion of the target gene by replacement with the *pac* cassette. Integration of the plasmids is selected by resistance to puromycin.

Several mutants have been constructed using this technique. For example, Stathopoulos *et al.* (2001) used the plasmid pIJ03-*cysS* to disrupt the gene *cysS* that encode cysteinyl-tRNA synthetase, proving that this gene is not essential for viability of *M. maripaludis*. Lin and Whitman (2004) cloned regions immediately upstream and downstream of the genes *porE* and *porF*, respectively, and used the plasmid pIJ03 + CR to replace genes *porE* and *porF* and demonstrate their importance in the anabolic pyruvate oxidoreductase of *M. maripaludis*. Recently, Liu *et al.* (2010) developed

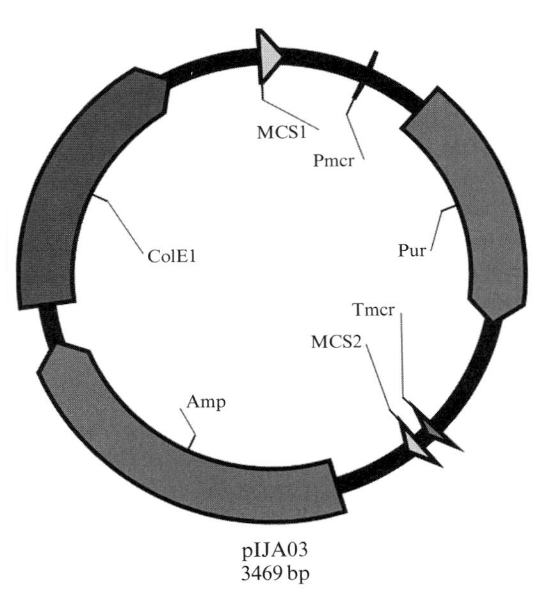

Figure 3.3 *Map of the methanococcal integration vector pIJA03* used for gene replacement. P*mcr*, methyl coenzyme M reductase promoter from *M. voltae*; Pur, puromycin transacetylase gene from *S. alboniger*; T*mcr*, methyl coenzyme M reductase terminator from *M. voltae*; MCS1, multicloning site 1(*Bam*HI, *Eco*RI, *Afl*III, *Mlu*I, *Nde*I, *Afl*II, *Bgl*II, *Xba*I, *Xba*I); MCS2, multicloning site 2 (*Eco*0109I, *Cla*I, *Afe*I, *Bmt*I, *Nhe*I, *Spe*I, *Kpn*I) Amp, ampicillin resistance cassette for *E. coli*; ColE1, origin of replication for *E. coli*.

M. maripaludis mmp1527::pac mutant, a lysine auxotroph, using the plasmid pIJA03-MMP1527. This work demonstrated that methanococci use the diaminopimelate aminotransferase (*dapL*) pathway for lysine biosynthesis.

4.4. Reporter genes

Two fusion reporter genes have been described to monitor gene expression in hydrogenotrophic methanogens, *uidA* and *lacZ*. These reporters can be placed under the control of different promoters to assay their strength. However, it must be remembered that reporter genes differ from the original genes that they are replacing, which could affect their expression levels.

Beneke *et al.* (1995) developed a reporter gene system for *M. voltae* based on *uidA*, which encodes for β-glucuronidase. This reporter gene was placed under the control of the intergenic region of the genes encoding the divergently transcribed selenium-free [NiFe] hydrogenases. The complete expression cassette was introduced into the integration vector Mip1, which carried the *pac* cassette and the gene *hisA* for homologous recombination into the genome. After transformation to *M. voltae*, β-glucuronidase expression became dependent on the depletion of selenium in the growth medium.

β-glucuronidase activity assay:

1. Cells are harvested by centrifugation, and the pellet is resuspended in test buffer (20 m*M* potassium phosphate pH 7, 100 m*M* β-mercaptoethanol), which lyses the cells.
2. Centrifuge the cell lysate for 15 min at 10,000 × g at room temperature and use the supernatant to test enzyme activity.
3. Mix 2 ml of reaction buffer with 1 mM 4-nitrophenyl-β-1,4-glucuronide and initiate the reaction by adding up to 100 μl of cell extract. The reaction is performed at 25 °C and is stopped with 100 μl of 1 *M* sodium carbonate.
4. The reaction product (4-nitrophenol) is detected by its absorption at 405 nm (molar extinction $\varepsilon = 18{,}500\ M^{-1}\ \mathrm{cm}^{-1}$).

Cohen-Kupiec *et al.* (1997) developed a series of promoter-*lacZ* fusions to demonstrate a repressor binding site in the *nifH* promoter that regulates gene expression of the nitrogenase reductase component of the nitrogenase complex. The fusions were constructed using a promoterless *lacZYA* operon, which was cloned after a 1.2-kb DNA fragment containing the *nifH* promoter region from *M. maripaludis*. The resulting cassette was cloned into an integration plasmid, which carried the *pac* cassette for puromycin resistance, and introduced into *M. maripaludis* by transformation. Similar constructs were developed to study the regulatory response of *M. maripaludis* to alanine (Lie and Leigh, 2002). In addition, an X-Gal colony screen was devised and used to identify superrepressor mutants of the transcriptional regulator NrpR (Lie and Leigh, 2007).

β-galactosidase activity assay: is carried out using the original procedure (Miller, 1972)

X-Gal colony screen:

1. After anaerobic growth of the colonies (see Section 3.2), plates are exposed for 30 min to air or until the resazurin in the medium is oxidized and has turned pink in color.
2. Two milliliters of X-Gal (25 mg/ml diluted in 100% dimethylformamide) are mixed with 3 ml of growth media and sprayed over the plates to completely cover the agar surface.
3. After 3–5 s, remove the excess solution and let the plates dry.
4. Color development is apparent after 30–60 min. Viable cells could be recovered from colonies exposed to air for as long as 2 h.

5. Transformation Methods

Over the years, different methods for transformation in methanogens have been tested. The first successfully approaches were developed for *M. voltae*. Bertani and Baresi (1987) discovered a natural transformation method that yielded between 8 and 9 transformants/μg of plasmid DNA. Four years later, the transformation frequency was increased 50- to 80-fold using chromosomal DNA through an anaerobic electroporation method, but this method was still inefficient for plasmid DNA (Micheletti *et al.*, 1991). An important modification in the transformation methods used protoplasts of *M. voltae* (Patel *et al.*, 1993). Protoplasting increased the natural transformation efficiency of plasmid DNA to 705 transformants/μg of DNA and the electroporation-mediated transformation efficiency to 177 transformants per microgram (Patel *et al.*, 1994). While these methods greatly improved the natural transformation efficiency, none of them reached the high number of transformants common with bacteria. However, in 1994, a polyethylene glycol (PEG)-mediated transformation method was developed for *M. maripaludis* (Tumbula *et al.*, 1994). This optimized procedure gave a frequency of transformation of 1.8×10^5 transfomants/μg of plasmid DNA, using 0.8 μg of plasmid and 3×10^9 cells, representing an increment of four orders of magnitude to the natural transformation method. Finally in 2001, a liposome delivery transformation method that was originally developed for *Methanosarcina acetivorans* (Metcalf *et al.*, 1997) was adapted for use in *M. voltae* (Thomas *et al.*, 2001).

In this chapter, the methodology for the natural- and electroporation-mediated transformation of protoplasts of *M. voltae* and the PEG-mediated transformation method of *M. maripaludis* are described. For more details about the liposome delivery transformation method, please refer to the chapter 2.

5.1. Transformation of *M. voltae* by protoplast regeneration

In many Archaea, the S-layer forms an external barrier outside of the cytoplasmic membrane and prevents the uptake of exogenous DNA. In methanococci, the S-layer is a paracrystalline array built by repetitions of a 76-kDa protein (Koval and Jarrell, 1987). The present transformation method is based on the disruption of the S-layer and the formation of protoplasts to allow an easier entrance of DNA into the cell.

Anaerobic protoplasting buffer (APB): Prepare solutions A, B, and C in 50-ml serum bottles, degassing with N_2 gas for 1 h to remove O_2 and autoclave.

- *Solution A*: 0.1 *M* Tris–HCl (pH 7.3) containing resazurin at 1 mg/l.
- *Solution B*: 2 M sucrose in 0.1 M Tris–HCl containing resazurin at 1 mg/l.
- *Solution C*: 1 *M* NaCl in 0.1 *M* Tris–HCl containing resazurin at 1 mg/l.

Under anaerobic conditions, add 4 ml of solution B and 1 ml of solution C to 15 ml of solution A in a 50-ml serum bottle. Inject 0.2 ml of H_2S gas and shake the solution until the resazurin turns colorless. In a fume hood, flush the vial headspace for 5 min with N_2. Leave a slight overpressure and store the solution inside the chamber at room temperature.

Protoplast formation and regeneration on agar medium: Cultivate *M. voltae* in Balch tubes with 5 ml of McN + vitamins solution and the suggested medium modification for this microorganism (Table 3.2) or in BD medium (Patel *et al.*, 1994). Incubate at 35 °C to midexponential growth phase (A_{660} of 0.5–0.8). Pressurize the culture tubes to 137 kPa with H_2/CO_2 and centrifuge them at 4000 rpm for 20 min at room temperature. Gently invert each tube and expel the supernatant using a PrecisionGlide™ Vacutainer needle (Becton, Dickinson and Company) that possesses two needles. Keeping the same Vacutainer needle inserted in the tube, invert the tube again and insert the other needle into a prepressurize tube with 5 ml of APB. The pressure in the APB tube forces the buffer through the needle to the side of the culture tube. Resuspend cell pellets into APB at no more than five times the initial cell concentration. Immediately, pressurize the culture tubes to 137 kPa using H_2/CO_2, centrifuge at 4000 rpm for 20 min at room temperature, and expel the supernatant fraction that possesses the lysed cell content, membranes, and cell wall material. Transfer the tubes inside the chamber, gently resuspend in anaerobic and fresh McCV medium supplemented with 1% of bovine serum albumin and dispense 0.2 ml aliquots of the appropriate dilution onto surface of McCV agar (or BD agar) in plates. Spread the inoculum by gently swirling the plates or using a sterile bent glass rod. Culture the plates in a canister pressurized with H_2/CO_2 (80:20, v/v) to 100 kPa at 30 °C, as described above.

Colonies develop after 7–10 days of incubation.

Side notes:

- Resuspension of whole cells of *M. voltae* into APB results in ~50% lysis, but 99% of the remaining cells are converted into protoplasts.
- Leaving protoplasts in the APB over extended periods of time results in additional cell lysis. Once the protoplast is formed (around 1 min), resuspend them in growth medium supplemented with BSA as soon as possible.
- Protoplasting procedure can be applied to other *Methanococcus* species with different degrees of success, particularly to those species that tend to lyse in hypotonic solutions. For example, *M. maripaludis* has a 75% conversion to protoplasts.
- During the 7–10 days of incubation of the plates, add H_2/CO_2 to maintain the canister pressure at 100 kPa.

Natural transformation of M. voltae *protoplasts method:*

1. Pressurize Balch tubes containing protoplasts in APB (see above) for 10 s with 137 kPa of H_2/CO_2 and centrifuge (4000 rpm) for 20 min at room temperature. Resuspend to $(0.5–1.5) \times 10^9$ protoplasts/ml in McCV medium-BSA.
2. Anaerobically add 15 μg of plasmid DNA in TE buffer to 3.6 ml of the protoplast suspension. To avoid oxygen, it is recommended to transfer DNA inside the chamber and allow it to exchange gases for 2 h prior to mixing with protoplasts.
3. Incubate without shaking for 2 h at 30 °C.
4. Pressurize the tube with 137 kPa of H_2/CO_2, centrifuge (4000 rpm) for 20 min at room temperature, discard the supernatant and resuspend the pellet in 6 ml of McCV medium-BSA.
5. Incubate the cell suspension at 30 °C in a water bath with gentle agitation. Pressurize the tube with H_2/CO_2 to 275 kPa after 2, 4, and 22 h of incubation.
6. After 24 h of incubation, centrifuge for 20 min at room temperature, discard the supernatant, and resuspend the pellet in 0.6 ml of McCV-BSA.
7. Inside the chamber, prepare serial dilutions of the transformed protoplast suspension. Plate in agar McCV supplemented with 5 μg/ml puromycin. For plating and incubation procedures, refer to the protoplast regeneration protocol.

Side notes:

- Prior to and after the 24-h incubation, withdraw 0.1 ml aliquot of protoplasts to determine the cell number by direct counting.
- From 10^9 protoplasts and 15 μg of plasmid DNA, expect to obtain 705 ± 4 transformants/μg of DNA.
- The frequency of spontaneous mutations to puromycin resistance in *M. voltae* is less than 10^{-7} per cell (Possot *et al.* 1988).
- Cells are collected by centrifugation directly in the Balch tubes in a J2-21 centrifuge (Beckman), using a JA-14 rotor and specific adaptors for these tubes. These glass tubes break under high centrifugation. For that reason, it is recommended not to exceed a rotational speed of 4000 rpm.

Anaerobic electroporation buffer (AEB) preparation: The AEB was originally proposed by Micheletti *et al.* (1991). Here, the same buffer is described with one modification, the solution is reduced using H_2S instead of Ti(III) citrate. In a 50-ml serum vial, prepare 10 ml aliquots of AEB (0.1 *M* HEPES (pH 6.5), 0.4 *M* sucrose, 0.05 *M* NaCl, 0.05 *M* KCl, 0.05 *M* $MgCl_2$, and resazurin (1 mg/l)) under N_2 and autoclave for 15 min at 121 °C. Inject 0.5 ml of H_2S gas and shake the contents. When resazurin turns colorless, flush the vial headspace with N_2. *M. voltae* is stable in AEB for at least 15 min.

Transformation by electroporation of M. voltae *protoplasts:*

1. Wash protoplasts once in APB (see above) and resuspend to 400× their initial concentration in AEB under an H_2/CO_2 gas phase.
2. Add 50 μl of the protoplast suspension (2–5 × 10^9 protoplasts) in AEB to 5 μg of plasmid DNA contained in a small 5-ml serum bottle kept under a continuous flow of N_2/CO_2. Add AEB to a final volume of 300 μl and seal the serum bottle under a headspace of N_2/CO_2 (137 kPa). Place the serum bottle and an electroporation cuvette in ice for 10 min.
3. Simultaneously, flush the serum bottle and the electroporation cuvette with N_2/CO_2. Transfer the complete contents of the serum bottle to the electroporation cuvette. Cap the cuvette (under N_2/CO_2 gas phase) and electroporate with a 2.9-ms time constant (400 V, 125 μF, 0.2-cm electrode gap, and 5-Ω resistor in-line).
4. Remove the cap and immediately flush the cuvette with N_2/CO_2.
5. Using a sterile 1-ml glass syringe, add 0.9 ml of McCV-BSA to the cuvette. Add the complete contents of the cuvette to a Balch tube with 9 ml of McCV-BSA and pressurize with 137 kPa H_2/CO_2.

6. Centrifuge at 4000 rpm for 20 min at room temperature and expel the supernatant. Resuspend the pellet in 2 ml of McCV-BSA and pressurize the tube with 137 kPa of H_2/CO_2.
7. Incubate the tube at 30 °C for 24 h and pressurize with H_2/CO_2 to 275 kPa after 2, 4, and 22 h of incubation.
8. Centrifuge the tube at 4000 rpm for 20 min at room temperature and resuspend the pellet in 0.2 ml of McCV-BSA.
9. Dilute with McCV-BSA and plate in McCV-BSA agar in the presence and absence of puromycin as described above.

Side notes:

- Electroporation cuvettes should have a headspace of N_2/CO_2 instead of H_2/CO_2 to avoid an explosion hazard.
- Prior to and after incubation for 24 h, withdraw 0.1 ml aliquot of protoplasts to determine the cell number by direct counting.
- From 10^9 protoplasts and 5 μg of plasmid DNA, expect to obtain 177 ± 74 transformants/μg of DNA.

5.2. Polyethylene glycol (PEG)-mediated transformation of *M. maripaludis*

Preparation of transformation buffer (TB) and TB+PEG: In a 100-ml beaker, mix the components for TB (50 m*M* Tris Base, 0.35 *M* sucrose, 0.38 *M* NaCl, 0.00001% resazurin, and 1 m*M* $MgCl_2$. For TB + PEG, also add 40% (wt/vol) PEG8000). Adjust pH to 7.5 using HCl and transfer to a 50-ml serum bottle.

In a small serum bottle, prepare a 50× cysteine/DTT solution (2.5% cysteine–HCl, 50 m*M* DTT). After adjusting the pH to 7.5 with Tris base, take into the anaerobic chamber.

Exchange gases by sparging with N_2 for 1 or 3 h for TB and TB + PEG, respectively. Transfer the serum bottles inside the chamber and add 1 ml of 50× cysteine/DTT solution to 50 ml TB or TB + PEG. Incubate the solutions unclosed overnight or until the resazurin turns colorless. Filter–sterilize TB (or TB + PEG) using disposable 0.2 μm filters and transfer 5 ml aliquots into sterile Balch tubes. Seal the tubes with a serum stoppers and secure with aluminum seals. Pressurize them for 10 s with 137 kPa of H_2/CO_2 (80:20, vol/vol).

Side notes:

- PEG takes several hours to dissolve. Mixing of the TB + PEG solution overnight is recommended.

- It is recommended to filter–sterilize TB + PEG using a vacuum-driven filtration system. The TB + PEG solution may take several hours to filter.

PEG-mediated transformation of M. maripaludis:

1. In a Balch tube, grow a 5-ml *M. maripaludis* culture in McCV medium to an A_{600} of 0.7–1.0.
2. Pressurize the Balch tube for 10 s with 137 kPa of H_2/CO_2 and centrifuge at 2300 rpm for 10 min at room temperature.
3. Gently invert each tube and discard the supernatant. Using a Vacutainer needle, add 5 ml of TB buffer (see section 5.1).
4. Pressurize the Balch tube with 137 kPa of H_2/CO_2 and centrifuge at 4000 rpm for 20 min at room temperature.
5. Discard the supernatant and add anaerobically 0.375 ml of TB. Resuspend the pellet by swirling.
6. Transfer into the chamber, add 0.8–1.5 μg of plasmid DNA, mix, and take out from the chamber.
7. Flush for 1 min with 100% N_2 and pressurize for 10 s using the same gas.
8. Carefully add 0.225 ml of TB-PEG with a sterile syringe. Allow the TB-PEG to fall directly onto the cell suspension without touching the sides of the tube. Mix thoroughly by swirling.
9. Incubate without shaking at 37 °C for 1 h.
10. Prepare two Balch tubes with 5 ml of McCV medium plus 100 μl of 2.5% $Na_2S \cdot 9H_2O$. Pressurize them with 137 kPa of H_2/CO_2.
11. Using a Vacutainer needle, add 5 ml of McCV medium to the transformants. Resuspend the pellet by swirling.
12. Pressurize the serum tube for 10 s with 137 kPa of H_2/CO_2 and centrifuge at 4000 rpm for 20 min at room temperature.
13. Using a Vacutainer needle, discard the supernatant of the transformation tube and add 5 ml of McCV medium. Resuspend the pellet.
14. Flush the serum tube with H_2/CO_2 and pressurize it at 275 kPa for 10 s. Incubate with shaking at 37 °C overnight.
15. Introduce the sample inside the chamber and make serial dilutions with McCV medium. Plate in McCV agar in the presence and absence of puromycin as described (see section 3.2).

Side notes:

- Check absorbance (600 nm) of the transformation tubes during the procedure to ensure viability.

6. Other Genetic Techniques

6.1. Markerless mutagenesis

Efficient markerless mutation requires a positive and negative selection. Positive selection is used to create a merodiploid from the homologous recombination of the recombinant plasmid into the genome. Negative selection is used to select for the removal of the plasmid by a second homologous recombination event. Normally, both selections are provided in the plasmid constructed for mutagenesis. Moore and Leigh (2005) developed a markerless mutation procedure for *M. maripaludis* with a positive selection for neomycin resistance and negative selection for sensitivity to the base analog 8-azahypoxanthine. This system was based on a markerless mutation system designed for *M. acetivorans* (Pritchett *et al.*, 2004).

In this section, we describe a modification of the method for markerless mutagenesis in *M. maripaludis*, which was developed to be used in strain Mm901. This strain was derived from strain S2 by deletion of the gene *upt*, which encodes for uracil phosphoribosyltransferase and confers 6-azauracil sensitivity. The plasmid pCRUPTNEO carries the positive selection cassette (Neo^R), the gene for negative selection (*upt*) under the control of the promoter P_{hmv}, and a multicloning site for construction of the deletion (Fig. 3.4). The complete process for markerless mutagenesis for the strain Mm901 of *M. maripaludis* is schematically represented in Fig. 3.5. The original method for markerless mutations, which was based in the susceptibility of *M. maripaludis* to the base analog 8-azahypoxanthine (Moore and Leigh, 2005), required the use of the strain Mm900. This strain was derived from strain S2 of *M. maripaludis* by deletion of the gene *hpt*, which encodes for the enzyme hypoxanthine phosphoribosyltransferase. Recently, it was discovered that this mutation has a polar effect on hydrogen metabolism. For this reason, this alternative method was developed.

Markerless mutagénesis for M. maripaludis *method:*

1. Amplify the gene of interest with more than 500 bp of flanking DNA to allow for homologous recombination and clone the resulting fragment in a desired plasmid.
2. Make an *in-frame* deletion of the gene of interest, amplifying by PCR the edges of the gene and the flanking DNA without the internal portion of the gene and clone in the plasmid pCRUPTNEO at any of the two multicloning sites (Fig. 3.4).
3. Transform the plasmid in *M. maripaludis* strain Mm901 and select for neomycin resistance (see Section 5).

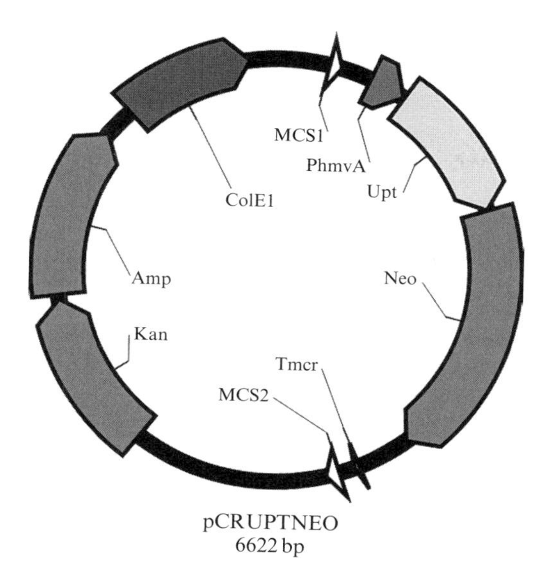

Figure 3.4 *Map of the methanococcal vector pCRUPTNEO* used for markerless mutagenesis. P*hmva*, histone gene promoter from *M. voltae*; Upt, uracil phosphoribosyltransferase gene; Neo, neomycin resistance cassette; T*mcr*, methyl coenzyme M reductase terminator from *M. voltae*; Kan, kanamycin resistance cassette for *E. coli*; Amp, ampicillin resistance cassette for *E. coli*; ColE1, origin of replication for *E. coli*; MCS1, multicloning site 1 (*Hind*III, *Acc*651, *Kpn*I, *Bam*HI, *Avr*II); MCS2, multicloning site 2 (*Apa*I, *Xba*I, *Not*I, *Afl*II).

4. Pick one colony and grow it overnight in 5 ml of McCas medium (see side notes) in the presence of neomycin.
5. From this culture, inoculate 0.05 ml into 5 ml of McCas medium without neomycin.
6. Take aliquots of 0.1 ml and plate in McCas agar with 6-azauracil (0.25 μg/ml).
7. Pick several colonies and inoculate them in McCas medium. Screen by Southern blotting to distinguish mutant colonies from wild type.

Side notes:

- McCas media refer to McCV medium but without yeast extract. Yeast extract contains nucleobases and reduce the sensitivity to base analogs.
- 6-azauracil is added to agar medium before pouring plates from a stock of 10 mg/ml in 0.25 *M* NaOH.

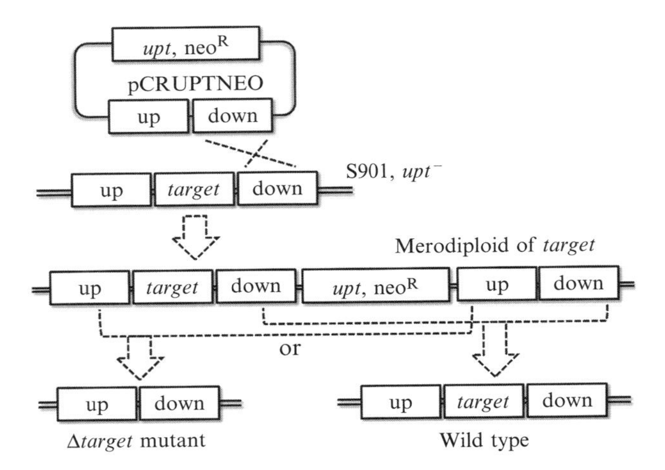

Figure 3.5 *Schematic representation of the markerless mutagenesis technique for M. maripaludis strain Mm901.* The first homologous recombination between the genomic DNA of the strain Mm901 of *M. maripaludis* and the cloned downstream (or upstream) DNA region in the plasmid pCRUPTNEO results in the formation of a merodiploid of the target gene. A second recombination can produce two different outcomes, a mutant strain for the target gene or restoration of the wild-type strain. In the absence of selection, these should occur with equal frequency and are identified by screening with PCR or Southern blotting. *upt,* uracil phosphoribosyltransferase gene; neo^R, neomycin resistance cassette; *target*, gene of interest to be deleted.

6.2. Random mutagenesis with ethylmethanesulfonate

Ethylmethanesulfonate (EMS, $C_3H_8O_3S$) is a volatile organic solvent that is mutagenic and carcinogenic. It produces random mutations in DNA and RNA by nucleotide substitution, specifically by guanine alkylation. This dangerous compound has been successfully used to randomly mutagenize *Methanobacterium ivanovii* (Jain and Zeikus, 1987) and methanococci (Ladapo and Whitman, 1990). The method used in this latter work is described here.

Ethylmethanesulfonate random mutagenesis:

1. Prepare a 0.2 M EMS stock solution in distilled water. Incubate it unstoppered for 24 h in the anaerobic chamber. Inside the chamber, filter–sterilize the solution by passage through a 0.2-μm filter.
2. Grow methanococci in 5 ml of the appropriate medium under 200 kPa H_2/CO_2 (80:20, v/v) at 37 °C (see Section 3.1) until an A_{600} of 0.5 (4–5 × 10^8 cells/ml) is obtained.
3. Mutagenize the cells by adding 0.1 ml of 0.2 M EMS to 5 ml of culture and pressurize it with H_2/CO_2 to 200 kPa. Incubate for 1 h at 37 °C.

4. Centrifuge cells in the culture tube at 1500 g for 20 min at room temperature and gently invert the tube. Discard the supernatant using a Vacutainer needle and resuspend the cell pellet in 5 ml of fresh medium without EMS (see section 5.1).
5. Repeat step 4 two additional times to remove residual EMS.
6. Inoculate 1 ml of the suspension into 5 ml of fresh medium. Incubate the cells under H_2/CO_2 (275 kPa) at 37 °C (A_{600} of 0.5). These cells can then be used for desired selection or screening procedure (see Section 3.2).

Side notes:

- EMS is a mutagen and carcinogen, use extreme caution when handling. It is harmful if swallowed, inhaled, or absorbed through the skin.
- After 1 h of treatment with EMS, the viable cell number is reduced to 25%.
- For effective mutagenesis, a killing curve should be prepared when the sensitivity to EMS is not known.

6.3. Selection for auxotrophic mutants

Certain analogs of purines and pyrimidines are lethal for microorganisms when they are incorporated by growing cells via the salvage pathways. *M. voltae* and *M. maripaludis* are susceptible to base analogs (Bowen and Whitman, 1987; Ladapo and Whitman, 1990). Indeed, a strategy to enrich for auxotrophic mutants based on this characteristic was developed for *M. maripaludis* (see below). In this section, a technique is described for the isolation of auxotrophs mutants using base analogs. This method was first developed for the isolation of acetate auxotrophs after EMS mutagenesis (Ladapo and Whitman, 1990). In the protocol below, McCV is a rich permissive medium and McN medium lacks the nutrient essential for growth of the auxotrophy.

Selection for auxotrophic mutants:

1. Transfer 1 ml of the mutagenized culture to a fresh tube of 5 ml McCV medium, pressurize with H_2/CO_2 (275 kPa) and let it grow until the A_{660} has reached ~0.4 at 37 °C with shaking.
2. Transfer 0.2 ml of culture to a fresh tube of 5-ml McN medium tube that contains 5 mg of each of 6-azauracil and 8-azahypoxanthin. Pressurize the tube with H_2/CO_2 (275 kPa) and incubate for 48 h at 37 °C with shaking.

3. Harvest cells by centrifugation and wash twice with fresh McN medium. Resuspend the pellet in 5-ml McCV medium and incubate to midexponential phase.
4. Repeat steps 2 and 3 twice.
5. Plate the culture on McCV agar plates (see Section 3.2). Replica plate colonies (using sterile toothpicks) onto McN and McCV agar plates. After growth, pick colonies that grow on McCV but not in McN plates with a sterile syringe. Inoculate into a tube with 5 ml of McCV medium. This last step should be done inside the chamber.

Side notes:

- Composition of McN medium varies depending on the auxotroph of interest.
- Base analogs have low solubility in McCV medium. Two analogs are used to prevent selection of spontaneous resistance mutants.

6.4. Transposon approaches in *Methanococcus*

Over the years, there have been two reported trials to establish transposon mutagenesis systems in hydrogenotrophic methanogens. In 1995, an *in vitro* transposon insertion mutagenesis technique was developed to study the *nif* operon of *M. maripaludis*. In that study, a 15.6-kb *nif* region from *M. maripaludis* was cloned in a λ vector and was used as a target for a transposon insertion using a Mudpur transposon that encoded the puromycin transacetylase gene for puromycin resistance. *M. maripaludis* was transformed using different mutagenized versions of the vector DNA, where each was found to replace the wild-type DNA (Blank *et al.*, 1995). This methodology represents the first successful attempt of *in vitro* transposon mutagenesis in *M. maripaludis*.

In 2009, an *in vitro* random transposon system was reported for a rapid knockout of the tryptophan operon in *M. maripaludis*. In that study, a plasmid pKJ331-KAN, bearing a transposon derived from the Tn5 transposable element and the kanamycin and puromycin resistance markers, was used in the mutagenesis of a plasmid that contained the tryptophan operon of *M. maripaludis* in an *in vitro* reaction with the hyperactive Tn5 transposase (*tnp*). The resulting plasmids were transformed into *E. coli* using the kanamycin selection. After selection, enrichment, and purification from *E. coli*, they were transformed into *M. maripaludis* using the puromycin selection (Porat and Whitman, 2009). Using this newly developed method, different tryptophan auxotrophs were obtained, which demonstrated the function of various genes on the biosynthesis of tryptophan.

ACKNOWLEDGMENTS

We thank Boguslaw Lupa for help in preparation of the figures. This work was supported in part by a grant from NIH.

REFERENCES

Agha-Amiri, K., and Klein, A. (1993). Nucleotide sequence of a gene encoding a histone-like protein in the archaeon *Methanococcus voltae*. *Nucleic Acids Res.* **21,** 1491.

Anderson, I., Sieprawska-Lupa, M., Goltsman, E., Lapidus, A., Copeland, A., Del Rio, T. G., Tice, H., Dalin, E., Barry, K., Pitluck, S., Hauser, L., Land, M., *et al.* (2009a). Complete genome sequence of *Methanocorpusculum labreanum* type strain Z. *Stand. Genomic Sci.* **1,** 197–203.

Anderson, I., Ulrich, L. E., Lupa, B., Susanti, D., Porat, I., Hooper, S. D., Lykidis, A., Sieprawska-Lupa, M., Dharmarajan, L., Goltsman, E., Lapidus, A., Saunders, E., *et al.* (2009b). Genomic characterization of methanomicrobiales reveals three classes of methanogens. *PLoS One* **4,** e5797.

Argyle, J. L., Tumbula, D. L., and Leigh, J. A. (1995). Neomycin resistance as a selectable marker in *Methanococcus maripaludis*. *Appl. Environ. Microbiol.* **62,** 4233–4237.

Balch, W. E., and Wolfe, R. S. (1976). New approach to the cultivation of methanogenic bacteria: 2-mercaptoethanesulfonic acid (HS-CoM)-dependent growth of *Methanobacterium ruminantium* in a pressurized atmosphere. *Appl. Environ. Microbiol.* **32,** 781–791.

Beckler, G. S., *et al.* (1984). Chloramphenicol acetyltransferase should not provide methanogens with resistance to chloramphenicol. *Appl. Environ. Microbiol.* **47,** 868–869.

Beneke, S., *et al.* (1995). Use of the *Escherichia coli* uidA gene as a reporter in *Methanococcus voltae* for the analysis of the regulatory function of the intergenic region between the operons encoding selenium-free hydrogenases. *Mol. Gen. Genet.* **248,** 225–228.

Bertani, G., and Baresi, L. (1987). Genetic transformation in the methanogen *Methanococcus voltae* PS. *J. Bacteriol.* **169,** 2730–2738.

Blank, C. E., *et al.* (1995). Genetics in methanogens: Transposon insertion mutagenesis of a *Methanococcus maripaludis* nifH gene. *J. Bacteriol.* **177,** 5773–5777.

Böck, A., and Kandler, O. (1985). Antibiotic sensitivity of archaebacteria. *In* "The Bacteria", Vol. 8, pp. 525–544. Academic Press, New York.

Bokranz, M., *et al.* (1990). Complete nucleotide sequence of plasmid pME2001 of *Methanobacterium thermoautotrophicum* (Marburg). *Nucleic Acids Res.* **18,** 363.

Bowen, T. L., and Whitman, W. B. (1987). Incorporation of exogenous purines and pyrimidines by *Methanococcus voltae* and isolation of analog-resistant mutants. *Appl. Environ. Microbiol.* **53,** 1822–1826.

Bult, C. J., White, O., Olsen, G. J., Zhou, L., Fleischmann, R. D., Sutton, G. G., Blake, J. A., FitzGerald, L. M., Clayton, R. A., Gocayne, J. D., Kerlavage, A. R., Dougherty, B. A., *et al.* (1996). Complete genome sequence of the methanogenic archaeon, *Methanococcus jannaschii*. *Science* **273,** 1058–1073.

Cohen-Kupiec, R., *et al.* (1997). Transcriptional regulation in Archaea: In vivo demonstration of a repressor binding site in a methanogen. *Proc. Natl. Acad. Sci. USA* **94,** 1316–1320.

Erkel, C., Kube, M., Reinhardt, R., and Liesack, W. (2006). Genome of rice cluster I archaea—The key methane producers in the rice rhizosphere. *Science* **313,** 370–372.

Fricke, W. F., Seedorf, H., Henne, A., Kruer, M., Liesegang, H., Hedderich, R., Gottschalk, G., and Thauer, R. K. (2006). The genome sequence of *Methanosphaera stadtmanae* reveals why this human intestinal archaeon is restricted to methanol and H2 for methane formation and ATP synthesis. *J. Bacteriol.* **188,** 642–658.

Gardner, W. L., and Whitman, W. B. (1999). Expression vectors for *Methanococcus maripaludis*: Overexpression of acetohydroxyacid synthase and beta-galactosidase. *Genetics* **152,** 1439–1447.

Gernhardt, P., *et al.* (1990). Construction of an integration vector for use in the archaebacterium *Methanococcus voltae* and expression of a eubacterial resistance gene. *Mol. Gen. Genet.* **221,** 273–279.

Hedderich, R., and Whitman, W. B. (2006). Physiology and biochemistry of the methane-producing archaea. *In* "The Prokaryotes", Vol. 2, pp. 1050–1079. Springer-Verlag, New York.

Hendrickson, E. L., Kaul, R., Zhou, Y., Bovee, D., Chapman, P., Chung, J., Conway de Macario, E., Dodsworth, J. A., Gillett, W., Graham, D. E., Hackett, M., Haydock, A. K., *et al.* (2004). Complete genome sequence of the genetically tractable hydrogenotrophic methanogen *Methanococcus maripaludis*. *J. Bacteriol.* **186,** 6956–6969.

Hippe, H. (1984). Maintenance of methanogenic bacteria. *In* "Maintenance of Microorganisms: A Manual of Laboratory Methods", pp. 69–81. Academic Press, London.

Hungate, R. E. (1950). The anaerobic mesophilic cellulolytic bacteria. *Bacteriol. Rev.* **14,** 1–49.

Jain, M. K., and Zeikus, J. G. (1987). Methods for isolation of auxotrophic mutants of *Methanobacterium ivanovii* and initial characterization of acetate auxotrophs. *Appl. Environ. Microbiol.* **53,** 1387–1390.

Jones, W. J., *et al.* (1983). Growth and plating efficiency of methanococci on agar media. *Appl. Environ. Microbiol.* **46,** 220–226.

Koval, S. F., and Jarrell, K. F. (1987). Ultrastructure and biochemistry of the cell wall of *Methanococcus voltae*. *J. Bacteriol.* **169,** 1298–1306.

Ladapo, J., and Whitman, W. B. (1990). Method for isolation of auxotrophs in the methanogenic archaebacteria: Role of the acetyl-CoA pathway of autotrophic CO2 fixation in *Methanococcus maripaludis*. *Proc. Natl. Acad. Sci. USA* **87,** 5598–5602.

Lie, T. J., and Leigh, J. A. (2002). Regulatory response of *Methanococcus maripaludis* to alanine, an intermediate nitrogen source. *J. Bacteriol.* **184,** 5301–5306.

Lie, T. J., and Leigh, J. A. (2007). Genetic screen for regulatory mutations in *Methanococcus maripaludis* and its use in identification of induction-deficient mutants of the euryarchaeal repressor NrpR. *Appl. Environ. Microbiol.* **73,** 6595–6600.

Lin, W., and Whitman, W. B. (2004). The importance of porE and porF in the anabolic pyruvate oxidoreductase of *Methanococcus maripaludis*. *Arch. Microbiol.* **181,** 68–73.

Liu, Y. (2010). Taxonomy of methanogens. *In* "Handbook of Hydrocarbon and Lipid Microbiology", Vol. 1, pp. 550–558. Sprimger-Verlag, Berlin Heidelberg, New York.

Liu, Y., and Whitman, W. B. (2008). Metabolic, phylogenetic and ecological diversity of the methanogenic archaea. *In* "Incredible Anaerobes", pp. 171–189. New York Academy of Sciences, New York.

Liu, Y., *et al.* (2010). Methanococci use the diaminopimelate aminotransferase (DapL) pathway for lysine biosynthesis. *J. Bacteriol.* **192,** 3304–3310.

Metcalf, W. W., *et al.* (1997). A genetic system for Archaea of the genus *Methanosarcina*: Liposome-mediated transformation and construction of shuttle vectors. *Proc. Natl. Acad. Sci. USA* **94,** 2626–2631.

Micheletti, P. A., *et al.* (1991). Isolation of a coenzyme M-auxotrophic mutant and transformation by electroporation in *Methanococcus voltae*. *J. Bacteriol.* **173,** 3414–3418.

Miller, J. H. (1972). Experiments in molecular genetics. Cold Spring Harbor Laboratory, Cold Spring Harbor, New York.

Miller, T. L. (1989). Methanobrevibacter. *In* "Bergey's Manual of Systematic Bacteriology", Vol. 3, pp. 2178–2183. Williams & Wilkins, Baltimore.

Moore, B. C., and Leigh, J. A. (2005). Markerless mutagenesis in *Methanococcus maripaludis* demonstrates roles for alanine dehydrogenase, alanine racemase, and alanine permease. *J. Bacteriol.* **187,** 972–979.

Nolling, J., *et al.* (1992). Modular organization of related Archaeal plasmids encoding different restriction-modification systems in *Methanobacterium thermoformicicum*. *Nucleic Acids Res.* **20,** 6501–6507.

Patel, G. B., *et al.* (1993). Formation and regeneration of *Methanococcus voltae* protoplasts. *Appl. Environ. Microbiol.* **59,** 27–33.

Patel, G. B., *et al.* (1994). Natural and electroporation-mediated transformation of *Methanococcus voltae* protoplasts. *Appl. Environ. Microbiol.* **60,** 903–907.

Porat, I., and Whitman, W. B. (2009). Tryptophan auxotrophs were obtained by random transposon insertions in the *Methanococcus maripaludis* tryptophan operon. *FEMS Microbiol. Lett.* **297,** 250–254.

Possot, O., *et al.* (1988). Analysis of drug resistance in the archaebacterium *Methanococcus voltae* with respect to potential use in genetic engineering. *Appl. Environ. Microbiol.* **54,** 734–740.

Pritchett, M. A., *et al.* (2004). Development of a markerless genetic exchange method for *Methanosarcina acetivorans* C2A and its use in construction of new genetic tools for methanogenic archaea. *Appl. Environ. Microbiol.* **70,** 1425–1433.

Sakai, S., *et al.* (2008). *Methanocella paludicola* gen. nov., sp. nov., a methane-producing archaeon, the first isolate of the lineage 'Rice Cluster I', and proposal of the new archaeal order Methanocellales ord. nov. *Int. J. Syst. Evol. Microbiol.* **58,** 929–936.

Samuel, B. S., Hansen, E. E., Manchester, J. K., Coutinho, P. M., Henrissat, B., Fulton, R., Latreille, P., Kim, K., Wilson, R. K., and Gordon, J. I. (2007). Genomic and metabolic adaptations of *Methanobrevibacter smithii* to the human gut. *Proc. Natl. Acad. Sci. USA* **104,** 10643–10648.

Slesarev, A. I., Mezhevaya, K. V., Makarova, K. S., Polushin, N. N., Shcherbinina, O. V., Shakhova, V. V., Belova, G. I., Aravind, L., Natale, D. A., Rogozin, I. B., Tatusov, R. L., Wolf, Y. I., *et al.* (2002). The complete genome of hyperthermophile *Methanopyrus kandleri* AV19 and monophyly of archaeal methanogens. *Proc. Natl. Acad. Sci. USA* **99,** 4644–4649.

Smith, D. R., Doucette-Stamm, L. A., Deloughery, C., Lee, H., Dubois, J., Aldredge, T., Bashirzadeh, R., Blakely, D., Cook, R., Gilbert, K., Harrison, D., Hoang, L., *et al.* (1997). Complete genome sequence of *Methanobacterium thermoautotrophicum* deltaH: Functional analysis and comparative genomics. *J. Bacteriol.* **179,** 7135–7155.

Stathopoulos, C., *et al.* (2001). Cysteinyl-tRNA synthetase is not essential for viability of the archaeon *Methanococcus maripaludis*. *Proc. Natl. Acad. Sci. USA* **98,** 14292–14297.

Stettler, R., Pfister, P., and Leisinger, T. (1994). Characterization of a plasmid carried by *Methanobacterium thermoautotrophicum* ZH3, a methanogens closely related to *Methanobacterium thermoautotrophicum* Marburg. *Syst. Appl. Microbiol.* **17,** 484–491.

Thomas, N. A., *et al.* (2001). Insertional inactivation of the flaH gene in the archaeon *Methanococcus voltae* results in non-flagellated cells. *Mol. Genet. Genomics* **265,** 596–603.

Tumbula, D. L., and Whitman, W. B. (1999). Genetics of Methanococcus: Possibilities for functional genomics in Archaea. *Mol. Microbiol.* **33,** 1–7.

Tumbula, D. L., Makula, R. A., and Whitman, W. B. (1994). Transformation of *Methanococcus maripaludis* and identification of a PstI-like restriction system. *FEMS Microbiol. Lett.* **121,** 309–314.

Tumbula, D. L., *et al.* (1997). Characterization of pURB500 from the archaeon *Methanococcus maripaludis* and construction of a shuttle vector. *J. Bacteriol.* **179,** 2976–2986.

Uffen, R. L., and Wolfe, R. S. (1970). Anaerobic growth of purple nonsulfur bacteria under dark conditions. *J. Bacteriol.* **104,** 462–472.

Weisberg, W. G., and Tanner, R. S. (1982). Aminoglycoside sensitivity of archaebacteria. *FEMS Microbiol. Lett.* **14,** 307–310.

Whitman, W. B., *et al.* (1986). Isolation and characterization of 22 mesophilic methanococci. *Syst. Appl. Microbiol.* **7,** 235–240.

CHAPTER FOUR

Molecular Tools for Investigating ANME Community Structure and Function

Steven J. Hallam,[*,†] Antoine P. Pagé,[*] Lea Constan,[*] Young C. Song,[†] Angela D. Norbeck,[‡] Heather Brewer,[§] *and* Ljiljana Pasa-Tolic[§]

Contents

Abstract

Methane production and consumption in anaerobic marine sediments is catalyzed by a series of reversible tetrahydromethanopterin (H_4MPT)-linked C1 transfer reactions. Although many of these reactions are conserved between one-carbon compound utilizing microorganisms, two remain diagnostic for archaeal methane metabolism. These include reactions catalyzed by N5-methyltetrahydromethanopterin: coenzyme M methyltransferase and methyl-coenzyme M reductase (MCR). The latter enzyme is central to C–H bond formation and cleavage underlying methanogenic and reverse methanogenic phenotypes.

[*] Department of Microbiology and Immunology, University of British Columbia, Vancouver, British Columbia, Canada
[†] Graduate Program in Bioinformatics, University of British Columbia, Vancouver, British Columbia, Canada
[‡] Pacific Northwest National Laboratory, Richland, Washington, USA
[§] Environmental and Molecular Sciences Laboratory, Pacific Northwest National Laboratory, Richland, Washington, USA

Methods in Enzymology, Volume 494
ISSN 0076-6879, DOI: 10.1016/B978-0-12-385112-3.00004-4

Here, we describe a set of novel tools for the detection and quantification of H4MPT-linked C1 transfer reactions mediated by uncultivated anaerobic methane-oxidizing archaea (ANME). These tools include polymerase chain reaction primers targeting ANME MCR subunit A subgroups and protein extraction methods from marine sediments compatible with high-resolution mass spectrometry for profiling community structure and functional dynamics.

1. Introduction

In marine sediments, microbial-mediated anaerobic oxidation of methane (AOM) acts as a biological filter limiting greenhouse gas exchange between the ocean and the atmosphere (reviewed in Knittel and Boetius, 2009). Several uncultivated anaerobic methane-oxidizing archaea (ANME) lineages (ANME-1, -2, and -3) affiliated with methanogenic archaea have been described using one or more molecular taxonomic, lipid biomarker or fluorescent *in situ* hybridization (FISH) approaches (Boetius *et al.*, 2000; Hinrichs *et al.*, 1999; Niemann *et al.*, 2006; Orphan *et al.*, 2001, 2002). Incubation studies suggest that ANME-1 and ANME-2 exhibit contrasting methane oxidation rates (Kruger *et al.*, 2008; Nauhaus *et al.*, 2005), and FISH studies have revealed specific physical associations between ANME-2 and bacterial partners affiliated with the Delta- (Boetius *et al.*, 2000; Orphan *et al.*, 2002) and Betaproteobacteria (Pernthaler *et al.*, 2008). Combined observations suggest that ANME partition along environmental gradients giving rise to ecologically differentiated populations adapted to alternative nutrient and energy demands (reviewed in Taupp *et al.*, 2010).

ANME utilize H_4MPT-linked C1 transfer reactions including the terminal step of methanogenesis mediated by methyl-coenzyme M reductase (MCR) consistent with a reversal of one or more canonical methanogenic pathways (Fig. 4.1; Hallam *et al.*, 2003, 2004; Meyerdierks *et al.*, 2005, 2010; Pernthaler *et al.*, 2008). Biochemical studies have purified alternative MCR subunits and cofactors expressed by ANME-1 subgroups (Kruger *et al.*, 2003; Mayr *et al.*, 2008) and provided functional evidence for MCR-mediated reverse methanogenesis (Scheller *et al.*, 2010). Despite an abundance of taxonomic and functional information, environmental and genetic regulation of ANME-encoded C1 transfer reactions remains poorly understood (reviewed in Taupp and Hallam, 2010). Here, we describe novel polymerase chain reaction (PCR) primers targeting ANME-encoded *mcrA* subgroups for use in quantitative PCR (qPCR) assays and sediment protein extraction methods compatible with high-resolution mass spectrometry for expression profiling of H_4MPT-linked C1 transfer reactions.

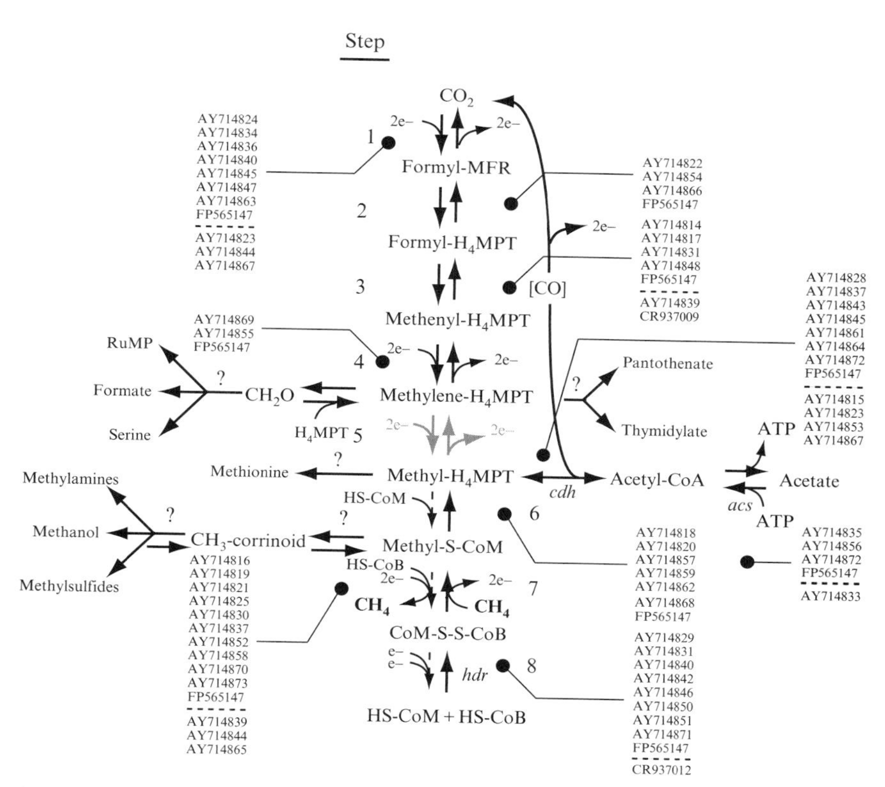

Figure 4.1 A conceptual map of H_4MPT-linked C1 transfer reactions mediating methanogenic and reverse methanogenic phenotypes. For each step, accession numbers indicate large insert clone sequences affiliated with ANME lineages for use in primer design and peptide matching. The *mer* gene encoding step 5 in the pathway has not yet been definitively linked to ANME genotypes and is therefore depicted in gray. The dashed line demarcates ANME-1 (top) and ANME-2 (bottom) encoded genes. A series of five reversible enzymatic steps, (1) formylmethanofuran dehydrogenase (FMD), (2) formylmethanofuran-tetrahydromethanopterin formyltransferase (FTR), (3) *N*5,*N*10-methenyltetrahydromethanopterin cyclohydrolase (MCH), (4) F420-dependent methylenetetrahydromethanopterin dehydrogenase (MTD), or H2-forming *N*5,*N*10-methylene-tetrahydromethanopterin dehydrogenase (HMD), and (5) F420-dependent *N*5, *N*10-methylenetetrahydromethanopterin reductase (MER), respectively, is involved in the transfer of C1 moieties between the carrier molecules methanofuran (MFR) and tetrahydromethanopterin (H_4MPT). Intermediate products have the potential to provide precursors for anabolic processes, including the biosynthesis of methionine and acetate, serine, thymidylate, or pantothenate (reviewed in Maden, 2000), or used as a source of reducing equivalents. In the aceticlastic direction, acetate is converted to acetyl-CoA in one step by ADP-forming acetyl-CoA synthetase (ACS) or in two steps by acetate kinase (ACK) and acetyl-Pi: acetyl phosphate phosphotransacetylase (ACK; not shown), followed by the reversible oxidation of the carbonyl moiety of acetyl-CoA and transfer of the methyl group to H_4MPT by carbon monoxide (CO) dehydrogenase/acetyl-CoA synthetase (CDH), resulting in the formation of methyl-H_4MPT. All reaction series converge on the activities of (6) *N*5-methyltetrahydromethanotperin coenzyme M methyltransferase (MTR), (7) methyl-coenzyme M reductase (MCR), and (8) heterodisulfide reductase (HDR; reviewed in Thauer, 1998).

2. Exploring ANME Population Structure

Current tools for investigating ANME community structure and function are limited in sensitivity, resolution, or throughput. FISH probes targeting the small subunit ribosomal RNA (SSU rRNA) have been used to image and count ANME from diverse environments (Boetius *et al.*, 2000; Knittel *et al.*, 2005; Losekann *et al.*, 2007; Orphan *et al.*, 2002; Pernthaler *et al.*, 2008). However, accurate detection and quantification in marine sediments is confounded by high background fluorescence, aggregate or biofilm formation, and differential SSU rRNA gene expression patterns. Moreover, FISH remains a time- and labor-intensive process reliant on high-end microscopy infrastructure with limited automation. qPCR using 5′endonuclease probe-based chemistry (Taqman) or dye assay chemistry (SYBR® Green) have been successfully adapted for rapid and high-throughput quantification of microbial populations in seawater (Suzuki *et al.*, 2000; Takai and Horikoshi, 2000), soil (Kolb *et al.*, 2003; Stubner 2002), and marine sediments (Stults *et al.*, 2001). Primers targeting ANME-1 and ANME-2c SSU rRNA genes have been used in SYBR® Green-based qPCR assays to monitor ANME growth rates in continuous flow bioreactors (Girguis *et al.*, 2003, 2005), and Taqman probes developed for the detection of genes encoding *mcrA* have been used to investigate the vertical distribution of ANME-1 and ANME-2 in sediments underlying hydrocarbon seeps (Nunoura *et al.*, 2006). The use of *mcr* subunits as functional markers for archaeal methane metabolism provides an exceptional adjunct or alternative to traditional phylogenetic screening with increased subgroup resolution along environmental gradients (reviewed in Friedrich, 2005).

2.1. Sample processing and DNA extraction

Sediment samples vary in pore water content, granularity, and chemical composition, and no single method will reliably provide ultraclean DNA in sufficient quantity and quality to support all molecular applications. Here, we provide a generic template used successfully to recover DNA from Eel River and Santa Barbara Basin sediments.

1. Bulk-frozen sediment is ground to a fine powder in a prechilled mortar under liquid nitrogen (*Note*: Start with 2–5 g and scale up as needed. Homogenate can be transferred to −80 °C for long-term storage).
2. Between 0.5 and 1.0 g of homogenate is processed in triplicate using open source methods (Lee and Hallam, 2009) or a commercial kit such as the PowerSoil® DNA isolation kit (Mobio, Carlsbad, CA) for total community DNA isolation. The resulting extract is purified on a CsCl gradient to remove inhibitory substances, as previously described

(Wright *et al.*, 2009), and concentrated using an Amicon® Ultra filter with 10 K nominal molecular weight limit cutoff (Millipore, Carrigtwohill, Co. Cork). The resulting DNA is diluted to a fixed volume of 500 μl prior to quantification.

3. DNA is quantified with the Quant-iTTM PicoGreen® dsDNA reagent (Molecular Probes, Eugene, OR) in 96- or 384-well microtiter plates on a Varioskan Flash spectral scanning plate reader (Thermo Scientific, Ontario) according to the manufacturer's instructions.

2.2. Methyl-coenzyme M reductase primer design

A total of 2216 nucleotide sequences for cultured methanogen and environmental *mcrA* sequences were downloaded from the NCBI nonredundant database and clustered using DOTUR (Schloss and Handelsman, 2005), resulting in 1672 unique OTUs (100% identity threshold). Deduced amino acid sequences for each OTU were subsequently aligned using the multi-sequence alignment tool MUSCLE (Edgar, 2004a,b), and alignments were imported into MacClade for manual refinement. MCR subunit A gene (not shown) and protein (Fig. 4.2) trees were inferred with PHYML (Guindon and Gascuel, 2003). ANME subgroups were identified based on the original classification scheme described by Hallam *et al.* (2003). Primers were manually selected from the nucleotide alignment for each ANME subgroup, and primer specificity and coverage were determined with blastn queries against the nonredundant database (Table 4.1). While predicted specificity was high for each primer combination, sequence coverage varied between subgroups from different environments. Sequences recovered from hydrocarbon seeps and deep-sea sediments exhibited the highest subgroup coverage, while mud volcanoes and hydrothermal vents exhibited the lowest subgroup coverage (data not shown). Coverage variation in a conserved protein-encoding gene is not unexpected, as new sequences from diverse environments are added to the database, thus reinforcing the need for periodic reevaluation of primer specificity and subgroup coverage.

2.3. Quantifying *mcrA* copy number using dye assay chemistry

Candidate primers are screened against a standard template panel representing ANME *mcrA* subgroups derived from Eel River Basin sediments (Hallam *et al.*, 2003) using end-point PCR to optimize reaction conditions.

1. ANME subgroup-specific *mcrA* copy number is determined by qPCR. Each 25 μl reaction contains 12.5 μl iQ™ SYBR® Green Supermix (Biorad, Hercules, CA), 300 n*M* final concentration of each primer (Table 4.1), 1 μl of template, and 10 μl of sterile, nuclease-free water.

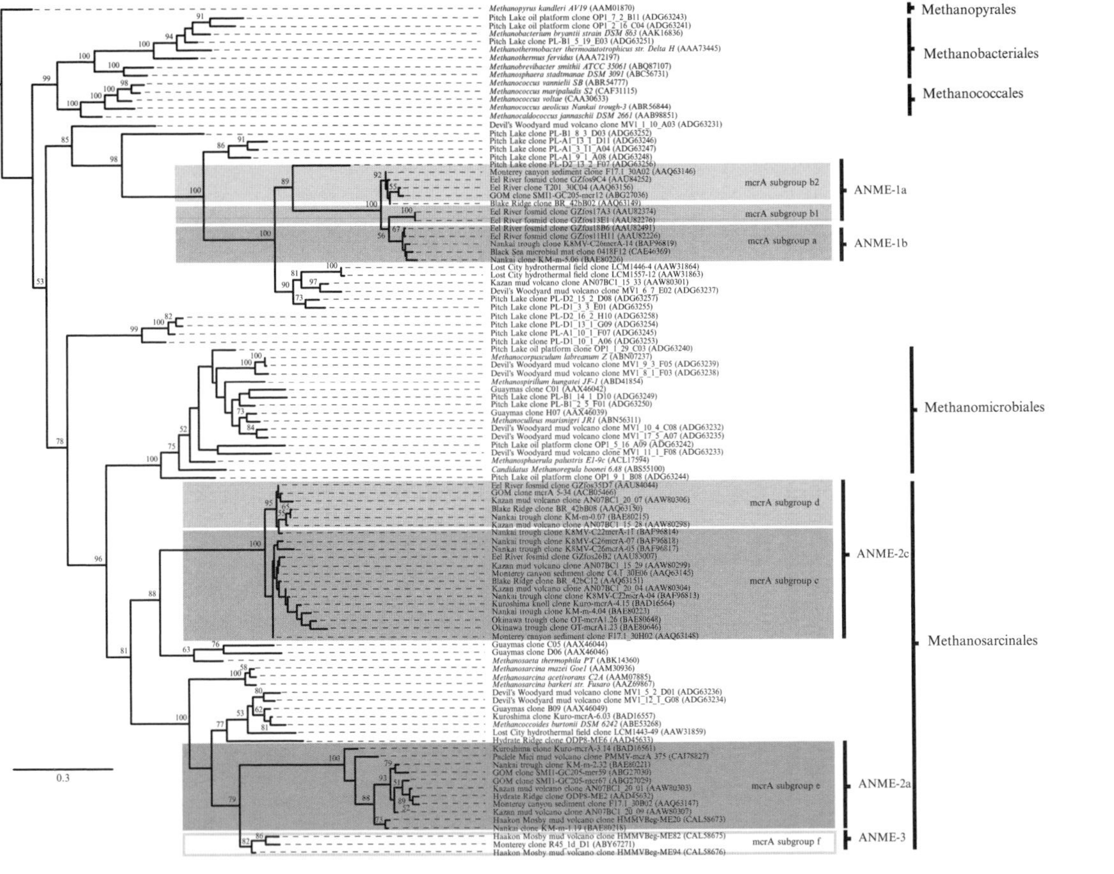

Figure 4.2 Distance tree of inferred environmental McrA amino acid sequences in relation to cultivated methanogenic reference strains. ANME McrA subgroups a–f are indicated in relation to taxonomic assignments derived from SSU rRNA gene trees. The tree was inferred with PHYML (Guindon and Gascuel, 2003) using a JTT model of protein evolution where the Alpha parameter of the Gamma distribution

Table 4.1 ANME *mcrA* subgroup-specific primer sequences and usage information

McrA subgroup	SSU subgroup	Name	Sequence (5′–3′)	Range	Annealing (°C)	Amplicon (bp)	qPCR standard
ab	ANME-1	ANME1abF	GACCAGTTGTGGTTCGGAACGTACATGTC	1021–1203	62	182	ERB T201 30c04
		ANME1abR	CTTGTCGCCCTTTA**KM**GA**HS**CCAT				
a	ANME-1b	ANME1aF	GGGACACTACGAGCGTAAAG	1434–1508	69	74	GZfos11H11
		ANME1aR	TGAATACGTCGTCGCTCTGGTA				
b1	ANME-1a	ANME1b1F	GAGGCGTACCCGACAGTA	1292–1500	67	208	GZfos13E1
		ANME1b1R	CTCTGGTATGAATACGAGCAGTTA				
b2	ANME-1a	ANME1b2F	AAACACTCCGCTGTGGTCAGGC	865–1035	71	170	ERB T201 30c04
		ANME1b2R	CGAACCACAACTGGTCATAGACCATT				
cd	ANME-2	ANME2cdF	CAGATGGCTGAGATGCTTCC	859–925	67	66	MC F17.1 30H02
		ANME2cdR	CATGTCAGCGCAGTAACC				
c	ANME-2c	ANME2cF	GGTTACTGCGCTGACATGGTCC	925–1046	69	121	MC F17.1 30H02
		ANME2cR	GAATCCGACACCACCTGACA				
d	ANME-2c	ANME2dF	CAGATGGCTGAGATGCTTCC	859–941	62	82	BR 42b B08
		ANME2dR	AGCGATTCTCTGTGTCTGA				
e	ANME-2a	ANME2eF	AAGAT**Y**GCTCT**K**GATACCTGT	979–1430	62	451	MC F17.1 30B02
		ANME2eR	GTG**KB**CCTCTT**Y**ATGGAGGTA				
f	ANME-3	Not developed					

R = AG, **Y** = CT, **M** = AC, **K** = GT, W = AT, **S** = CG, **B** = CGT, D = AGT, **H** = ACT, V = ACG, N = ACGT.

Reactions are carried out in low tube strips or 96-well white qPCR plates and run on an Opticon® 2 DNA Engine Real-Time PCR detection system (Biorad) under the following PCR conditions: initial denaturation at 95 °C for 3 min, followed by 45 cycles of 95 °C for 30 s, and extension at 72 °C for 30 s. Annealing temperatures will vary depending on the primer set in use (Table 4.1). (*Note*: 200 μ*M* primer concentrations were chosen to minimize primer dimmer formation in low copy samples. Optimal primer concentrations and annealing temperatures should be empirically determined by running a primer dilution series across a range of annealing temperatures and template concentrations.)

2. Following the sample run, a melting curve from 55 to 95 °C, held at each 0.5 °C increment for 1 s, is performed to check for reaction specificity. Real-time data is analyzed with the MJ Opticon Monitor™ Analysis Software version 3.1 (Biorad).
3. Standards used for ANME subgroup-specific *mcrA* quantification are prepared using the Qiagen® Plasmid Midi kit (Qiagen, Mississauga, ON), followed by a plasmid-safe DNase™ treatment (Epicenter® Biotechnologies, Madison, WI) to remove residual *Escherichia coli* genomic DNA according to the manufacturer's instructions.
4. After enzyme treatment and heat inactivation, the standard solution is extracted with phenol–chloroform–isoamyl alcohol (25:24:1), followed by buffer exchange (3×) with Tris–ethylenediaminetetraacetic (EDTA), pH 8.0, using an Amicon Ultra 4 10K filter (Millipore).
5. The resulting concentrated plasmid DNA is quantified using the Quant-iTTM PicoGreen® dsDNA reagent (Molecular Probes) as described above. A 10-fold dilution series for each standard ranging from 10^2 to 10^8 copies is used in real-time analysis. For the purposes of *mcrA* subgroup primers, the limit of detection was established based on a comparison of dissociation curves and threshold cycle (Ct) number for all sample replicates to be $\sim 2 \times 10^2$ copies/μl. Each assay produces logarithmic amplification curves with 90–100% amplification efficiency over the standard range.
6. To determine specificity, each primer pair is tested with nontarget *mcrA* standards and *Menthanosarcina mazei* genomic DNA at the end point of the standard range (3×10^8 copies). For each ANME *mcrA* subgroup primer set, nontarget amplification is negligible (data not shown).
7. Sample inhibition of SYBR® Green qPCR assays is assessed by "spiking" template DNA with $\sim 10^5$ copies of cognate plasmid standard. A template minus control is run in parallel. The ratio of SYBR® Green quantification for (template + standard)/(standard alone) is calculated to determine inhibitory effects.
8. Following optimization, each primer pair is tested in triplicate on sediment DNA extracts using ~1 μl template DNA per reaction (Fig. 4.3).

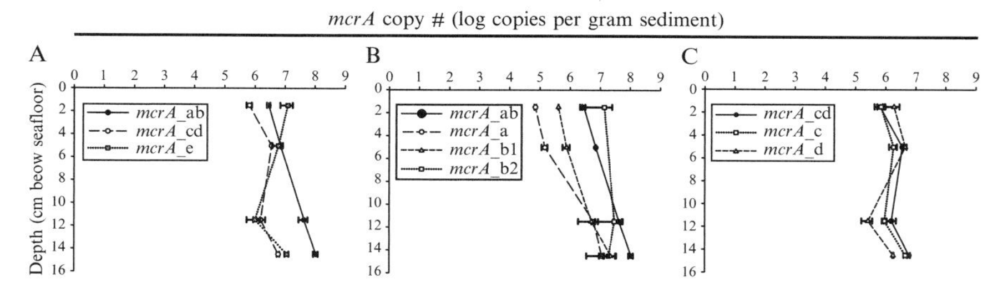

Figure 4.3 Quantification of *mcrA* copy number from AOM sediments using ANME subgroup-specific primers. Consult Fig. 4.1 for subgroup affiliations and Table 4.1 for primer information. (A) Total ANME-1 and ANME-2 *mcrA*; (B) ANME-1 *mcrA* subgroups a, b1, and b2; (C) ANME-2 *mcrA* subgroups c and d. Error bars represent standard error of triplicate DNA extractions from homogenized core sections.

9. Group-specific *mcrA* copy numbers are determined using Ct values from the cognate standards. To calculate *mcrA* copy numbers per gram of sediment, raw numbers are multiplied by the DNA extraction volume, for example, 50 μl, and divided by the amount of sediment from which the DNA was extracted, for example, 1.0 g. To correct for inefficient DNA extraction, *mcrA* copy numbers obtained from extracts producing suboptimal recovery are multiplied by the ratio of DNA yield in replicate divided by highest DNA yield. Average copy numbers per gram of sediment and standard deviations are then calculated from the corrected values.
10. (*Note*: Although developed for quantifying *mcrA* copy number at the DNA level, ANME *mcrA* subgroup primers should be extensible to reverse transcription (RT) qPCR assays to quantify environmental gene expression.)

3. Detecting ANME Proteins in Marine Sediments

While the detection and quantification of ANME *mcrA* subgroups opens a functional genomic window on ANME community structure and dynamics, additional methods for profiling ANME gene expression in the environment are needed for pathway validation. The application of tandem mass spectrometry (MS/MS) to identify expressed protein sequences from AOM sediments offers a rapid and high-throughput profiling solution. Effective peptide matching currently relies on the availability of reference datasets derived from AOM environments (Hallam *et al.*, 2003, 2004; Meyerdierks *et al.*, 2005, 2010; Pernthaler *et al.*, 2008). For targeted analysis of ANME-encoded H4MPT-linked C1 transfer reactions, refer to Fig. 4.1 for pathway reference sequences and Fig. 4.4 for an example of peptide recruitment, coverage, and specificity.

3.1. Sample processing and protein extraction

1. Bulk-frozen sediment is ground to a fine powder in a prechilled mortar under liquid nitrogen (*Note*: Start with 1–5 g and scale up as needed).
2. The resulting homogenate is diluted with three volumes of lysis buffer composed of 50 m*M* Tris–HCl (pH 7.5), 2% sodium deoxycholate w/v, 1 m*M* dithiothreitol (DTT), and 1 m*M* EDTA at 4 °C.
3. The suspension is then vortexed at maximum speed for 10 min at 4 °C, followed by probe sonication at 100 W for 5 min on ice with a 1-min cooling step between each minute pulse.
4. Clarify by centrifugation in a microfuge for 10 min at 4 °C and 13,000×*g* and collect the supernatant. (*Note*: The more dilute the solution, the more difficult it is to precipitate proteins in the concentration step below.)
5. Concentrate the sample by diluting the supernatant four times with absolute ethanol supplemented with a final concentration of 50 m*M* sodium acetate ($NaCH_3COO$), pH 5, and 20 μg glycogen.
6. Precipitate at room temperature for 2 h, followed by centrifugation in a microfuge for 10 min at 13,000×*g*.
7. Decant the supernatant and resuspend the pellet in a minimum of 25 μl 1% sodium deoxycholate in a 50-m*M* ammonium bicarbonate (NH_4HCO_3) buffer. (*Note*: A pellet should be visible although sometimes the proteins are smeared up the side of the tube. Let the tube sit for a few more minutes before removing the last bit of solvent. If the pellet does not dissolve, incubate sample for 5 min at 70–80 °C.)
8. Determine protein concentration with bicinchoninic acid (BCA) or equivalent assay.

3.2. Protein digestion and HPLC separation

1. Heat to denature sample at 99 °C for 5 min.
2. Reduce with 1 μg DTT per 50 μg protein for 30 min at 37 °C.
3. Cool to room temperature. Alkylate with 2.5 μg iodoacetamide (IAA) per 50 μg protein for 30 min at 37 °C in the dark.
4. Digest with 12.5 ng/μl trypsin (sequence grade) at 37 °C for 12 h. (*Note*: Samples may appear discolored due to the presence of residual humic substances.)
5. To reduce residual humic substances in the sample, acidify by diluting 1:1 with a solution containing 3% acetonitrile, 1% trifluoroacetic acid (TFA), and 0.5% acetic acid.

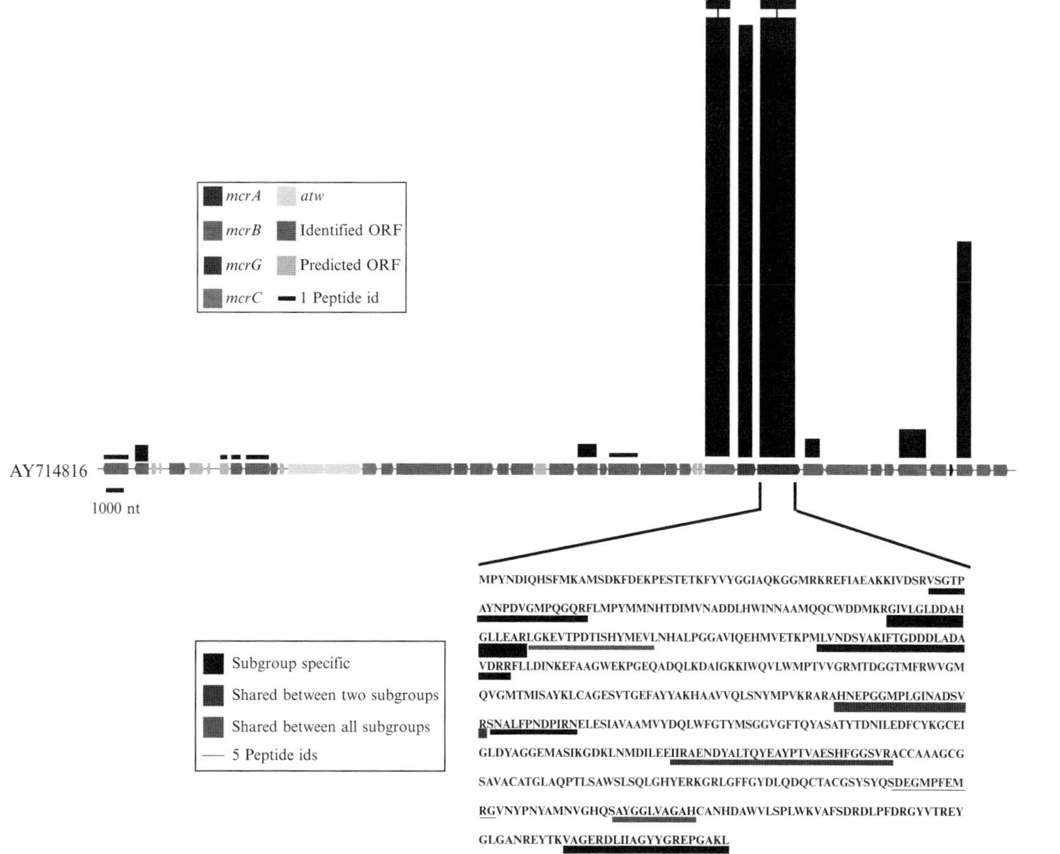

Figure 4.4 Peptide recruitment plot for ANME-1 *mcrA* operon containing large insert clone recovered from AOM sediments (Hallam *et al.*, 2004). Peptide coverage over the length of the clone indicates patterns of correlated gene expression (top). A closer look at peptide coverage over the *mcrA* locus reveals common and unique peptide signatures associated with different ANME *mcrA* subgroups.

6. Clarify by centrifugation in a microfuge for 10 min at 13,000×*g* and collect the supernatant.
7. Peptides are purified and concentrated using OMIX C18 reverse-phase resin (Varian, Palo Alto, CA) according to the manufacturer's specifications. Briefly, columns are activated with 20 μl methanol, followed by rinsing with 0.5% acetic acid. Samples are then passed through the column, washed with 0.5% acetic acid, and eluted with 0.5% acetic acid in 80% acetonitrile. (*Note*: The trypsin used in digestion remains in solution. To remove trypsin, prior to peptide purification, pass the sample through an Amicon® Ultra filter with 10K nominal molecular weight limit cutoff (Millipore) and recover the flow-through.)
9. Concentrate the peptides on a speedvac system and quantify with BCA or equivalent assay. (*Note*: Residual DTT or thiourea will reduce the copper in the protein assay, artificially increasing protein concentration measurements.)
10. Following quantification, the sample is passed through a strong cation exchange (SCX) precolumn prior to loading onto a reverse-phase column for liquid chromatographic (LC) MS/MS analysis.
11. A high-pressure capillary LC (HPLC) system consisting of a pair of Model 100DM 100-ml syringe pumps and a Series D controller (Isco, Inc., Lincoln, NE): a custom stir-bar style mobile-phase mixer (2.5-ml volume); two 4-port, 2-position valves (Valco Instruments Co., Houston, TX) for mobile-phase and capillary column selection; and a 6-port, 2-position Valco valve equipped with a 10 μl sample loop for automated injections are used.
12. The mixer and valves are mounted on a custom rack assembly fitted to a PAL autosampler (Leap Technologies, Carrboro, NC) for unattended routine analysis. Reversed-phase capillary HPLC columns are prepared on demand by slurry packing 5-μm Jupiter phase capillary C18 stationary phase (Phenomenex, Torrence, CA) into a 60-cm length of 360 μm o.d. × 150 μm i.d. fused silica capillary tubing (Polymicro Technologies Inc., Phoenix, AZ) incorporating a 2 μm retaining screen in a 1/16-in. capillary-bore union (Valco Instruments Co.).
13. Samples are then loaded onto the capillary reverse-phase C18 column using the autosampler (~10 μg in a 10 μl volume per sample).
14. The mobile phase consists of 0.2% acetic acid and 0.05% TFA in water (A) and 0.1% TFA in 90% acetonitrile/10%water (B) degassed with an in-line Alltech vacuum degasser (Alltech Associates, Inc., Deerfield, IL).
15. The HPLC system is initially equilibrated at 5000 psi with 100% mobile phase A. The mobile-phase selection valve is then switched from position A to B 20 min after injection, creating an exponential gradient.
16. A ~5 cm length of 360 o.d. × 150 i.d. fused silica tubing packed with 5 μm C18 is used to split ~25 μl/min of flow before the injection

valve. This split flow controls the gradient speed under conditions of constant pressure operation. The flow through the capillary HPLC column is ~1.8 μl/min when equilibrated to 100% mobile phase A.

3.3. Tandem mass spectrometry and peptide identification

A Thermo Electron LTQ linear ion trap (ThermoElectron Corp., San Jose, CA) with electrospray ionization (ESI) is used for the MS/MS acquisition.

1. The capillary LC column is directly infused into the mass spectrometer using a custom interface. No sheath gas or make-up liquid is used. The heated capillary temperature and ESU voltage are 200 °C and 2.2 kV, respectively.
2. Samples are analyzed over a mass to charge (*m/z*) range of 400–2000. The instrument is operated with a duty cycle consisting of one MS event, followed by 10 MS/MS events on the most intense ions using the data-dependent setting and a collision energy setting of 45%. A dynamic exclusion window of −0.5 *m/z* units and +1.5 *m/z* units is used with a residence time of 1 min. Fragmentation occurs on ions within an isolation width of 3 *m/z* units and an activation time of 30 ms within the trap with a repeat count set to 1.
3. Peptides are identified using SEQUEST™ to search mass spectra from the LC–MS/MS run. Each search is performed using an environmental database of predicted protein sequences from the location of interest. In addition, a standard parameter file with no modifications to amino acid residues and a mass error window of 3 *m/z* units for precursor mass and 0 *m/z* units for fragmentation mass is used. Identifications are allowed for all possible peptide termini, that is, not limited by tryptic only termini.
4. Peptide identifications by SEQUEST™ are filtered using a combination of scores provided in the SEQUEST™ output files. Minimum threshold filters include those proposed by Washburn and Yates in 2001 (Washburn *et al.*, 2001). Additional filter thresholds included a minimum discriminant value (Purohit and Rocke, 2003) and a minimum Peptide Prophet Probability (Keller *et al.*, 2002) to reduce the occurrence of false positive identifications.
5. Both discriminant score and peptide prophet probability are based on Xcorr, delCN, and Sp values provided by SEQUEST™ to calculate a single score used as a global confidence measurement for each peptide sequence. Both scores are normalized to a scale of 0–1, with 1 being the most confident identification. This method of determining the score value differs conceptually between the two in that the discriminant score emphasizes separation of right and wrong answers, and Peptide Prophet Probability uses score distributions and random chance calculations to determine the likelihood of match.

6. The number of peptide observations from each protein is used as a rough measure of relative abundance. Multiple charge states of a single peptide are considered as individual observations, as are the same peptides detected in different mass spectral analyses. (*Note*: MCR is highly expressed in AOM sediments, contributing up to 40% of the total identified peptides in some samples (Fig. 4.4).)

ACKNOWLEDGMENTS

We thank the Canadian Foundation for Innovation, the British Columbia Knowledge Development Fund, the National Sciences and Engineering Research Council (NSERC) of Canada, and Genome Canada for supporting ongoing studies on the AOM. A. P. was supported by a fellowship from Le Fonds québécois de la recherche sur la nature et les technologies (FQRNT). L. C. was supported by a fellowship from NSERC. Proteomic methods were performed using EMSL, a national scientific user facility sponsored by the Department of Energy's Office of Biological and Environmental Research and located at Pacific Northwest National Laboratory.

REFERENCES

Boetius, A., Ravenschlag, K., Schubert, C. J., Rickert, D., Widdel, F., Gieseke, A., Amann, R., Jorgensen, B. B., Witte, U., and Pfannkuche, O. (2000). A marine microbial consortium apparently mediating anaerobic oxidation of methane. *Nature* **407,** 623–626.

Edgar, R. C. (2004a). MUSCLE: A multiple sequence alignment method with reduced time and space complexity. *BMC Bioinform.* **5,** 113.

Edgar, R. C. (2004b). MUSCLE: Multiple sequence alignment with high accuracy and high throughput. *Nucleic Acids Res.* **32,** 1792–1797.

Friedrich, M. W. (2005). Methyl-coenzyme M reductase genes: Unique functional markers for methanogenic and anaerobic methane-oxidizing Archaea. *Methods Enzymol.* **397,** 428–442.

Girguis, P. R., Orphan, V. J., Hallam, S. J., and DeLong, E. F. (2003). Growth and methane oxidation rates of anaerobic methanotrophic archaea in a continuous-flow bioreactor. *Appl. Environ. Microbiol.* **69,** 5472–5482.

Girguis, P. R., Cozen, A. E., and DeLong, E. F. (2005). Growth and population dynamics of anaerobic methane-oxidizing archaea and sulfate-reducing bacteria in a continuous-flow bioreactor. *Appl. Environ. Microbiol.* **71,** 3725–3733.

Guindon, S., and Gascuel, O. (2003). A simple, fast, and accurate algorithm to estimate large phylogenies by maximum likelihood. *Syst. Biol.* **52,** 696–704.

Hallam, S. J., Girguis, P. R., Preston, C. M., Richardson, P. M., and DeLong, E. F. (2003). Identification of methyl coenzyme M reductase A (*mcrA*) genes associated with methane-oxidizing archaea. *Appl. Environ. Microbiol.* **69,** 5483–5491.

Hallam, S. J., Putnam, N., Preston, C. M., Detter, J. C., Rokhsar, D., Richardson, P. M., and DeLong, E. F. (2004). Reverse methanogenesis: Testing the hypothesis with environmental genomics. *Science* **305,** 1457–1462.

Hinrichs, K. U., Hayes, J. M., Sylva, S. P., Brewer, P. G., and DeLong, E. F. (1999). Methane-consuming archaebacteria in marine sediments. *Nature* **398,** 802–805.

Keller, A., Nesvizhskii, A. I., Kolker, E., and Aebersold, R. (2002). Empirical statistical model to estimate the accuracy of peptide identifications made by MS/MS and database search. *Anal. Chem.* **74,** 5383–5392.

Knittel, K., and Boetius, A. (2009). Anaerobic oxidation of methane: Progress with an unknown process. *Annu. Rev. Microbiol.* **63,** 311–334.

Knittel, K., Losekann, T., Boetius, A., Kort, R., and Amann, R. (2005). Diversity and distribution of methanotrophic archaea at cold seeps. *Appl. Environ. Microbiol.* **71,** 467–479.

Kolb, S., Knief, C., Stubner, S., and Conrad, R. (2003). Quantitative detection of methanotrophs in soil by novel pmoA-targeted real-time PCR assays. *Appl. Environ. Microbiol.* **69,** 2423–2429.

Kruger, M., Meyerdierks, A., Glockner, F. O., Amann, R., Widdel, F., Kube, M., Reinhardt, R., Kahnt, J., Bocher, R., Thauer, R. K., and Shima, S. (2003). A conspicuous nickel protein in microbial mats that oxidize methane anaerobically. *Nature* **426,** 878–881.

Kruger, M., Blumenberg, M., Kasten, S., Wieland, A., Kanel, L., Klock, J. H., Michaelis, W., and Seifert, R. (2008). A novel, multi-layered methanotrophic microbial mat system growing on the sediment of the Black Sea. *Environ. Microbiol.* **10,** 1934–1947.

Lee, S., and Hallam, S. J. (2009). Extraction of high molecular weight genomic DNA from soils and sediments. *J. Vis. Exp.* **33**10.3791/1569, Pii: 1569.

Losekann, T., Knittel, K., Nadalig, T., Fuchs, B., Niemann, H., Boetius, A., and Amann, R. (2007). Diversity and abundance of aerobic and anaerobic methane oxidizers at the Haakon Mosby Mud Volcano, Barents Sea. *Appl. Environ. Microbiol.* **73,** 3348–3362.

Maden, B. E. (2000). Tetrahydrofolate and tetrahydromethanopterin compared: Functionally distinct carriers in C1 metabolism. *Biochem. J.* **350,** 609–629.

Mayr, S., Latkoczy, C., Kruger, M., Gunther, D., Shima, S., Thauer, R. K., Widdel, F., and Jaun, B. (2008). Structure of an F430 variant from archaea associated with anaerobic oxidation of methane. *J. Am. Chem. Soc.* **130,** 10758–10767.

Meyerdierks, A., Kube, M., Lombardot, T., Knittel, K., Bauer, M., Glockner, F. O., Reinhardt, R., and Amann, R. (2005). Insights into the genomes of archaea mediating the anaerobic oxidation of methane. *Environ. Microbiol.* **7,** 1937–1951.

Meyerdierks, A., Kube, M., Kostadinov, I., Teeling, H., Glockner, F. O., Reinhardt, R., and Amann, R. (2010). Metagenome and mRNA expression analyses of anaerobic methanotrophic archaea of the ANME-1 group. *Environ. Microbiol.* **12,** 422–439.

Nauhaus, K., Treude, T., Boetius, A., and Kruger, M. (2005). Environmental regulation of the anaerobic oxidation of methane: A comparison of ANME-I and ANME-II communities. *Environ. Microbiol.* **7,** 98–106.

Niemann, H., Losekann, T., de Beer, D., Elvert, M., Nadalig, T., Knittel, K., Amann, R., Sauter, E. J., Schluter, M., Klages, M., Foucher, J. P., and Boetius, A. (2006). Novel microbial communities of the Haakon Mosby mud volcano and their role as a methane sink. *Nature* **443,** 854–858.

Nunoura, T., Oida, H., Toki, T., Ashi, J., Takai, K., and Horikoshi, K. (2006). Quantification of mcrA by quantitative fluorescent PCR in sediments from methane seep of the Nankai Trough. *FEMS Microbiol. Ecol.* **57,** 149–157.

Orphan, V. J., Hinrichs, K. U., Ussler, W., III, Paull, C. K., Taylor, L. T., Sylva, S. P., Hayes, J. M., and Delong, E. F. (2001). Comparative analysis of methane-oxidizing archaea and sulfate-reducing bacteria in anoxic marine sediments. *Appl. Environ. Microbiol.* **67,** 1922–1934.

Orphan, V. J., House, C. H., Hinrichs, K. U., McKeegan, K. D., and DeLong, E. F. (2002). Multiple archaeal groups mediate methane oxidation in anoxic cold seep sediments. *Proc. Natl. Acad. Sci. USA* **99,** 7663–7668.

Pernthaler, A., Dekas, A. E., Brown, C. T., Goffredi, S. K., Embaye, T., and Orphan, V. J. (2008). Diverse syntrophic partnerships from deep-sea methane vents revealed by direct cell capture and metagenomics. *Proc. Natl. Acad. Sci. USA* **105,** 7052–7057.

Purohit, P. V., and Rocke, D. M. (2003). Discriminant models for high-throughput proteomics mass spectrometer data. *Proteomics* **3,** 1699–1703.

Scheller, S., Goenrich, M., Boecher, R., Thauer, R. K., and Jaun, B. (2010). The key nickel enzyme of methanogenesis catalyses the anaerobic oxidation of methane. *Nature* **465,** 606–608.

Schloss, P. D., and Handelsman, J. (2005). Introducing DOTUR, a computer program for defining operational taxonomic units and estimating species richness. *Appl. Environ. Microbiol.* **71,** 1501–1506.

Stubner, S. (2002). Enumeration of 16S rDNA of Desulfotomaculum lineage 1 in rice field soil by real-time PCR with SybrGreen detection. *J. Microbiol. Methods* **50,** 155–164.

Stults, J. R., Snoeyenbos-West, O., Methe, B., Lovley, D. R., and Chandler, D. P. (2001). Application of the 5′ fluorogenic exonuclease assay (TaqMan) for quantitative ribosomal DNA and rRNA analysis in sediments. *Appl. Environ. Microbiol.* **67,** 2781–2789.

Suzuki, M. T., Taylor, L. T., and DeLong, E. F. (2000). Quantitative analysis of small-subunit rRNA genes in mixed microbial populations via 5′-nuclease assays. *Appl. Environ. Microbiol.* **66,** 4605–4614.

Takai, K., and Horikoshi, K. (2000). Rapid detection and quantification of members of the archaeal community by quantitative PCR using fluorogenic probes. *Appl. Environ. Microbiol.* **66,** 5066–5072.

Taupp, M., Hallam, S. J., Taupp, M., and Hallam, S. J. (2010a). The meta-methanoxgenome. *In* "Handbook of Hydrocarbon and Lipid Microbiology," (K. Timmis, ed.), Vol. 2, pp. 2232–2244. Springer-Verlag, Berlin.

Taupp, M., Constan, L., and Hallam, S. J. (2010b). The biochemistry of anaerobic methane oxidation. *In* "Handbook of Hydrocarbon and Lipid Microbiology," (K. Timmis, ed.), Vol. 2, pp. 890–907. Springer-Verlag, Berlin.

Thauer, R. K. (1998). Biochemistry of methanogenesis: A tribute to Marjory Stephenson. 1998 Marjory Stephenson Prize Lecture. *Microbiology* **144,** 2377–2406.

Washburn, M. P., Wolters, D., and Yates, J. R., III (2001). Large-scale analysis of the yeast proteome by multidimensional protein identification technology. *Nat. Biotechnol.* **19,** 242–247.

Wright, J. J., Lee, S., Zaikova, E., Walsh, D. A., and Hallam, S. J. (2009). DNA extraction from 0.22 microM Sterivex filters and cesium chloride density gradient centrifugation. *J. Vis. Exp.* **31,** 135210.3791/1352, pii.

CHAPTER FIVE

Studying Gene Regulation in Methanogenic Archaea

Michael Rother, Christian Sattler, *and* Tilmann Stock

Contents

Abstract

Methanogenic archaea are a unique group of strictly anaerobic microorganisms characterized by their ability, and dependence, to convert simple C1 and C2 compounds to methane for growth. The major models for studying the biology of methanogens are members of the *Methanococcus* and *Methanosarcina* species. Recent development of sophisticated tools for molecular analysis and for genetic manipulation allows investigating not only their metabolism but also their cell cycle, and their interaction with the environment in great detail. One

Institut für Molekulare Biowissenschaften, Molekulare Mikrobiologie & Bioenergetik, Johann Wolfgang Goethe-Universität, Frankfurt am Main, Germany

Methods in Enzymology, Volume 494
ISSN 0076-6879, DOI: 10.1016/B978-0-12-385112-3.00005-6

aspect of such analyses is assessment and dissection of methanoarchaeal gene regulation, for which, at present, only a handful of cases have been investigated thoroughly, partly due to the great methodological effort required. However, it becomes more and more evident that many new regulatory paradigms can be unraveled in this unique archaeal group. Here, we report both molecular and physiological/genetic methods to assess gene regulation in *Methanococcus maripaludis* and *Methanosarcina acetivorans*, which should, however, be applicable for other methanogens as well.

1. Introduction

Methanogens are a monophyletic group of anaerobic microorganisms belonging to the domain Archaea. As the name implies they are unique in that their sole means to conserve energy relies on the process of methanogenesis, the biological formation of methane. This process is of profound ecological importance because it is the last step of combustion of organic matter in anaerobic environments lacking abundant electron acceptors such as sulfate or nitrate (Breas *et al.*, 2001; Thauer *et al.*, 2008). Thus, methanogenesis is an important link in the carbon cycle connecting anaerobic with aerobic environments. Studies on methanogenesis have produced a wealth of novel physiological and biochemical knowledge due to the unique nature of methanogenic metabolism. Although the basic biochemistry of methane production has been elucidated, relatively little is known about other aspects of the methanogens, many of which have direct relevance to the process of methanogenesis. One reason for this lack of understanding is the sensitivity of methanogens toward oxygen, which necessitates elaborate methods and equipment for their cultivation and manipulation. Another significant factor was the constraint in experimentation resulting from the limited genetic methods that could be employed. However, this has changed within the last decade (see, e.g., Chapters 2 and 3). The model organisms for genetic analysis developed to date are members of the two genera *Methanococcus* and *Methanosarcina*.

Like all other organisms, methanogens have to react, and adapt, to changing environmental conditions, and, thus, expression of genes and the consequential composition of the protein inventory of methanogens is regulated in response to external stimuli, such as nature and availability of supplements (reviewed in, e.g., Rother, 2010), stress (Spanheimer and Müller, 2008), surfaces or other cells (Bellack *et al.*, 2010; Schopf *et al.*, 2008), and so on, as well as "internal" stimuli (which are always a consequence of external), such as the cell cycle (Walters and Chong, 2009). Thus far, gene regulation in methanogens has been studied mostly at the phenomenological level, that is, by determining (relative) mRNA or protein abundances under different growth conditions or in different strains (wild type vs. mutant). Molecular methods employed include Northern analyses (Eggen *et al.*, 1996; Ehlers *et al.*, 2002; Morgan *et al.*, 1997; Pihl *et al.*, 1994; Sowers *et al.*, 1993b), quantitative reverse transcription PCR (qRT-PCR) analyses (Deppenmeier, 1995; Saum *et al.*, 2009), and transcriptome and/or proteome analyses of *Methanosarcina mazei* Gö1 (Hovey *et al.*, 2005; Veit *et al.*, 2005, 2006), *Methanosarcina acetivorans* (Lessner *et al.*, 2006; Li *et al.*, 2006; Rohlin and Gunsalus, 2010; Rother *et al.*, 2007), and *Methanococcus maripaludis* (Hendrickson *et al.*, 2007, 2008; Xia *et al.*, 2006). Some major findings regarding methanogenesis-related genes were recently summarized (Rother, 2010). In most cases, where regulation of a gene, of an operon, or coordinated regulation of a scattered set of genes (a regulon) is observed in a methanogen, the underlying mechanism of regulation is still unknown (Kennelly, 2007). This lack of knowledge is also illustrated by the fact that promoters of genes from methanogens are usually not well defined. Instead, "promoter regions," simply 1000 bp upstream of the (often putative) translational start site, are often analyzed (see Section 3). A notable-exception is the promoter of *mcrB*, encoding a subunit of methyl-coenzyme M reductase, which was one of the first to be characterized (Allmansberger *et al.*, 1988; Cram *et al.*, 1987; Thomm *et al.*, 1988). The "classical" regulation via two-component systems appears to be the exception rather than the rule, because only very few of these systems are encoded in methanoarchaeal genomes (Ashby, 2006). In cases where effectors of gene expression are known, their mode of action mostly follows the bacterial paradigm, that is, as activators or repressors of transcription, via binding to an operator sequence in the proximity of the transcription start point and either blocking access of RNA polymerase or modulating its activity (Dodsworth and Leigh, 2006; Lie *et al.*, 2005, 2009; Sun and Klein, 2004). Notable exceptions are the regulation of genes encoding CODH/ACS of *Methanosarcina thermophila* (Anderson *et al.*, 2009) and of genes encoding the methyltransferases required for funneling methanol into the central methanogenic pathway of *M. acetivorans* (Bose and Metcalf, 2008; Opulencia *et al.*, 2009).

Recent progress in the phenomenological analysis of gene regulation was made by applying advanced sequencing/probing technology to determine transcriptional start sites (Bensing *et al.*, 1996; Bose and Metcalf, 2008; Jäger *et al.*, 2009) and the development of means to quantify gene expression via analysis of reporter gene fusions. For *Methanosarcina*, the most widely used reporter gene is *uidA*, which encodes β-glucuronidase from *Escherichia coli*. For *M. maripaludis*, the best-developed reporter system is *lacZ*, which encodes β-galactosidase from *E. coli* (Cohen-Kupiec *et al.*, 1997; Wood *et al.*, 2003).

In this chapter, we describe two established procedures to assess regulation of genes of interest in *M. maripaludis* and in *M. acetivorans*, respectively, qRT-PCR and *uidA* fusion analysis. Additionally, we show how to apply the latter to identify groups of genes controlled by one regulatory mechanism (regulon) in *M. acetivorans*. Lastly, a procedure for creating random, stable chromosomal transposon insertions in *M. maripaludis* that can be used to analyze regulatory principles and networks is described.

2. Quantitative Reverse Transcription PCR for *M. maripaludis*

qRT-PCR, also called quantitative real-time RT-PCR, is a powerful tool to determine both relative and absolute amounts of a given mRNA at a certain time point under certain experimental conditions. It is both cost-effective (provided one has access to a qPCR-compatible cycler; see Section 2.4) and comparably simple to employ. By varying the conditions or by using different strains such as a mutant with a known genotype, and by comparing the outcome of the two data sets, an idea can be derived as to whether and to what extent the gene of interest might be regulated. The method described here was successfully applied to quantify abundances of mRNAs encoding both selenocysteine-containing enzymes, as well as their cysteine-containing homologs, involved in the hydrogenotrophic pathway of methanogenesis in various mutant backgrounds of *M. maripaludis* strain JJ (T. Stock and M. Rother, unpublished data; see Stock and Rother, 2009 for further details). It is principally similar to a TaqMan method recently employed to confirm omics data for *M. maripaludis* strain S2 (Xia *et al.*, 2006), but SYBR Green based, which is less sensitive but cheaper.

2.1. *M. maripaludis* culture conditions

M. maripaludis strain JJ (DSMZ 2067; Jones *et al.*, 1983) and its derivatives are grown in McSe medium (Rother *et al.*, 2003), which is based on McN medium with all metal sulfates replaced by the respective chlorides without changing the concentration of the corresponding cations, containing 2 g/l casamino acids, 10 m*M* acetate (added from a 4 *M* anaerobic stock solution; Whitman *et al.*, 1986), and 1 μ*M* sodium selenite (added from a 0.1 m*M* anaerobic stock solution). The medium is prepared as described (Whitman *et al.*, 1986) and 25 ml are dispensed after reduction with cysteine–HCl (ad 0.5 g/l) and Na_2S (ad 0.1 g/l) into 120 ml serum vials. After 1:50 inoculation with a preculture, the vials are pressurized in a sterile fashion with 2×10^5 Pa of H_2:CO_2 (80:20), which serves as the energy source. Cultures may be repressurized after 12 h to expedite reaching the final cell density. Alternatively, for formate-dependent growth 2%, (w/v) sodium formate (from a 50% (w/v) anaerobic stock solution) and 80 m*M* morpholinepropanesulfonic acid, pH 6.8 (to keep the pH constant, from a 2 *M* anaerobic stock solution), are added and 0.5×10^5 Pa of N_2:CO_2 (80:20) is applied. The cultures are incubated at 35–37 °C with gentle agitation (e.g., on a rotary shaker at 40 rpm) to avoid sedimentation of the cells and to facilitate gas/liquid phase mass transfer. Growth of the cultures can be monitored by following their optical density (OD) at wavelengths from 578 to 600 nm. Adding a small amount of sodium dithionite prior to the measurement reduces the resazurin oxidized from withdrawal of the sample.

2.2. Cell harvest and mRNA isolation

Cultures are harvested by centrifugation for 10 min at $5000 \times g$ and the cells are washed in sucrose buffer (800 m*M* sucrose, 50 m*M* $NaHCO_3$, pH 7.4). RNase-free disposable plasticware should be used as well as glassware oven baked at 240 °C for at least 4 h to prevent RNA degradation. Furthermore, all solutions should be made with RNase-free compounds and diethyl pyrocarbonate (Sigma, Steinheim, Germany)-treated water. RNA is isolated from the cells using the High Pure RNA isolation Kit (Roche, Mannheim, Germany) and following the manufacturer's instructions but leaving out the on-column DNase digestion step. Instead, the eluted RNA is treated three times (for 2 h each) with RQ1 RNase-free DNase (Promega, Mannheim, Germany) according to the manufacturer's recommendation, to remove contaminating DNA. Subsequently, the RNA preparation is again purified using the Nucleospin RNA Clean-up Kit (Macherey-Nagel, Düren) to remove digested DNA and DNase. To confirm absence of DNA, the RNA preparation is used as template in a conventional PCR. If no DNA appears to be amplified, judged by conventional agarose gel electrophoresis and staining with ethidium bromide, qPCR (see below) using the same constituents is performed to increase the sensitivity to detect contaminating DNA. If no DNA is present, the RNA preparation is used or can be stored at −80 °C for later use.

2.3. Reverse transcription (cDNA synthesis)

For the synthesis of a specific cDNA from one particular mRNA, one gene-specific oligonucleotide is derived (Fig. 5.1). We found this approach to be more successful in this system than using statistical (hexa)nucleotides for random priming (Xia *et al.*, 2006), particularly if one wants to determine abundances of highly homologous genes, such as *fdhA1* and *fdhA2* (Wood *et al.*, 2003). Our criteria in deriving the primers for cDNA synthesis are primer length (±20 nt), a melting temperature of >55 °C (calculated at, e.g., www.cnr.berkeley.edu/~zimmer/oligoTMcalc.html), and to differ at least in the four 3′ bases from any homologous mRNA not to be amplified with this primer. A 50 μl reaction mixture contains 10 μl 5× Moloney

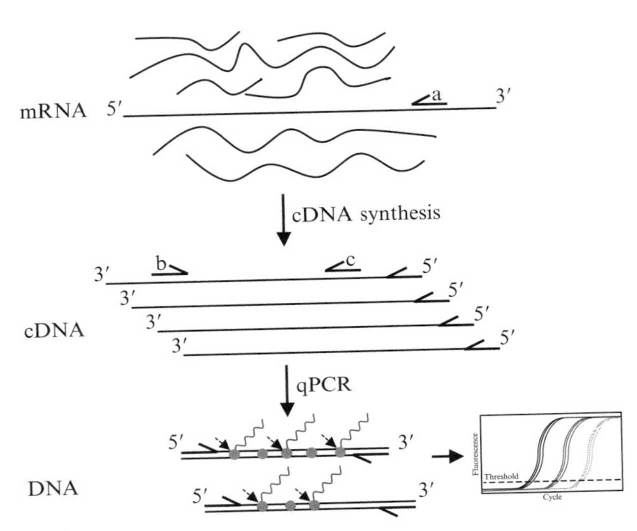

Figure 5.1 Schematic overview of qRT-PCR of *M. maripaludis*. From the RNA pool, one mRNA is specifically reverse transcribed (cDNA synthesis) with a primer a. From the cDNA, the second strand is synthesized and the double strand amplified in the qPCR reaction using primers b + c. Incorporation of SYBR Green (green spheres) allows the amplification to be followed in real time via the excitation/emission fluorescence (red wavy line); the box on the bottom right represents computational integration and presentation of the data. (For interpretation of the references to color in this figure legend, the reader is referred to the Web version of this chapter.)

murine leukemia virus (M-MLV) reaction buffer, 2 μg RNA, 10 pmol oligonucleotide (we reverse transcribe as many as 12 mRNAs per reaction), 0.5 m*M* dNTPs, and RNase-free water ad 48 μl. The sample is incubated at 70 °C for 5 min to reduce the amount of double-stranded RNA and secondary structures. To allow annealing of the oligonucleotide(s), the sample is then incubated at 55 °C for additional 5 min, before 400 U (2 μl) of M-MLV reverse transcriptase RNase H Minus, point mutant (Promega) is added to start cDNA synthesis. The reaction is incubated at 55 °C for 1 h before the enzyme is inactivated by incubation at 80 °C for 5 min. The cDNA can be directly used for quantitative PCR (qPCR, see below) or stored at −80 °C for later use.

2.4. Quantitative PCR and data analysis

For every mRNA to be analyzed, two additional oligonucleotides are derived, both annealing within the cDNA for second-strand synthesis and amplification (Fig. 5.1). This measure further increases the overall specificity of the analysis by reducing the amount of DNA amplified via unspecific priming. The cDNA (see Section 2.3) is diluted 10-fold before subjected to the qPCR analysis. For this purpose, Absolute QPCR SYBR Green Mastermix (Thermo Scientific, Dreieich, Germany) is used in a 25 μl reaction containing 1 μl of the diluted cDNA, 2 pmol of each primer, 12.5 μl of SYBR Green mastermix, and DNase-free water. The qPCR is run in a qPCR-compatible cycler (e.g., Rotor-Gene RG-3000 qPCR cycler, Corbett Research, Cambridge, UK) employing a "hot start" protocol to activate the thermostable DNA polymerase contained in the mastermix. During the following "conventional" PCR run, incorporation of the fluorescent dye into double-stranded DNA is monitored and the data are integrated and presented by suitable software (e.g., Rotor-Gene 6, Corbett Research). Interpretation and plotting of the data can be done by several methods (Pfaffl, 2001); we use the $2^{-\Delta\Delta C_T}$ method (Livak and Schmittgen, 2001). For normalization of the analytical data, reference data from a gene, which is constitutively expressed and not regulated under the conditions tested, are required. Ribosomal RNA, for example, 16S rRNA can serve this purpose (Xia *et al.*, 2006). We use mRNA from the *mcrB* gene encoding the β-subunit of methyl-coenzyme M reductase, which is not regulated in *M. maripaludis* S2 (Xia *et al.*, 2006).

3. Analyzing Gene Expression Using *uidA* Reporter Gene Fusions in *M. acetivorans*

A major leap forward in the functional analysis of expression signals in methanogens came when *uidA*, the *E. coli* gene for β-glucuronidase, was functionalized as an *in vivo* reporter for *Methanococcus voltae* in the group of Klein (Beneke *et al.*, 1995). The system was adapted to the developing genetic system for *Methanosarcina* species (Pritchett *et al.*, 2004) and is, today, a very important standard tool. One major reason for this is the high degree of genetic redundancy in this genus, with sequence identities of greater 70% for paralogous genes (Deppenmeier *et al.*, 2002; Galagan *et al.*, 2002; Maeder *et al.*, 2006), which makes qRT-PCR analysis a moderately tedious and Northern blot analysis an extremely tedious task. Second, once a promoter–*uidA* fusion construct has been inserted into the genome, many different conditions can be investigated without spending much more time and effort. Third, quantification of *uidA* gene product activity (GUS) is very easy using a commercially available chromogenic substrate. The method described here is the standard procedure for *M. acetivorans* first described by Pritchett *et al.* (2004) for stable insertion of a *uidA* fusion construct into the *hpt* locus (encoding hypoxanthine phosphoribosyl transferase, MA0717; Galagan *et al.*, 2002) of the organism. The *hpt* locus, which is also used as a negative selectable marker in the markerless deletion method (Pritchett *et al.*, 2004;

see Chapter 2), is permissive for insertion of foreign DNA. Mutants with deletions of *hpt*, or insertions into *hpt*, have no discernable phenotype under any conditions so far examined. After insertion of a nonreplicating plasmid carrying the reporter construct, partial diploids are selected as puromycin resistant via expression of the *pac* gene (see Section 3.1). Subsequently, clones that have resolved the partial diploid state by deletion of *hpt* (leaving the reporter construct on the chromosome) can be selected because a defective *hpt* gene confers resistance to the toxic base 8-aza-2,6-diamino-purine (8-ADP). However, other permissive chromosomal loci can also be used for integration: the *ssuBCA* locus (MA0063–MA0065), which encodes a putative sulfonate transporter, is useful as a permissive site because its loss can be selected on the basis of resistance to the methanogenesis-inhibitor 2-bromoethane sulfonic acid (Guss *et al.*, 2005). We also use MA2965, annotated as encoding a transposase (although no inverted repeats indicating the presence of a corresponding transposon can be found in its vicinity), which proved useful for integrating foreign DNA (C. Sattler and M. Rother, unpublished data).

3.1. Construction of the reporter plasmid

The starting plasmid is pMR51 (Fig. 5.2), a derivative of pMP42 (Pritchett *et al.*, 2004). It contains an origin of replication (*ori*R6K) and a selectable marker (*bla*, coding for β-lactamase which confers resistance to ampicillin) for *E. coli*, a selectable marker for *Methanosarcina* (*pac*, coding for puromycin acetyl transferase which confers resistance to puromycin), the promoter region of the *mtaCB1* operon (coding for one isoform of the methanol-specific methyltransferase MT1, Pritchett and Metcalf, 2005) from *M. acetivorans* (which will be replaced with the DNA fragment to be analyzed), *uidA*, and two 700–800 bp fragments corresponding to flanking regions of *hpt*. For isolation and manipulation of plasmid DNA from *E. coli*, standard methods are used throughout (Ausubel *et al.*, 1997; Sambrook *et al.*, 1989). As promoters are not well defined in *Methanosarcina* species, amplify by PCR approximately 1 kb of the DNA region upstream of the deduced translational start site of the gene, the expression of which is to be analyzed, thereby introducing suitable restriction sites for cloning into pMR51. At the 5′end, *Bgl*II, *Nsi*I, or *Nhe*I are convenient options, while the 3′end needs to be a *Nde*I site overlapping with translational start codon in order to create superimposed start codons at the *catATG*. Thus, a fusion is created in which the "promoter" to be analyzed controls transcription and translation initiation of *uidA*. The DNA fragment is ligated with pMR51 (restricted in the appropriate fashion), transformed into *E. coli* (note that the R6K origin requires the host strain to be *pir*$^+$; Filutowicz *et al.*, 1986) and plated onto LB agar (Bertani, 1951). Plasmid DNA from ampicillin-resistant transformants is isolated, subjected to appropriate restriction analysis, and the region obtained via PCR is sequenced to rule out base errors.

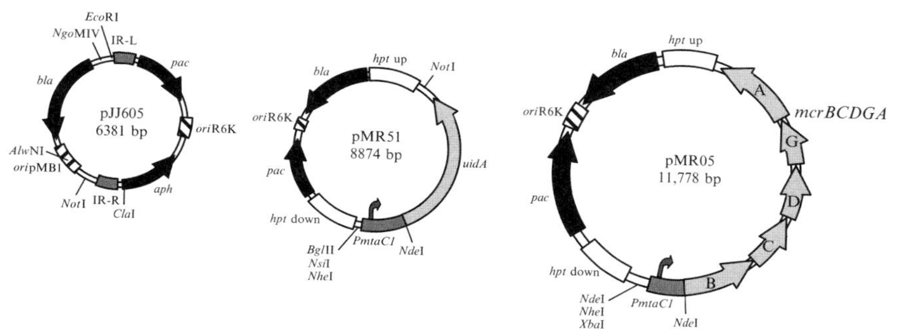

Figure 5.2 Plasmids for genetic manipulation of methanogenic archaea. The numbers indicate the sizes of the plasmids in base pairs (bp); black arrows indicate genes for factors conferring antibiotic resistance; *bla*: gene for β-lactamase conferring resistance to ampicillin; *pac*: gene for puromycin acetyltransferase conferring resistance to puromycin; *aph*: gene for aminoglycoside phosphotransferase conferring resistance to kanamycin; striped boxes represent replication origins (*ori*); light gray arrows indicate construct-specific structural genes; *uidA*: gene for β-glucuronidase; *mcrBCDGA*: genes encoding methyl-coenzyme M reductase from *M. barkeri* Fusaro; solid dark gray boxes indicate otherwise functional DNA; IR-L and IR-R: inverted repeats flanking the modified transposon; *PmtaC1*: *mtaC1* promoter region from *M. acetivorans*, transcription is indicated by the bent arrow; white boxes indicate homologous regions up- and downstream of *hpt*, encoding hypoxanthine phosphoribosyl transferase, the absence of which confers resistance to 8-ADP; the elements are not drawn to scale; see text for details.

3.2. Creation of the *M. acetivorans* reporter strain and determining GUS activity

M. acetivorans strain C2A (DSM 2834; Sowers *et al.*, 1984) is grown in HS medium as described (Sowers *et al.*, 1993a) with modifications (Boccazzi *et al.*, 2000). The most convenient (and least malodorant) growth substrate is methanol added to a final concentration of 60–150 m*M* from an anaerobic 25 *M* stock solution. Medium can be solidified with 1.5% (w/v) Bacto agar (Difco, Augsburg, Germany; the source of the agar has significant impact on plating efficiency). For selection of *Methanosarcina* strains carrying *pac*, puromycin (CalBiochem, San Diego, CA) is added from sterile, anaerobic stocks at a final concentration of 2 μg/ml. 8-ADP (Sigma) is added from a sterile, 2 mg/ml anaerobic solution to a final concentration of 20 μg/ml for selection against *hpt*. Growth of *M. acetivorans* can be monitored spectroscopically by following the OD at 578–600 nm. Transformation of *M. acetivorans* with the respective plasmid(s) (pMR50 should be included in the analysis as a positive control. It contains a fusion of *uidA* with *PmcrB*, the promoter of *mcrB* from *M. acetivorans* constitutively and highly expressed and resulting in specific GUS activities of ca. 1000 mU/mg, Longstaff *et al.*, 2007) can be achieved by either a liposome-mediated (Metcalf *et al.*, 1997) or a polyethylene glycol-mediated (Oelgeschläger and Rother, 2009) transformation method. While the former requires less hands-on and results in somewhat higher transformation frequencies, the latter is considerably cheaper. Cells are plated onto HS agar containing puromycin and incubated in anaerobic jars with a N_2:CO_2:H_2S (79.9:20:0.1) atmosphere at 37 °C for ca. 10 days. Colonies are streak purified, and then grown in liquid HS media without selection for 1–5 transfers. Colonies arising from subsequent plating on medium containing 8-ADP will carry the reporter fusion while having lost the plasmid backbone.

GUS activity can now be readily determined in whole cell extracts prepared aerobically by pelleting mid-log-phase cultures (OD_{600} 0.3–0.5) by centrifugation (5000×*g* for 10 min), osmotic lysis of the cell pellet by resuspension in 50 m*M* Tris–HCl (pH 8.0) containing 0.1 μg/ml RNaseA and DNase I, respectively, and removal of cell debris by centrifugation (14,000×*g* for 15 min). Extracts should be held on ice until assayed, and it should be noted that specific GUS activity, either in whole cells stored at − 20 or − 80 °C, or in extracts, decays quite rapidly. Assays are performed at room temperature and consist of 0.85 ml 50 m*M* Tris–HCl (pH 8.0) and 0.05 ml cleared extract (diluted if necessary). The reaction is started by adding 0.1 ml 40 m*M* *p*-nitrophenyl-β-D-glucuronide (pNP-Gluc, Fluka, Steinheim, Germany), dissolved in 50 m*M* Tris–HCl, pH 8.0. Assays are continuously monitored at 415 nm to derive a reaction rate (due to hydrolysis of the chromogenic substrate absorbance at 415 nm increases). Protein concentration in the extracts can be determined by the Bradford assay (Bradford, 1976). The specific GUS activity $\mu mol \times min^{-1} \times mg\ protein^{-1}$ is calculated by applying the Lambert–Beer law with the molar extinction coefficient for pNP-Gluc ($12{,}4021 \times mol^{-1} \times cm^{-1}$), which was experimentally determined in the buffer conditions used in this assay (Pritchett *et al.*, 2004).

3.3. Isolation of mutants with defects in regulatory pathways of *M. acetivorans*

Having created a methanoarchaeal strain, which expresses a reporter in a clearly discernable fashion depending on different conditions, for example, high reporter activity on one growth substrate and very low activity on another sets the stage for investigating underlying mechanisms of this regulation. Random mutagenesis (see below) can be conducted if nothing is known about said mechanism, and mutants may be isolated with altered regulatory pattern, like expression of the reporter under nonexpressing conditions and vice versa. Color screening of mutant libraries with plates containing the substrate analog 5-bromo-4-chloro-3-indolyl-β-D-galactopyranoside (X-Gal) for β-galactosidase, or 5-bromo-4-chloro-3-indolyl glucuronide (X-Gluc) for β-glucuronidase, employed for decades in bacteria, is problematic in methanogens because formation of the respective chromogenic cleavage product (indigo dye) requires the presence of oxygen. However, this obstacle can be overcome, by either replica patching and sacrificing the replicate plate in air (Apolinario *et al.*, 2005; Pritchett *et al.*, 2004) or short-term exposure of plates to air, followed by color development, and salvaging clones of interest by reestablishing anaerobic conditions (Lie and Leigh, 2007).

There are some disadvantages of a reporter-based random mutagenesis approach as just described to characterize molecular mechanisms underlying apparent regulatory phenomena. One is that mostly effects resulting from loss-of-function mutation can be identified. Another is the potentially very high number of mutants that have to be screened. For *M. acetivorans*, this obstacle was overcome by making occurrence of a regulatory mutation selectable (Rother *et al.*, 2005): the methanol-regulated *PmtaC1* promoter was cloned in front of the *mcrBCDGA* operon from *Methanosarcina barkeri* and inserted into the *hpt* locus of *M. acetivorans*, thus creating a *mcr* allele, the expression of which depends on methanol. The respective plasmid pMR05 (Fig. 5.2) is a suitable starting point for such an approach. Because methyl-coenzyme M reductase, encoded by the *mcr* genes, is essential, disruption of the native *mcr* operon renders growth of the resulting strain methanol dependent. However, prolonged incubation under nonexpressing conditions (in the presence of trimethylamine) gives rise to suppressor mutants able to grow again, which have lost repression of *mtaC1* in the absence of methanol. Such an approach is suitable to identify members of regulons, that is, genes regulated by a common principle such as a master regulator, by subsequently analyzing either the proteome or the transcriptome of the suppressor mutant in comparison to the wild type.

Another disadvantage stems from false positive mutants where no *cis*- and *trans*-acting factor involved in the regulation of the promoter under investigation but the reporter itself is mutated. One possibility to solve this issue is to insert a second copy of the promoter under study in a way its expression can be distinctly assessed, for example, fused to a second reporter, into the chromosome. In the case described above (Rother *et al.*, 2005), presence of a *PmtaC1-uidA* fusion on the chromosome (in addition to the *PmtaC1–mcrBCDGA* fusion) enabled to distinguish between mutations affecting only *mcr* expression and true regulatory mutations because the latter affected expression of *uidA* as well.

4. Creating Random Disruption Mutants of *M. maripaludis* by *In Vitro* Transposon Mutagenesis

A premium tool to identify factors involved in exhibiting a certain phenotype, be it a naturally occurring one or the presence of a reporter gene product activity, is random mutagenesis. Methanogens can be randomly mutagenized with radiation (Jain and Zeikus, 1987), chemical agents such as ethyl methanesulfonate (Yang *et al.*, 1995), purine analogs (Bowen *et al.*, 1995), or less randomly using potentially toxic intermediates in the metabolic pathway under study (Cubonova *et al.*, 2007). The problem with following such strategies is the difficulty in identifying in a mutant exhibiting an interesting phenotype the locus that was originally mutated. The ability to tag mutations with known sequence elements or with selectable antibiotic resistance genes has made transposon mutagenesis one of the most powerful genetic techniques developed for *M. acetivorans* (Zhang *et al.*, 2000 and see below). For *Methanococcus*, a random plasmid insertion mutagenesis system was developed principally fulfilling said requirement (Whitman *et al.*, 1997); however, chromosomal insertion of circular DNA leads to rather unstable mutations. Transposon mutagenesis was applied previously in *Methanococcus* (Blank *et al.*, 1995; Porat and Whitman, 2009). However, there the targets for the transposons were defined chromosomal sequences, which render these systems not really random.

The random *in vitro* transposon mutagenesis system described in the following was derived from that for genome wide *in vivo* transposon insertion in *M. acetivorans* (Zhang *et al.*, 2000) which we tried to apply to *M. maripaludis* without much success (S. Scheuffele and M. Rother, unpublished data). Although modified from the "original," the system also makes use of a modified insect transposon of the *mariner* family, *Himar1* (Figs. 5.2 and 5.3). However, from one of the original transposon plasmids, pJK60 (Zhang *et al.*, 2000), the gene for the hyperactive *Himar1* transposase (Lampe *et al.*, 1999) was removed (resulting in pJJ605; Fig. 5.2) and instead cloned into the *E. coli* expression plasmid pT7-7 (Tabor and Richardson, 1985). The enzyme can now be heterologously overproduced and purified (Fig. 5.3) following an established protocol (Lampe *et al.*, 1996). When the transposase is incubated with the transposon delivery plasmid pJJ605 and fragmented chromosomal DNA from *M. maripaludis* JJ, transposition creates random insertions of the transposon into the chromosomal fragments, which can be isolated and recombined onto the *M. maripaludis* chromosome by double homologous recombination *in vivo* (Fig. 5.3). In addition to the transposable element, the transposon contains an *E. coli* origin of replication and the *aph* gene (encoding aminoglycoside phosphotransferase conferring kanamycin resistance) (Zhang *et al.*, 2000), which allows recovery of the transposon from *M. maripaludis*. Chromosomal DNA from clones exhibiting an interesting phenotype can be isolated, fragmented by restriction digest, ligated, and transformed into *E. coli*. There, the chromosomal region of *M. maripaludis* surrounding the transposon is replicated like a plasmid allowing its amplification and isolation for subsequent sequencing, which leads to the identification of the exact chromosomal location into which a transposon has integrated. We have created thousands of transposon mutants employing the following protocol with an apparent transposition

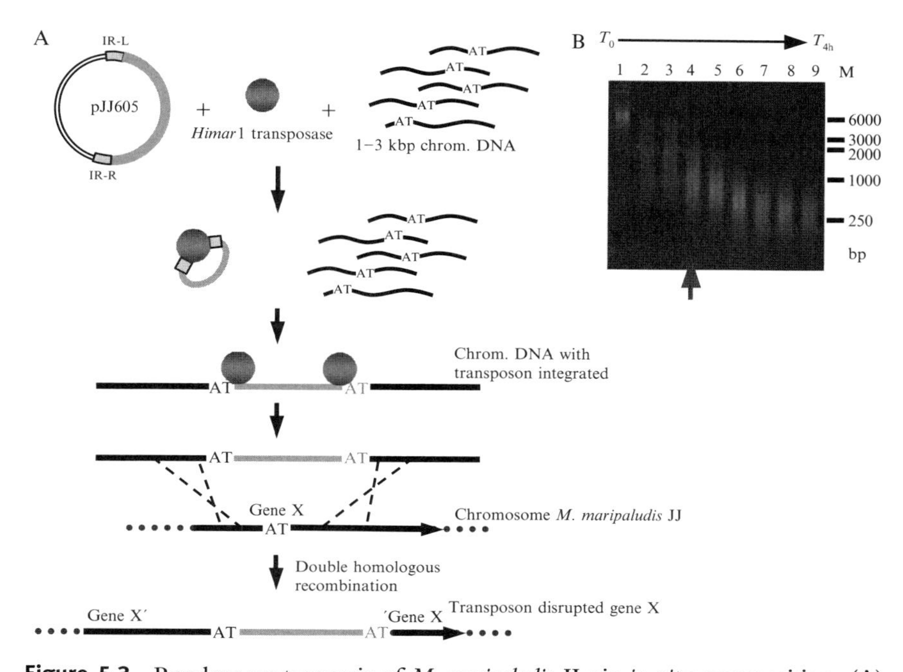

Figure 5.3 Random mutagenesis of *M. maripaludis* JJ via *in vitro* transposition. (A) Scheme of the procedure. Chromosomal DNA of *M. maripaludis* JJ is isolated and fragmented by digestion with *Alu*I and *Dra*I to average 1–3 kb (wavy black lines). *Himar*1 transposase (sphere) recognizes the transposon localized on the donor plasmid (pJJ605) via inverted repeats (IR-L, IR-R) and "activates" the recipient DNA at random AT sites for integration. *Himar*1 transposase inserts the transposon into the chromosomal fragment by a "cut-and-paste" mechanism. The gaps are repaired by the host cell thereby duplicating the AT-dinucleotide integration site (indicated by different shades of gray); dashed lines indicate homologous recombination events; dotted lines indicate chromosomal DNA of *M. maripaludis* JJ; solid gray lines indicate transposed/ transposable DNA fragments. (B) Partial digestion of *M. maripaludis* JJ chromosomal DNA with *Alu*I/*Dra*I to obtain 1–3 kb DNA fragments for transposition; the time course from 0 (T_0) to 4 h ($T_{4\ h}$) is shown from left to right; samples taken (lanes 1–9: 0, 3, 5, 15, 30, 60, 90, 120, and 240 min after addition of enzymes) were resolved on a 1% (w/v) agarose gel (in TAE buffer) at 100 V for 1 h and stained with ethidium bromide; the red arrow indicates the sample that should be used (in this example) for transposition; M, 1 kb DNA ladder (Fermentas, St. Leon-Roth). (For interpretation of the references to color in this figure legend, the reader is referred to the Web version of this chapter.)

frequency of 10–30 transposon mutants μg^{-1} recipient DNA; thus, a typical transformation leads to 50–150 puromycin-resistant *Methanococcus* clones on one agar plate. All mutants we have tested in this regard carried random transposon insertions (judged by Southern hybridization with a *pac*-specific probe) which were highly stable (judged from repeating the Southern analysis after nonselective growth of mutants for 50 generations; C. Sattler and M. Rother, unpublished data).

4.1. Preparation of reagents required for *in vitro* transposition

4.1.1. Transposon plasmid

The transposon delivery plasmid pJJ605 can be conveniently propagated in any ampicillin-sensitive cloning host and purified in large quantities by established procedures (Ausubel *et al.*, 1997).

4.1.2. Transposase

The overexpression plasmid carrying the gene for the hyperactive *Himar1* transposase (Lampe *et al.*, 1999), pT7tnp (C. Sattler and M. Rother, unpublished), is propagated in *E. coli* BL21 (DE3) in LB medium; T7 RNA polymerase-based overexpression of the gene (Tabor and Richardson, 1985) and protein production is induced at an OD_{600} of ca. 1 by addition of isopropyl-β-D-thiogalactopyranoside (ad 0.5 m*M*) and allowed to proceed for 3 h. *Himar1* transposase is purified from inclusion bodies as described (Lampe *et al.*, 1996), with the following modifications: in buffers, benzamidine is omitted and NP-40 is replaced by Nonidet P40; instead of DEAE-Sephacryl, DEAE-Sepharose is used.

4.1.3. Recipient DNA

Chromosomal DNA from *M. maripaludis* JJ (grown as described in Section 2.1) is isolated via a modified cetyltrimethylammonium bromide method (Metcalf *et al.*, 1996; Murray and Thompson, 1980), and partially restricted with *Alu*I and *Dra*I. For every large scale batch of isolated chromosomal DNA, the appropriate enzyme concentrations and incubation times should be determined (Fig. 5.3B). We digest for ca. 15 min preparations containing the restriction enzymes at ca. 0.08 U/μg DNA (Fig. 5.3B, red arrow). Fragments are preparatively separated on low melting agarose (USB, Cleveland, USA), and those of 1–3 kb excised from the gel. The gel is melted in 0.4 vol. of TE at 65 °C and the DNA extracted with phenol/chloroform, precipitated with ethanol, and redissolved in a small volume of water (Sambrook *et al.*, 1989).

4.2. *In vitro* transposition and transformation of *M. maripaludis*

For an *in vitro* transposition reaction, combine in 50 μl final volume: reaction buffer (25 m*M* 4-(2-hydroxyethyl)-1-piperazineethanesulfonic acid, 100 m*M* NaCl, 10 m*M* $MgCl_2$, 2 m*M* dithiothreitol, pH 7.9), 5 μg recipient DNA, 12.5 μg transposon plasmid pJJ605, and 12.5 μg bovine serum albumin. The reaction is started by adding transposase ad 10–20 n*M* (depending on the quality of the batch), incubated for 6 h at 25 °C, and stopped by shifting to 75 °C for 10 min. After ethanol precipitation, the whole sample is used to transform *M. maripaludis* JJ. Transformation is conducted as described (Tumbula *et al.*, 1994), except that McSe medium containing casamino acids and acetate (see Section 2.1) is used and that for each DNA to be transformed, two 5 ml cultures with OD_{578} of ca. 0.5 are used. For selection of *pac*, the plates contain 2.5 μg/ml puromycin. Phenotypic characteristics of mutants (e.g., the ability to utilize formate as a growth substrate or amino acid auxotrophy) can be assessed by replica patching on solid media containing/lacking the supplement under investigation. Those exhibiting the sought-for characteristics are streak purified, transferred to liquid media, and investigated further.

4.3. Localizing the site of transposon integration

To determine where on the chromosome of a *M. maripaludis Himar1* mutant with interesting phenotypic characteristics the transposon was inserted, isolate its chromosomal DNA (see Section 4.1) and digest it with *Eco*RI. No such recognition site is present in the transposon and the *M. maripaludis* JJ genome is cut to fragments of moderate lengths (that of strain S2 for which the genome sequence is available to fragments of ca. 2.4 kb). Self-ligate the DNA using T4 DNA ligase (Fermentas, St. Leon-Roth) according to the manufacturer's instructions. A pir^{+} *E. coli* strain (see Section 3.1) is subsequently transformed with the circularized DNA, which is replicated in *E. coli* like a plasmid and selected with kanamycin. The plasmid DNA isolated from the transformants can be directly subjected to sequencing analysis using transposon-specific, outward-directed primers (e.g., IR-5-Rev: 5′-CTCTTGAAGGGAACTATG-3′ and IR-3-For: 5′-GCATTTAATACTAGCGAC-3′) to determine the junction sequences of the transposon insertions.

ACKNOWLEDGMENTS

We thank V. Müller, University of Frankfurt, for his support. Work in M. R.'s laboratory is supported by grants from the Deutsche Forschungsgemeinschaft through SFB 579, SPP 1319, and RO 2445/6-1.

REFERENCES

Allmansberger, R., Knaub, S., and Klein, A. (1988). Conserved elements in the transcription initiation regions preceding highly expressed structural genes of methanogenic archaebacteria. *Nucleic Acids Res.* **16,** 7419–7436.

Anderson, K. L., Apolinario, E. E., MacAuley, S. R., and Sowers, K. R. (2009). A 5′ leader sequence regulates expression of methanosarcinal CO dehydrogenase/acetyl coenzyme A synthase. *J. Bacteriol.* **191,** 7123–7128.

Apolinario, E. E., Jackson, K. M., and Sowers, K. R. (2005). Development of a plasmid-mediated reporter system for *in vivo* monitoring of gene expression in the archaeon *Methanosarcina acetivorans*. *Appl. Environ. Microbiol.* **71,** 4914–4918.

Ashby, M. K. (2006). Distribution, structure and diversity of "bacterial" genes encoding two-component proteins in the Euryarchaeota. *Archaea* **2,** 11–30.

Ausubel, F. M., Brent, R., Kingston, R. E., Moore, D. D., Seidmann, J. G., Smith, J. A., and Struhl, K. (1997). Current Protocols in Molecular Biology. John Wiley & sons, Inc., New York.

Bellack, A., Huber, H., Rachel, R., Wanner, G., and Wirth, R. (2010). *Methanocaldococcus villosus* sp. nov., a heavily flagellated archaeon adhering to surfaces and forming cell-cell contacts. *Int. J. Syst. Evol. Microbiol.* 10.1099/ijs.1090.023663-023660.

Beneke, S., Bestgen, H., and Klein, A. (1995). Use of the *Escherichia coli uidA* gene as a reporter in *Methanococcus voltae* for the analysis of the regulatory function of the intergenic region between the operons encoding selenium-free hydrogenases. *Mol. Gen. Genet.* **248,** 225–228.

Bensing, B. A., Meyer, B. J., and Dunny, G. M. (1996). Sensitive detection of bacterial transcription initiation sites and differentiation from RNA processing sites in the pheromone-induced plasmid transfer system of *Enterococcus faecalis*. *Proc. Natl. Acad. Sci. USA* **93,** 7794–7799.

Bertani, G. (1951). Studies on lysogenesis. I. The mode of phage liberation by lysogenic *Escherichia coli*. *J. Bacteriol.* **62,** 293–300.

Blank, C. E., Kessler, P. S., and Leigh, J. A. (1995). Genetics in methanogens: Transposon insertion mutagenesis of a *Methanococcus maripaludis nifH* gene. *J. Bacteriol.* **177,** 5773–5777.

Boccazzi, P., Zhang, J. K., and Metcalf, W. W. (2000). Generation of dominant selectable markers for resistance to pseudomonic acid by cloning and mutagenesis of the *ileS* gene from the archaeon *Methanosarcina barkeri* Fusaro. *J. Bacteriol.* **182,** 2611–2618.

Bose, A., and Metcalf, W. W. (2008). Distinct regulators control the expression of methanol methyltransferase isozymes in *Methanosarcina acetivorans* C2A. *Mol. Microbiol.* **67,** 649–661.

Bowen, T. L., Ladapo, J. A., and Whitman, W. B. (1995). Selection for auxotrophic mutants of *Methanococcus*. *In* "Methanogens, Vol. 2," (K. R. Sowers and H. J. Schreier, eds.), pp. 405–407. Cold Spring Harbor Laboratory Press, Plainview, NY.

Bradford, M. M. (1976). A rapid and sensitive method for the quantitation of microgram quantities of protein utilizing the principle of protein-dye binding. *Anal. Biochem.* **72,** 248–254.

Breas, O., Guillou, C., Reniero, F., and Wada, E. (2001). The global methane cycle: Isotopes and mixing ratios, sources and sinks. *Isot. Environ. Health Stud.* **37,** 257–379.

Cohen-Kupiec, R., Blank, C., and Leigh, J. A. (1997). Transcriptional regulation in Archaea: *In vivo* demonstration of a repressor binding site in a methanogen. *Proc. Natl. Acad. Sci. USA* **94,** 1316–1320.

Cram, D. S., Sherf, B. A., Libby, R. T., Mattaliano, R. J., Ramachandran, K. L., and Reeve, J. N. (1987). Structure and expression of the genes, *mcrBDCGA*, which encode the subunits of component C of methyl coenzyme M reductase in *Methanococcus vannielii*. *Proc. Natl. Acad. Sci. USA* **84,** 3992–3996.

Cubonova, L., Sandman, K., Karr, E. A., Cochran, A. J., and Reeve, J. N. (2007). Spontaneous *trpY* mutants and mutational analysis of the TrpY archaeal transcription regulator. *J. Bacteriol.* **189,** 4338–4342.

Deppenmeier, U. (1995). Different structure and expression of the operons encoding the membrane-bound hydrogenases from *Methanosarcina mazei* Gö1. *Arch. Microbiol.* **164,** 370–376.

Deppenmeier, U., Johann, A., Hartsch, T., Merkl, R., Schmitz, R. A., Martinez-Arias, R., Henne, A., Wiezer, A., Bäumer, S., Jacobi, C., Brüggemann, H., Lienard, T., *et al.*

(2002). The genome of *Methanosarcina mazei*: Evidence for lateral gene transfer between bacteria and archaea. *J. Mol. Microbiol. Biotechnol.* **4,** 453–461.

Dodsworth, J. A., and Leigh, J. A. (2006). Regulation of nitrogenase by 2-oxoglutarate-reversible, direct binding of a PII-like nitrogen sensor protein to dinitrogenase. *Proc. Natl. Acad. Sci. USA* **103,** 9779–9784.

Eggen, R. I., van Kranenburg, R., Vriesema, A. J., Geerling, A. C., Verhagen, M. F., Hagen, W. R., and de Vos, W. M. (1996). Carbon monoxide dehydrogenase from *Methanosarcina frisia* Gö1: Characterization of the enzyme and the regulated expression of two operon-like *cdh* gene clusters. *J. Biol. Chem.* **271,** 14256–14263.

Ehlers, C., Veit, K., Gottschalk, G., and Schmitz, R. A. (2002). Functional organization of a single nif cluster in the mesophilic archaeon *Methanosarcina mazei* strain Gö1. *Archaea* **1,** 143–150.

Filutowicz, M., McEachern, M. J., and Helinski, D. R. (1986). Positive and negative roles of an initiator protein at an origin of replication. *Proc. Natl. Acad. Sci. USA* **83,** 9645–9649.

Galagan, J. E., Nusbaum, C., Roy, A., Endrizzi, M. G., Macdonald, P., FitzHugh, W., Calvo, S., Engels, R., Smirnov, S., Atnoor, D., Brown, A., Allen, N., *et al.* (2002). The genome of *M. acetivorans* reveals extensive metabolic and physiological diversity. *Genome Res.* **12,** 532–542.

Guss, A. M., Mukhopadhyay, B., Zhang, J. K., and Metcalf, W. W. (2005). Genetic analysis of *mch* mutants in two *Methanosarcina* species demonstrates multiple roles for the methanopterin-dependent C-1 oxidation/reduction pathway and differences in H_2 metabolism between closely related species. *Mol. Microbiol.* **55,** 1671–1680.

Hendrickson, E. L., Haydock, A. K., Moore, B. C., Whitman, W. B., and Leigh, J. A. (2007). Functionally distinct genes regulated by hydrogen limitation and growth rate in methanogenic Archaea. *Proc. Natl. Acad. Sci. USA* **104,** 8930–8934.

Hendrickson, E. L., Liu, Y., Rosas-Sandoval, G., Porat, I., Söll, D., Whitman, W. B., and Leigh, J. A. (2008). Global responses of *Methanococcus maripaludis* to specific nutrient limitations and growth rate. *J. Bacteriol.* **190,** 2198–2205.

Hovey, R., Lentes, S., Ehrenreich, A., Salmon, K., Saba, K., Gottschalk, G., Gunsalus, R. P., and Deppenmeier, U. (2005). DNA microarray analysis of *Methanosarcina mazei* Gö1 reveals adaptation to different methanogenic substrates. *Mol. Genet. Genomics* **273,** 225–239.

Jäger, D., Sharma, C. M., Thomsen, J., Ehlers, C., Vogel, J., and Schmitz, R. A. (2009). Deep sequencing analysis of the *Methanosarcina mazei* Gö1 transcriptome in response to nitrogen availability. *Proc. Natl. Acad. Sci. USA* **106,** 21878–21882.

Jain, M. K., and Zeikus, G. J. (1987). Methods for isolation of auxotrophic mutants of *Methanobacterium ivanovii* and initial characterization of acetate auxotrophs. *Appl. Environ. Microbiol.* **53,** 1387–1390.

Jones, W. J., Paynter, M. J. B., and Gupta, R. (1983). Characterization of *Methanococcus maripaludis* sp. nov., a new methanogen isolated from salt marsh sediment. *Arch. Microbiol.* **135,** 91–97.

Kennelly, P. J. (2007). Sensing, signal transduction, and posttranslational modification. *In* "Archaea—Molecular and Cellular Biology," (R. Cavicchioli, ed.), pp. 224–259. ASM Press, Washington, DC.

Lampe, D. J., Churchill, M. E., and Robertson, H. M. (1996). A purified *mariner* transposase is sufficient to mediate transposition *in vitro*. *EMBO J.* **15,** 5470–5479.

Lampe, D. J., Akerley, B. J., Rubin, E. J., Mekalanos, J. J., and Robertson, H. M. (1999). Hyperactive transposase mutants of the *Himar1 mariner* transposon. *Proc. Natl. Acad. Sci. USA* **96,** 11428–11433.

Lessner, D. J., Li, L., Li, Q., Rejtar, T., Andreev, V. P., Reichlen, M., Hill, K., Moran, J. J., Karger, B. L., and Ferry, J. G. (2006). An unconventional pathway for reduction of CO_2

to methane in CO-grown *Methanosarcina acetivorans* revealed by proteomics. *Proc. Natl. Acad. Sci. USA* **103,** 17921–17926.

Li, Q., Li, L., Rejtar, T., Lessner, D. J., Karger, B. L., and Ferry, J. G. (2006). Electron transport in the pathway of acetate conversion to methane in the marine archaeon *Methanosarcina acetivorans*. *J. Bacteriol.* **188,** 702–710.

Lie, T. J., and Leigh, J. A. (2007). Genetic screen for regulatory mutations in *Methanococcus maripaludis* and its use in identification of induction-deficient mutants of the euryarchaeal repressor NrpR. *Appl. Environ. Microbiol.* **73,** 6595–6600.

Lie, T. J., Wood, G. E., and Leigh, J. A. (2005). Regulation of *nif* expression in *Methanococcus maripaludis*: Roles of the euryarchaeal repressor NrpR, 2-oxoglutarate, and two operators. *J. Biol. Chem.* **280,** 5236–5241.

Lie, T. J., Hendrickson, E. L., Niess, U. M., Moore, B. C., Haydock, A. K., and Leigh, J. A. (2009). Overlapping repressor binding sites regulate expression of the *Methanococcus maripaludis glnK* operon. *Mol. Microbiol.* **75,** 755–762.

Livak, K. J., and Schmittgen, T. D. (2001). Analysis of relative gene expression data using real-time quantitative PCR and the $2^{-\Delta\Delta C}T$ method. *Methods* **25,** 402–408.

Longstaff, D. G., Blight, S. K., Zhang, L., Green-Church, K. B., and Krzycki, J. A. (2007). *In vivo* contextual requirements for UAG translation as pyrrolysine. *Mol. Microbiol.* **63,** 229–241.

Maeder, D. L., Anderson, I., Brettin, T. S., Bruce, D. C., Gilna, P., Han, C. S., Lapidus, A., Metcalf, W. W., Saunders, E., Tapia, R., and Sowers, K. R. (2006). The *Methanosarcina barkeri* genome: Comparative analysis with *Methanosarcina acetivorans* and *Methanosarcina mazei* reveals extensive rearrangement within methanosarcinal genomes. *J. Bacteriol.* **188,** 7922–7931.

Metcalf, W. W., Zhang, J. K., Shi, X., and Wolfe, R. S. (1996). Molecular, genetic, and biochemical characterization of the *serC* gene of *Methanosarcina barkeri* Fusaro. *J. Bacteriol.* **178,** 5797–5802.

Metcalf, W. W., Zhang, J. K., Apolinario, E., Sowers, K. R., and Wolfe, R. S. (1997). A genetic system for Archaea of the genus *Methanosarcina*: Liposome-mediated transformation and construction of shuttle vectors. *Proc. Natl. Acad. Sci. USA* **94,** 2626–2631.

Morgan, R. M., Pihl, T. D., Nölling, J., and Reeve, J. N. (1997). Hydrogen regulation of growth, growth yields, and methane gene transcription in *Methanobacterium thermoautotrophicum* DH. *J. Bacteriol.* **179,** 889–898.

Murray, M. G., and Thompson, W. F. (1980). Rapid isolation of high molecular weight plant DNA. *Nucleic Acids Res.* **8,** 4321–4325.

Oelgeschläger, E., and Rother, M. (2009). *In vivo* role of three fused corrinoid/methyl transfer proteins in *Methanosarcina acetivorans*. *Mol. Microbiol.* **72,** 1260–1272.

Opulencia, R. B., Bose, A., and Metcalf, W. W. (2009). Physiology and posttranscriptional regulation of methanol:coenzyme M methyltransferase isozymes in *Methanosarcina acetivorans* C2A. *J. Bacteriol.* **191,** 6928–6935.

Pfaffl, M. W. (2001). A new mathematical model for relative quantification in real-time RT-PCR. *Nucleic Acids Res.* **29,** e45.

Pihl, T. D., Sharma, S., and Reeve, J. N. (1994). Growth phase-dependent transcription of the genes that encode the two methyl coenzyme M reductase isoenzymes and N^5-methyltetrahydromethanopterin:coenzyme M methyltransferase in *Methanobacterium thermoautotrophicum* DH. *J. Bacteriol.* **176,** 6384–6391.

Porat, I., and Whitman, W. B. (2009). Tryptophan auxotrophs were obtained by random transposon insertions in the *Methanococcus maripaludis* tryptophan operon. *FEMS Microbiol. Lett.* **297,** 250–254.

Pritchett, M. A., and Metcalf, W. W. (2005). Genetic, physiological and biochemical characterization of multiple methanol methyltransferase isozymes in *Methanosarcina acetivorans* C2A. *Mol. Microbiol.* **56,** 1183–1194.

Pritchett, M. A., Zhang, J. K., and Metcalf, W. W. (2004). Development of a markerless genetic exchange method for *Methanosarcina acetivorans* C2A and its use in construction of new genetic tools for methanogenic archaea. *Appl. Environ. Microbiol.* **70,** 1425–1433.

Rohlin, L., and Gunsalus, R. P. (2010). Carbon-dependent control of electron transfer and central carbon pathway genes for methane biosynthesis in the Archaean, *Methanosarcina acetivorans* strain C2A. *BMC Microbiol.* **10,** 62.

Rother, M. (2010). Methanogenesis. *In* "Handbook of Hydrocarbon and Lipid Microbiology, Vol. 1," (K. N. Timmis, ed.), pp. 483–499. Springer, Berlin, Heidelberg.

Rother, M., Mathes, I., Lottspeich, F., and Böck, A. (2003). Inactivation of the *selB* gene in *Methanococcus maripaludis*: Effect on synthesis of selenoproteins and their sulfur-containing homologs. *J. Bacteriol.* **185,** 107–114.

Rother, M., Boccazzi, P., Bose, A., Pritchett, M. A., and Metcalf, W. W. (2005). Methanol-dependent gene expression demonstrates that methyl-CoM reductase is essential in *Methanosarcina acetivorans* C2A and allows isolation of mutants with defects in regulation of the methanol utilization pathway. *J. Bacteriol.* **187,** 5552–5559.

Rother, M., Oelgeschläger, E., and Metcalf, W. W. (2007). Genetic and proteomic analyses of CO utilization by *Methanosarcina acetivorans*. *Arch. Microbiol.* **188,** 463–472.

Sambrook, J., Fritsch, E. F., and Maniatis, T. (1989). Molecular Cloning: A Laboratory Manual. Cold Spring Harbor Laboratory Press, Plainview, NY.

Saum, R., Mingote, A., Santos, H., and Müller, V. (2009). Genetic analysis of the role of the ABC transporter Ota and Otb in glycine betaine transport in *Methanosarcina mazei* Gö1. *Arch. Microbiol.* **191,** 291–301.

Schopf, S., Wanner, G., Rachel, R., and Wirth, R. (2008). An archaeal bi-species biofilm formed by *Pyrococcus furiosus* and *Methanopyrus kandleri*. *Arch. Microbiol.* **190,** 371–377.

Sowers, K. R., Baron, S. F., and Ferry, J. G. (1984). *Methanosarcina acetivorans* sp. nov., an acetotrophic methane-producing bacterium isolated from marine sediments. *Appl. Environ. Microbiol.* **47,** 971–978.

Sowers, K. R., Boone, J. E., and Gunsalus, R. P. (1993a). Disaggregation of *Methanosarcina* spp. and growth as single cells at elevated osmolarity. *Appl. Environ. Microbiol.* **59,** 3832–3839.

Sowers, K. R., Thai, T. T., and Gunsalus, R. P. (1993b). Transcriptional regulation of the carbon monoxide dehydrogenase gene (*cdhA*) in *Methanosarcina thermophila*. *J. Biol. Chem.* **268,** 23172–32178.

Spanheimer, R., and Müller, V. (2008). The molecular basis of salt adaptation in *Methanosarcina mazei* Gö1. *Arch. Microbiol.* **190,** 271–279.

Stock, T., and Rother, M. (2009). Selenoproteins in Archaea and Gram-positive bacteria. *Biochim. Biophys. Acta* **1790,** 1520–1532.

Sun, J., and Klein, A. (2004). A LysR-type regulator is involved in the negative regulation of genes encoding selenium-free hydrogenases in the archaeon *Methanococcus voltae*. *Mol. Microbiol.* **52,** 563–571.

Tabor, S., and Richardson, C. C. (1985). A bacteriophage T7 RNA polymerase/promoter system for controlled exclusive expression of specific genes. *Proc. Natl. Acad. Sci. USA* **82,** 1074–1078.

Thauer, R. K., Kaster, A. K., Seedorf, H., Buckel, W., and Hedderich, R. (2008). Methanogenic archaea: Ecologically relevant differences in energy conservation. *Nat. Rev. Microbiol.* **6,** 579–591.

Thomm, M., Sherf, B. A., and Reeve, J. N. (1988). RNA polymerase-binding and transcription initiation sites upstream of the methyl reductase operon of *Methanococcus vannielii*. *J. Bacteriol.* **170,** 1958–1961.

Tumbula, D. L., Bowen, T. L., and Whitman, W. B. (1994). Transformation of *Methanococcus maripaludis* and identification of a *Pst*I-like restriction system. *FEMS Microbiol. Lett.* **121,** 309–314.

Veit, K., Ehlers, C., and Schmitz, R. A. (2005). Effects of nitrogen and carbon sources on transcription of soluble methyltransferases in *Methanosarcina mazei* strain Gö1. *J. Bacteriol.* **187,** 6147–6154.

Veit, K., Ehlers, C., Ehrenreich, A., Salmon, K., Hovey, R., Gunsalus, R. P., Deppenmeier, U., and Schmitz, R. A. (2006). Global transcriptional analysis of *Methanosarcina mazei* strain Gö1 under different nitrogen availabilities. *Mol. Genet. Genomics* **276,** 41–55.

Walters, A. D., and Chong, J. P. (2009). *Methanococcus maripaludis*: An archaeon with multiple functional MCM proteins? *Biochem. Soc. Trans.* **37,** 1–6.

Whitman, W. B., Shieh, J., Sohn, S., Caras, D. S., and Premachandran, U. (1986). Isolation and characterisation of 22 mesophilic methanococci. *Syst. Appl. Microbiol.* **7,** 235–240.

Whitman, W. B., Tumbula, D. L., Yu, J. P., and Kim, W. (1997). Development of genetic approaches for the methane-producing archaebacterium *Methanococcus maripaludis*. *Biofactors* **6,** 37–46.

Wood, G. E., Haydock, A. K., and Leigh, J. A. (2003). Function and regulation of the formate dehydrogenase genes of the methanogenic archaeon *Methanococcus maripaludis*. *J. Bacteriol.* **185,** 2548–2554.

Xia, Q., Hendrickson, E. L., Zhang, Y., Wang, T., Taub, F., Moore, B. C., Porat, I., Whitman, W. B., Hackett, M., and Leigh, J. A. (2006). Quantitative proteomics of the archaeon *Methanococcus maripaludis* validated by microarray analysis and real time PCR. *Mol. Cell. Proteomics* **5,** 868–881.

Yang, Y.-L., Ladapo, J. A., and Whitman, W. B. (1995). Mutagenesis of *Methanococcus* spp. with ethylmethanesulfonate. *In* "Methanogens, Vol. 2," (K. R. Sowers and H. J. Schreier, eds.), pp. 403–404. Cold Spring Harbor Laboratory Press, Plainview, NY.

Zhang, J. K., Pritchett, M. A., Lampe, D. J., Robertson, H. M., and Metcalf, W. W. (2000). *In vivo* transposon mutagenesis of the methanogenic archaeon *Methanosarcina acetivorans* C2A using a modified version of the insect *mariner*-family transposable element *Himar1*. *Proc. Natl. Acad. Sci. USA* **97,** 9665–9670.

CHAPTER SIX

Growth of Methanogens Under Defined Hydrogen Conditions

John A. Leigh

Contents

Abstract

Hydrogen (H_2) is a primary electron donor for methanogenesis and its availability can have profound effects on gene expression and the physiology of energy conservation. The rigorous evaluation of the effects of hydrogen conditions requires the comparison of cultures that are grown under hydrogen limitation and hydrogen excess. The growth of methanogens under defined hydrogen conditions is complicated by the dynamics of hydrogen dissolution and its utilization by the cells. In batch culture, gassing and agitation conditions must be carefully calibrated, and even then variations in growth rate and cell density are hard to avoid. Using chemostats, continuous cultures can be achieved whose nutritional states are known, while growth rate and cell density are invariant. Cultures whose growth is limited by hydrogen can be compared to cultures whose growth is limited by some other nutrient and are therefore under hydrogen excess.

Department of Microbiology, University of Washington, Seattle, Washington, USA

Methods in Enzymology, Volume 494
ISSN 0076-6879, DOI: 10.1016/B978-0-12-385112-3.00006-8

1. Introduction

Hydrogen (H_2) is a primary electron donor for methanogenesis and its availability has profound effects on gene expression and the physiology of energy conservation. Key steps in methanogenesis are regulated by hydrogen availability. In *Methanococcus maripaludis*, for example, mRNA abundances for coenzyme F_{420} oxidoreductases were shown to be especially responsive to hydrogen, with increases of 4- to 22-fold when hydrogen was limiting (Hendrickson *et al.*, 2007). Similar effects on mRNA abundances were observed in *Methanothermobacter thermautotrophicus* (Morgan *et al.*, 1997), and in *Methanocaldococcus jannaschii*, flagella increased with a low supply of hydrogen (Mukhopadhyay *et al.*, 2000). Measurements of relative protein abundance in *M. maripaludis* showed that many enzymes of methanogenesis increased moderately with hydrogen limitation, and a few proteins decreased (Xia *et al.*, 2009). Hydrogen availability also affects the physiology of methanogenesis and energy conservation. Growth yields (Y_{CH4}, or biomass produced per methane generated) can increase markedly when hydrogen is limiting, reflecting a more efficient use of methanogenesis from hydrogen to generate ATP (de Poorter *et al.*, 2007; Schonheit *et al.*, 1980).

The accurate assessment of the response of methanogens to hydrogen availability is complicated by the dynamics of hydrogen dissolution and its utilization by the cells. Hydrogen has limited solubility (0.8 m*M* at 1 bar at 20 °C) and the concentration of hydrogen experienced by the cells is a function of the hydrogen partial pressure, the volume of the culture, the surface area exposed to the gas atmosphere, the cellular rate of hydrogen utilization, and cell density. During growth, a low-density culture agitated vigorously tends to experience hydrogen excess, while a high-density culture typically reaches hydrogen limitation. Experiments testing the effects of hydrogen availability have been done in batch culture and continuous culture, and both approaches are described here.

In batch culture, one approach that has been used to vary hydrogen availability has been to change the rate of impeller rotation in a fermenter (e.g., Morgan *et al.*, 1997). Another approach has been to vary the percentage of hydrogen in the gas mixture delivered to the fermenter vessel, while holding the impeller speed constant (e.g., Mukhopadhyay *et al.*, 2000). With both approaches, growth rate and/or rate of methane production varied along with the change in agitation or percentage of hydrogen. One must sound a cautionary note when parameters such as agitation or partial pressure of hydrogen are used to control hydrogen availability in batch culture—the dependence of growth and methanogenesis on these parameters indicates that hydrogen is limiting at low settings, but does not always guarantee that it is in excess at high settings. Also, studies employing batch culture inevitably compare cultures at two growth rates and often at different cell densities, so physiological effects could stem from these parameters rather than directly from hydrogen availability.

Another approach to control hydrogen availability uses continuous culture, where a growing culture is constantly supplied with fresh medium and excess volume is removed. Once the culture is stabilized, its growth rate, density, and nutritional state are constant. In a chemostat, the growth rate equals the dilution rate, which equals the rate at which fresh medium is pumped in and excess volume is removed divided by the culture volume. Cell density is determined by the concentration of the limiting nutrient. My laboratory worked out a continuous culture system for the model species *M. maripaludis* (Haydock *et al.*, 2004). We imposed hydrogen limitation and two additional nutrient limitations, for example, nitrogen limitation and phosphorus limitation. By comparing the response to hydrogen limitation with each of the other two nutrient limitations, we were able to infer the effects of hydrogen availability (Hendrickson *et al.*, 2007; Xia *et al.*, 2009). One advantage of the continuous culture approach is that both hydrogen limitation and hydrogen excess are assured. The culture is hydrogen-limited when the hydrogen supply limits the cell density. It is under hydrogen excess when the supply of another nutrient, such as nitrogen or phosphorus, limits the cell density. Another advantage is that growth rate and cell density are invariant.

This chapter describes batch and continuous culture methods to achieve hydrogen-limited and hydrogen excess growth conditions. The determination of the *response* to hydrogen conditions, employing general techniques for measuring gene expression, mRNA abundance, or protein abundance, is not covered here. The response of growth yield (Y_{CH4}) involves standard measurements of biomass and methane, and is also not covered in this chapter.

2. Hydrogen Limitation in Batch Culture

2.1. General considerations

Differing hydrogen conditions can be achieved in batch culture using a fermenter equipped with an impeller and modified for use of anoxic gas mixtures. For many purposes, relatively small volumes of culture provide enough material for analysis and a small fermenter vessel is adequate. The New Brunswick BioFlo 110 Benchtop Fermenter with a 1.3-l vessel capacity is in common use, and the newer BioFlo 115 model is designed for the same purpose. However, other volumes may also be suitable, and the choice of equipment will depend on the purpose and what is available in the laboratory. The trick is to determine the rates of hydrogen delivery and/or the rates of impeller rotation that achieve hydrogen limitation and excess. Hydrogen limitation presumably is the case at a given setting if an increase in hydrogen or impeller speed results in a concurrent increase in the rate of growth or methanogenesis. If growth or methanogenesis persists at the same rate after an increase, then hydrogen is presumably in excess. The parameters will depend on the species, the incubation temperature, and the equipment specifications, so the following examples serve only as general guidelines. For *M. thermautotrophicus* at 55 °C, one study (Morgan *et al.*, 1997) used a 1.5-l culture volume and fed the culture at 200 ml/min with a gas mixture containing 89% H_2. Varying the impeller speed between 120 and 600 rpm resulted in changes in growth rate and methanogenesis as well as in gene expression that were attributed to hydrogen regulation. Alternatively, at a constant impeller speed of 600 rpm, varying the percentage of H_2 between 18% and 89% had similar effects. Another study (Mukhopadhyay *et al.*, 2000) investigated the effects of hydrogen partial pressure in *M. jannaschii* at 85 °C, using a 12-l culture volume. Gas was supplied at 24,000 ml/min, the vessel was maintained at a slight positive pressure of 1.24×10^5 Pa, the impeller speed was 600 rpm, and the proportion of hydrogen was varied between 80% and 0.3%. Flagellum synthesis was induced when hydrogen became limiting either at high cell density under high partial pressure of H_2 or at low cell density under low partial pressure of H_2.

2.2. The gas delivery system

The primary modification of a commercially purchased fermenter will be the installation of the gas delivery system (Fig. 6.1). Gas mixtures of hydrogen and carbon dioxide, and often also including nitrogen, argon, or hydrogen sulfide, can be achieved using metering valve-floating ball-type flow meter (rotameter) assemblies (e.g., Cole-Parmer® multitube flowmeter systems with direct reading flow tubes for specialty gasses). It is often critical to be able to accurately set the flow of hydrogen at markedly different rates for hydrogen excess and hydrogen availability. This requires two different metering valve–flow tube combinations, and the gas can be directed through one or the other from a three-way ball valve (e.g., Swagelok®). After passing through the flow meters, the individual gasses enter a common stream that flows toward the vessel. The gas mixture may be passed through a copper-reducing furnace to remove traces of oxygen, but if hydrogen sulfide is included, this gas must join the stream after the furnace. Copper or stainless steel tubing with Swagelok® fittings is convenient, although gas-impermeable flexible tubing such as Norprene® or butyl rubber may be used. Where hydrogen sulfide is present, stainless steel tubing and fittings should be used in place of copper and brass.

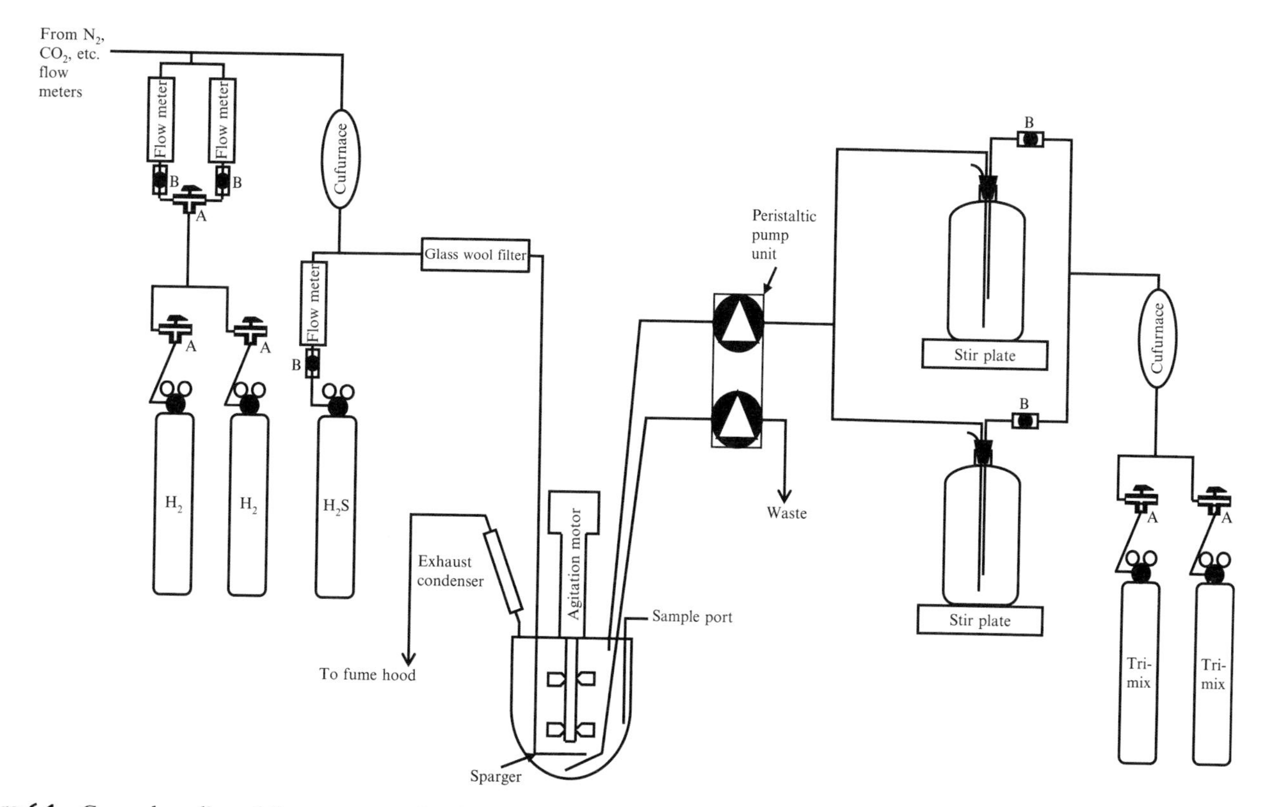

Figure 6.1 Gas and medium delivery systems for the chemostat. The gas delivery system may also be used for batch culture in fermenters. (A) Three-way ball valves. (B) Metering valves. Modified from Haydock *et al.* (2004).

It is convenient to install an on–off toggle valve (e.g., Swagelok®) in the gas stream for each individual gas as well as in the pooled gas stream. Before entering the vessel, the gas may be sterilized by passage through a glass wool filter, which can be made from a stainless steel hex long nipple (Swagelok®) filled with glass wool and closed at the ends with tubing-pipe thread adapters.

When setting up a vessel with fresh medium, gas flow (excluding hydrogen sulfide) may be started (and vented into a fume hood) and the copper-reducing furnace turned on while the vessel is being autoclaved. To prevent oxygenation, gas is directed through the vessel immediately after autoclaving and moving into place. Hydrogen sulfide is turned on subsequently.

3. Hydrogen Limitation in Continuous Culture

3.1. General considerations

Continuous culture systems have been described for *M. maripaludis* (Haydock *et al.*, 2004) and *M. thermautotrophicus* (de Poorter *et al.*, 2007). For *M. maripaludis*, we adopted an approach, detailed in Haydock *et al.*, 2004, whereby a single known nutrient limits growth. We generally hold the dilution rate (and hence the growth rate) constant and limit the cell density by lowering the level of a given nutrient. Hydrogen becomes the limiting nutrient when its proportion in the gas mixture delivered to the chemostat vessel is lowered to the point where it limits the cell density at steady state. Table 6.1 shows an example of gas mixtures that are used for hydrogen limitation and hydrogen excess. A nutrient such as phosphate or ammonia is limiting when its concentration in the growth medium is

Table 6.1 Example of gas flow rates (ml/min) for growth of *M. maripaludis* in a chemostat under hydrogen limitation and hydrogen excess[a]

	Hydrogen limitation[b]	Hydrogen excess[b]
H_2	20	110
N_2 or Ar	125	35
CO_2	40	40
H_2S (1% in N_2 or Ar)	15	15

[a] These gas flow rates were determined for continuous culture of *M. maripaludis* at a specific growth rate of 0.083 h^{-1} at a cell density (OD_{660}) of 0.6 in a 1-l volume with an agitation rate of 1000 rpm. Appropriate values will vary with different organisms, different conditions, and different equipment.

[b] *Hydrogen limitation*: This level of H_2 holds the cell density (OD_{660}) at about 0.6. *Hydrogen excess*: A limiting level of some nutrient in the medium (e.g., nitrogen or phosphate) holds the OD_{660} at about 0.6.

lowered to the point where it similarly determines cell density. If the cell density is the same for each limiting nutrient, and the growth rate is also the same, differences between cultures result solely from which nutrient is limiting. In our studies, we regard a culture as experiencing hydrogen excess if its density is limited by some other nutrient such as phosphate or ammonia (Hendrickson *et al.*, 2007; Xia *et al.*, 2009). For the continuous culture of *M. thermautotrophicus* (de Poorter *et al.*, 2007), the dilution rate and gassing rate were varied, resulting in cultures of different densities. Two kinds of culture were obtained: cultures whose density remained generally constant over a range of dilution rates and gassing rates were apparently limited by a nutrient in the growth medium, and cultures whose density decreased with increasing dilution rates, or increased with increasing gassing rates, were hydrogen-limited.

3.2. The gas delivery system

The gas delivery system to the chemostat vessel is similar to that for a batch fermenter (above and Fig. 6.1). Additional measures can be taken to avoid changes to the gas flow that would otherwise perturb the steady state of the culture. The use of two-stage gas regulators minimizes changes in gas delivery pressure as a tank empties. In addition, it is a good idea to have two tanks of each gas on line, joined at a "T," and with regulator pressures equalized. An empty tank can be changed while the other tank continues to supply gas. The tanks can be vented through three-way ball valves to purge the lines.

3.3. The medium delivery system

The additional component needed for continuous culture is the system that feeds the medium from a medium reservoir to the vessel. A system that works well for *M. maripaludis* has been described (Haydock *et al.*, 2004) and is diagrammed in Fig. 6.1. Medium can be prepared in 10-l Pyrex bottles, which are a convenient size for autoclaving and moving into place. Caution must be exercised when moving the bottles of hot medium. Each bottle is equipped with a three-holed butyl rubber stopper penetrated by 1/8-in. stainless steel tubing for gas sparging, gas venting, and medium withdrawal. To keep the medium anoxic, a gas mixture is constantly bubbled slowly through the medium. The gas mixture for sparging the medium reservoir flows from a tank through copper tubing to a copper reducing furnace, then through a glass fiber filter assembly (see above), and into the reservoir bottle through the stainless steel tubing. We use a mixture ("trimix" in Fig. 6.1) of 90% N_2, 5% CO_2 (for pH buffering in a bicarbonate-containing medium), and 5% H_2 (to maintain the reduced state of the copper furnace). The medium is stirred gently by a magnetic stir bar. Gas sparging of the medium reservoir is begun immediately after autoclaving, after which cysteine and any other additions are made from anoxic stocks by temporary dislodging of the stopper. The medium is delivered to the chemostat vessel through stainless steel tubing or flexible tubing of a gas-impermeable material such as Norprene®. For passage through the peristaltic pump, Viton® tubing which is also gas-impermeable is recommended. Provision for two medium reservoirs allows preparation of one bottle, while the other is in use. The two reservoirs are joined at a "T," where attachment through short segments of Norprene® tubing allows one reservoir to be clamped off. Tubing can be obtained from Cole-Parmer®.

ACKNOWLEDGMENTS

Research in my laboratory on hydrogen regulation and hydrogen metabolism is supported by the U.S. Department of Energy Office of Science Grants DE-FG02-08ER64685 (BER) and DE-FG02-05ER15709 (BES, Basic Research for the Hydrogen Fuel Initiative).

REFERENCES

de Poorter, L. M., *et al.* (2007). Coupling of *Methanothermobacter thermautotrophicus* methane formation and growth in fed-batch and continuous cultures under different H_2 gassing regimens. *Appl. Environ. Microbiol.* **73,** 740–749.

Haydock, A. K., *et al.* (2004). Continuous culture of *Methanococcus maripaludis* under defined nutrient conditions. *FEMS Microbiol. Lett.* **238,** 85–91.

Hendrickson, E. L., *et al.* (2007). Functionally distinct genes regulated by hydrogen limitation and growth rate in methanogenic Archaea. *Proc. Natl. Acad. Sci. USA* **104,** 8930–8934.

Morgan, R. M., *et al.* (1997). Hydrogen regulation of growth, growth yields, and methane gene transcription in *Methanobacterium thermoautotrophicum* ΔH. *J. Bacteriol.* **179,** 889–898.

Mukhopadhyay, B., *et al.* (2000). A novel pH_2 control on the expression of flagella in the hyperthermophilic strictly hydrogenotrophic methanarchaeaon *Methanococcus jannaschii. Proc. Natl. Acad. Sci. USA* **97,** 11522–11527.

Schonheit, P., *et al.* (1980). Growth parameters (K_s, μ_{max}, Y_s) of *Methanobacterium thermoautotrophicum. Arch. Microbiol.* **127,** 59–65.

Xia, Q., *et al.* (2009). Quantitative proteomics of nutrient limitation in the hydrogenotrophic methanogen *Methanococcus maripaludis. BMC Microbiol.* **9,** 149.

CHAPTER SEVEN

Preparation of [Fe]-Hydrogenase from Methanogenic Archaea

Seigo Shima,*,† Michael Schick,* *and* Haruka Tamura*

Contents

Abstract

[Fe]-hydrogenase is one of the three types of hydrogenases. This enzyme is found in many hydrogenotrophic methanogenic archaea and catalyzes the reversible hydride transfer from H_2 to methenyl-H_4MPT^+ in methanogenesis from H_2 and CO_2. The enzyme harbors a unique iron-guanylyl pyridinol (FeGP)

* Max Planck Institute for Terrestrial Microbiology, Marburg, Germany
† PRESTO, Japan Science and Technology Agency (JST), Honcho, Kawaguchi, Saitama, Japan

Methods in Enzymology, Volume 494
ISSN 0076-6879, DOI: 10.1016/B978-0-12-385112-3.00007-X

cofactor as a prosthetic group. Here, we describe the purification of [Fe]-hydrogenase from *Methanothermobacter marburgensis*, the isolation of the FeGP cofactor from the native holoenzyme, and the reconstitution of [Fe]-hydrogenase from the isolated FeGP cofactor and the heterologously produced apoenzyme.

1. Introduction

Methane is an end product of the anaerobic degradation of organic materials in freshwater environments. Biological methane production is mediated by methanogenic archaea (Thauer, 1998). Most methanogenic archaea utilize H_2 as an electron donor for the eight-electron reduction of CO_2 to methane, which is coupled with energy conservation. The oxidation of H_2 in the methanogens is catalyzed by hydrogenases (Thauer *et al.*, 2010).

The three types of known hydrogenases are classified by the structure of the active-site metal center: [NiFe]-, [FeFe]-, and [Fe]-hydrogenases (Shima and Thauer, 2007). The metal centers of [NiFe]- and [FeFe]-hydrogenases consist of dinuclear metal ions composed of nickel and iron, and two irons, respectively. These two types of hydrogenases also contain iron–sulfur clusters. Three [NiFe]-hydrogenases operate in methanogenic archaea: F_{420}-reducing hydrogenases (Frh), integral-membrane energy-conserving hydrogenases (Ech), and heterodisulfide-reductase-associated hydrogenases (Mvh). In addition, many methanogens synthesize one [Fe]-hydrogenase (Thauer *et al.*, 2010). [Fe]-hydrogenase contains a mono-iron catalytic center and is devoid of iron–sulfur clusters. This enzyme has therefore also been referred to as an iron–sulfur-cluster-free hydrogenase.

The systematic name of [Fe]-hydrogenase is H_2-forming methylene-tetrahydromethanopterin (methylene-H_4MPT) dehydrogenase (Hmd), and the enzyme is encoded by the *hmd* gene. [Fe]-hydrogenase catalyzes the heterolytic cleavage of H_2 and the reversible hydride transfer to methenyl-H_4MPT^+, forming methylene-H_4MPT and a proton; the hydride ion is stereospecifically transferred to the *proR* position of methylene-H_4MPT (Fig. 7.1; Shima and Thauer, 2007). H_4MPT is a C1-carrier of methanogenic archaea, and a structural and functional analog of tetrahydrofolate (Thauer *et al.*, 1996). [Fe]-hydrogenase is found in many hydrogenotrophic methanogenic archaea belonging to the orders Methanopyrales, Methanobacteriales, and Methanococcales (Table 7.1; Thauer *et al.*, 1996). The *hmd* gene is not found in the genome of members of the orders Methanosarcinales and Methanomicrobiales, with one exception, in *Methanocorpusculum labreanum* in the order Methanomicrobiales (Table 7.1; Anderson *et al.*, 2009).

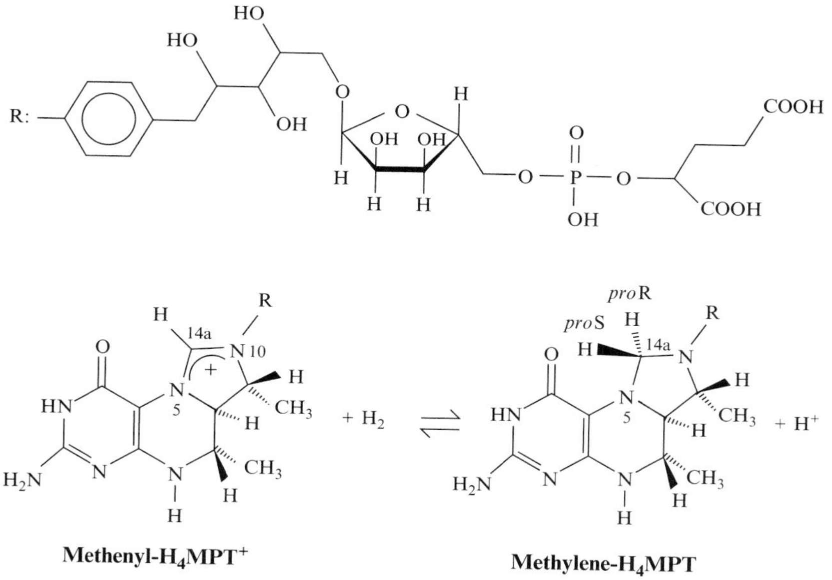

Figure 7.1 Reaction catalyzed by [Fe]-hydrogenase. A hydride is reversibly transferred from H_2 to the *proR* position of methenyl-H_4MPT^+ (Zirngibl *et al.*, 1990, 1992).

[Fe]-hydrogenase contains an iron-guanylyl pyridinol (FeGP) cofactor at the active site (Buurman *et al.*, 2000; Lyon *et al.*, 2004b). The FeGP cofactor can be extracted from the [Fe]-hydrogenase in an intact form, and the holoenzyme can be reconstituted from the extracted cofactor and the apoenzyme heterologously produced in *Escherichia coli* (Buurman *et al.*, 2000; Lyon *et al.*, 2004b). The crystal structures of the reconstituted [Fe]-hydrogenase have been solved, and a catalytic mechanism based on the enzyme–substrate complex has been proposed (Hiromoto *et al.*, 2009b; Shima *et al.*, 2008; Vogt *et al.*, 2008). Spectroscopic, chemical, and crystallographic analyses revealed its unique iron-complex structure (Korbas *et al.*, 2006; Lyon *et al.*, 2004a; Salomone-Stagni *et al.*, 2010; Shima *et al.*, 2004, 2005). The iron site is ligated with cysteine thiol, two CO, one acyl-carbon, and one sp^2-hybridized pyridinol nitrogen. The latter two ligands and the iron form a five-membered metallacycle (Fig. 7.2; Hiromoto *et al.*, 2009a; Shima *et al.*, 2008). The proposed H_2-binding site appears to be occupied by a solvent molecule in the crystal structure (Hiromoto *et al.*, 2009a).

This chapter describes the cultivation of *Methanothermobacter marburgensis* under nickel-limiting conditions and the purification of [Fe]-hydrogenase from the cells. We also present methods for the isolation of the FeGP

Table 7.1 Specific activity of [Fe]-hydrogenase in cell extract and catalytic properties of the purified enzyme from different methanogenic archaea

Methanogenic archaeon (optimum growth temperature)	Activity in cell extract[a] (U mg^{-1})	Apparent K_m (μM) for $CH_2{=}H_4MPT$	V_{max} (U mg^{-1})	Heterologous protein in *E. coli*[b]
Methanopyrus kandleri (98 °C)	7[c]	50[c]	2700[d]	Soluble[e]
Methanocaldococcus jannaschii (85 °C)	4[f]	10[g,h]	300[g,h]	Soluble[e]
Methanotorris igneus (85 °C)	9[i]	120[d]	500[d]	nd
Methanothermococcus thermolithotrophicus (65 °C)	18[j]	85[j]	760[j]	Soluble[k]
Methanothermus fervidus (83 °C)	5[i]	45[d]	860[d]	Inclusion body[k]
Methenothermobacter marburgensis (65 °C)	12[l]	40[l]	2000[l]	Inclusion[m] body
Methanothermobacter wolfeii (60 °C)	2[n]	20[l]	500[l]	nd
Methanothermobacter sp.[o] (55 °C)	3[p]	35[d]	400[d]	Inclusion body[q]
Methanobrevibacter smithii (37 °C)	0.03[r]	nd[s]	nd	Inclusion body[r]
Methanobrevibacter arboriphilus (37 °C)	0.04[n]	nd	nd	Inclusion body[k]

Methanococcus voltae (40 °C)	0.7[n]	55[d]	140[d]	Inclusion body[q]
Methanococcus maripaludis (37 °C)	0.05[f]	nd	nd	Partially soluble[f]
Methanocorpusculum labreanum (37 °C)	< 0.1[r]	10[r,g]	45[r,g]	Partially soluble[r]

[a] Under nickel-sufficient conditions.
[b] Expression in TP medium.
[c] Ma *et al.* (1991).
[d] Thauer *et al.* (1996).
[e] Buurman *et al.* (2000).
[f] Kaster *et al.* (unpublished results).
[g] Assay used the reconstituted enzyme.
[h] Vogt *et al.* (unpublished results).
[i] Klein *et al.* (unpublished results).
[j] Hartmann *et al.* (1996).
[k] Pilak *et al.* (unpublished results).
[l] Zirngibl *et al.* (1992).
[m] Codon usage and the mRNA structure optimized for *E. coli* production (Vogt *et al.*, unpublished results).
[n] Schwörer and Thauer (1991).
[o] *Methanobacterium thermoformicicum.*
[p] Vaupel *et al.* (unpublished result).
[q] Shima *et al.* (unpublished results).
[r] Schick *et al.* (unpublished results).
[s] Not determined.

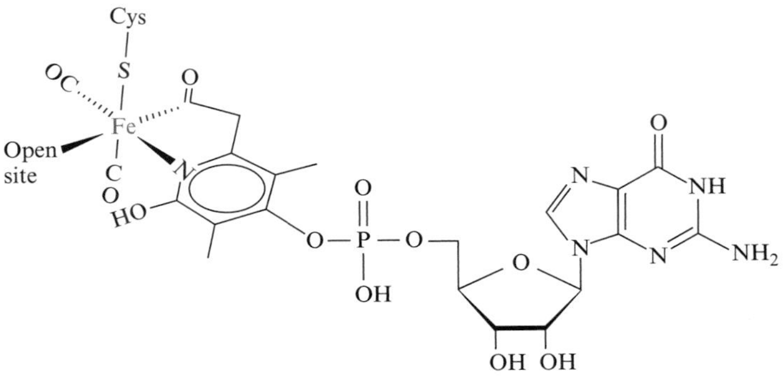

Figure 7.2 FeGP cofactor of [Fe]-hydrogenase (Hiromoto *et al.*, 2009a; Shima *et al.*, 2008).

cofactor, the purification of the apoenzyme heterologously overproduced in *E. coli,* and the reconstitution of the active holoenzyme.

2. Oxygen, Light, and Copper Sensitivity of [Fe]-Hydrogenase and FeGP Cofactor

[Fe]-hydrogenase is extracted and purified under strictly anoxic conditions under red light. The purified [Fe]-hydrogenase is stable for several hours in air, and the activity is not inhibited by air—a feature particular to [Fe]-hydrogenase that differs from the features of the other types of hydrogenases (Lyon *et al.*, 2004b). With longer exposure to air, [Fe]-hydrogenase is slowly inactivated ($t_{1/2}$ at 4 °C = 2–3 days). In cell extracts, the [Fe]-hydrogenase activity is sometimes quickly inactivated, probably owing to superoxide radical anions produced by the side reactions of other enzymes present in the cell extracts (Shima *et al.*, 2009). The FeGP cofactor in [Fe]-hydrogenase and in its protein-free state is sensitive to UV-A and blue light (Fig. 7.3). Therefore, samples of [Fe]-hydrogenase and FeGP cofactor must be stored under a N_2 atmosphere in amber-colored glass vials closed with butyl-rubber stoppers. The enzyme is purified in anoxic chambers set up in rooms under red or yellow lights at 18 °C. The anoxic chambers used in our laboratory are commercially available from Coy Laboratory Products Inc. The gas phase in the chamber is N_2/H_2 (95/5, v/v). Traces of oxygen are removed continuously by palladium catalysts (BASF). [Fe]-hydrogenase is inhibited with very low concentration of Cu^{2+}. Therefore, the enzyme assay solution has to contain 1 m*M* EDTA (Shima *et al.*, 2009).

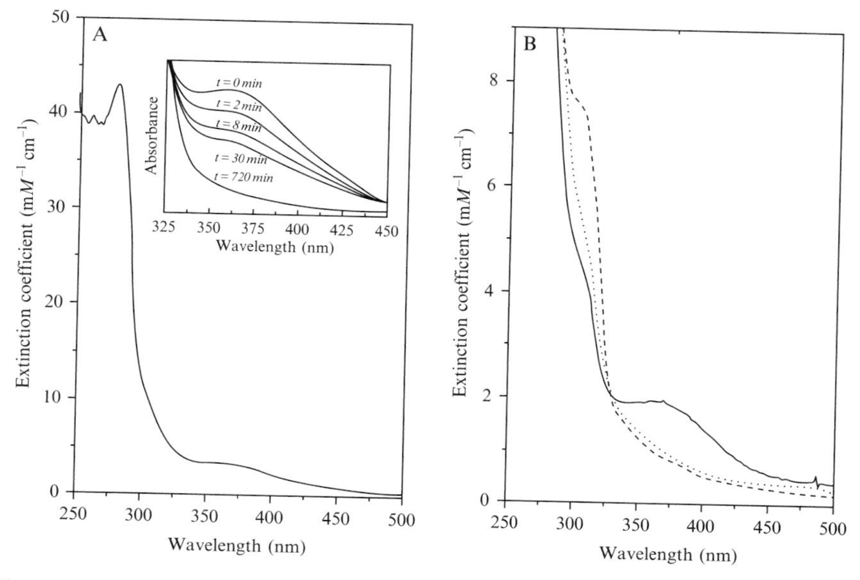

Figure 7.3 (A) UV–Vis spectrum of the [Fe]-hydrogenase. The inset shows the bleaching upon exposure of [Fe]-hydrogenase to white light. The protein concentration was determined using the Bradford method. (B) UV–Vis spectrum of the FeGP cofactor. Solid line, before irradiation with white light; dotted line, after 5-min irradiation; and dashed line, after 30-min irradiation. The cofactor concentration was determined using the absorption at 300 nm of the light-inactivated product.

3. Preparation of [Fe]-Hydrogenase

3.1. Cultivation of *M. marburgensis* under nickel-limiting conditions

The production of [Fe]-hydrogenase in *M. marburgensis*, *Methanocaldococcus jannaschii,* and *Methanococcus maripaludis* is four- to sevenfold upregulated in nickel-limited (<0.05 μ*M* nickel ions) medium, compared to the normal growth medium (5 μ*M* nickel ions; Afting *et al.*, 1998; Kaster *et al.*, unpublished results). Therefore, for purification of [Fe]-hydrogenase, *M. marburgensis* is cultivated in a medium with low nickel concentration. The growth medium at pH 7 contains 40 m*M* NH_4Cl, 50 m*M* KH_2PO_4, 24 m*M* Na_2CO_3, 0.5 m*M* nitrilotriacetic acid, 0.2 m*M* $MgCl_2{\cdot}6H_2O$, 1 μ*M* $CoCl_2{\cdot}6H_2O$, 1 μ*M* $Na_2MoO_4{\cdot}2H_2O$, 20 μ*M* resazurin, 50 μ*M* $FeCl_2$, and 0 (large-scale cultivation) or 0.65 μ*M* (precultivation) $NiCl_2$ (Schönheit *et al.*, 1980; Shima *et al.*, 2005). The nickel in the water used for the medium must be removed by scrubbing on a mixed-bed ion-exchange column. *M. marburgensis* (DSM2133) is precultivated in 360 ml medium in a

400-ml glass fermenter agitated with a plastic-coated magnetic stir bar at 400 rpm and gassed with 0.1% H_2S in H_2/CO_2 (80/20, v/v) at a rate of 100 ml min^{-1} at 65 °C. The overnight (~16 h) preculture (100 ml aliquot) is then used to anoxically inoculate 10 l of nickel-free medium in a fermenter using 50-ml plastic syringes. The fermenter consists of a glass vessel (11 l) with impellers and other parts made of stainless steel. The methanogens are incubated at 65 °C under the gas mixture described above at a flow rate of 1.5 l min^{-1} and at an agitation speed of 1000 rpm.

In the early growth phase ($\Delta OD_{578} < \sim 3$–4), *M. marburgensis* cells grow rapidly with a doubling time of ~2.5 h, because the amount of nickel leached from the stainless-steel parts of the fermenter appears to be enough to sustain the normal growth rate. However, when the culture reaches an optical density of ~3.0, the growth slows down to a doubling time of ~11 h, at which the *hmd* gene expression is upregulated. When the ΔOD_{578} reaches ~5–6 (~21 h cultivation), the culture is cooled down to ~10 °C. The cells (75 g wet cell mass per 10 l) are then harvested under H_2/CO_2 (80/20, v/v) by continuous-flow centrifugation. Because nickel-limited cultures are crucial for purification of the [Fe]-hydrogenase, the overproduction conditions of the enzyme in each fermenter should be tested carefully. The cells containing [Fe]-hydrogenase can be stored at −80 °C for several years without affecting the enzyme activity and the other properties.

3.2. Purification of [Fe]-hydrogenase from *M. marburgensis*

To purify the [Fe]-hydrogenase, ~100 g of *M. marburgensis* cells are suspended in 200 ml of 50 m*M* potassium phosphate, pH 7.0 and disrupted by sonication in ice water with a 50% cycle and 160 W (six times for 8 min, with 7-min pauses). The cell extract is ultracentrifuged at 140,000×*g* for 30 min at 4 °C to remove unbroken cells, cell debris, and the membrane fraction. Ammonium sulfate powder is then added to the supernatant to 60% saturation and the mixture is incubated on ice for 20 min; [Fe]-hydrogenase is highly soluble in the presence of ammonium sulfate at neutral pH, whereas most other proteins are precipitated. The precipitated proteins are removed by centrifugation at 13,000×*g* for 20 min at 4 °C. [Fe]-hydrogenase in the supernatant is precipitated by the addition of ammonium sulfate to 90% saturation and incubation on ice for 20 min; the mixture is then centrifuged. Through this fractionation, [Fe]-hydrogenase is enriched sixfold. The pellet is suspended in 15 ml of 50 m*M* Mops/KOH, pH 7.0 and then dialyzed against 50 m*M* citric acid/NaOH, pH 5.0 at 4 °C for 18 h. The turbid, dialyzed protein solution is centrifuged at 17,000×*g* for 20 min at 4 °C. The supernatant is applied to a Source 30Q column (5 cm diameter × 15 cm length) equilibrated with 50 m*M* citric acid/NaOH, pH 5.0. The column is washed with 250 ml of the equilibration buffer containing 200 mM NaCl, and the [Fe]-hydrogenase is eluted with an increasing linear gradient of NaCl from 200 to 500 mM in 500 ml. The flow rate is 7 ml min^{-1}, and 10 ml fractions are collected. The 10 fractions (~100 ml) of around 300 mM NaCl are combined and neutralized by the addition of 10 ml of 1 M Mops/KOH, pH 7.0 and 0.6 ml 1 M NaOH to avoid the precipitation of [Fe]-hydrogenase, and then concentrated to ~15 ml using ultrafiltration (30 kDa cutoff). The concentrated protein solutions are applied to a desalting Sephadex G-25 (HiPrep 26/10) column equilibrated with water. Following this purification procedure, one can obtain 100–150 mg of [Fe]-hydrogenase with a specific activity of 300–600 U mg^{-1} at 40 °C using 20 μM methylene-H_4MPT as a substrate (Table 7.2; Lyon *et al.*, 2004b).

The purified [Fe]-hydrogenase preparation can be stored at −80 °C for several years without any loss of activity. However, the enzyme activity tends to be partially inactivated by freeze–thaw cycles. Quick freezing of the enzyme solution in liquid nitrogen decreases the loss of activity.

3.3. Characterization of the purified [Fe]-hydrogenase

The purified enzyme preparation is routinely characterized by protein determination, enzyme activity assays, SDS-PAGE, and UV–Vis spectroscopy. The protein concentration is determined by the Bradford assay using bovine serum albumin as a standard (Bradford, 1976). The activity assay is described below in detail. In the preparation of the SDS-PAGE sample, boiling in the SDS sample buffer should be avoided because then, several extra protein bands are found in the SDS–polyacrylamide gel. The sample should instead be gently heated at 80 °C for 5 min prior to loading on the gel. The UV–Vis spectrum of purified [Fe]-hydrogenase is shown in Fig. 7.3

Table 7.2 Purification of [Fe]-hydrogenase from *Methanothermobacter marburgensis* grown under nickel-limiting conditions

	Activity (U)	Protein (mg)	Specific activity (U mg^{-1})	Purification (fold)	Yield (%)
Cell extract	110,000	5700	20	1	100
60–90% $(NH_4)_2SO_4$	80,000	750	110	6	73
Source 30Q	63,000	160	400	20	57
Sephadex G25	57,000	140	400	20	52

Protein was purified from 100 g cells (wet mass) according to the procedures described in the text. One unit (U) activity corresponds to the formation of 1 μmol of methenyl-H_4MPT^+ from 20 μM methylene-H_4MPT per minute under standard assay conditions (pH 6.0, 40 °C).

(Lyon *et al.*, 2004b). The small absorption peak at 360 nm ($\varepsilon_{360} \sim 3.5$ mM^{-1} cm^{-1}), a shoulder at 300 nm ($\varepsilon_{350} \sim 12$ mM^{-1} cm^{-1}), and a tiny but significant peak at 259 nm ($\varepsilon_{350} \sim 40$ mM^{-1} cm^{-1}) are attributed to the cofactor bound to the enzyme. Examination of the UV–Vis spectrum can reveal whether the preparation is contaminated with other substances and whether the [Fe]-hydrogenase is inactivated. If the absorption peak at 360 nm of the purified preparation is lower than usual ($< \varepsilon_{350} \sim 3.5$ mM^{-1} cm^{-1}), the enzyme is partially inactivated or contaminants are present. A typical contaminant is methyl-coenzyme M reductase, which absorbs at 420 nm.

4. Preparation of FeGP Cofactor

4.1. Extraction of FeGP cofactor from [Fe]-hydrogenase

The crystal structure of [Fe]-hydrogenase from *M. jannaschii* revealed that the guanosine monophosphate moiety of the FeGP cofactor is bound to the mononucleotide-binding site on the N-terminal peripheral domain of the enzyme by hydrogen-bonding, and the iron site of the cofactor is covalently anchored to the thiol of Cys176 (Shima *et al.*, 2008). These two types of bonds are cleaved by 60% methanol/1 m*M* 2-mercaptoethanol/1% ammonia (Fig. 7.4). Ammonia stabilizes the thiolate form of 2-mercaptoethanol (Buurman *et al.*, 2000; Lyon *et al.*, 2004b).

Approximately 150 mg purified [Fe]-hydrogenase (60,000 U) in 40 ml water is supplemented with methanol, 2-mercaptoethanol, and ammonia at the final concentrations mentioned above. The protein concentration in the extract has to be lower than 5 mg ml^{-1}. Previously, the cofactor was extracted from the hydrogenase at 4 °C overnight, but sometimes, very small amounts of the extracted cofactor were obtained. Recently, we found that incubation of the extract at 40 °C for 15 min suffices for complete extraction with minimum inactivation of the cofactor. Approximately 100,000 U of FeGP cofactor is extracted from 140 mg of the enzyme after this shorter incubation. Thus, the extracted cofactor activity (see below) is sometimes much higher than the total activity of [Fe]-hydrogenase extracted, which suggests that part of the protein with bound FeGP cofactor is inactive.

The extract contains not only the FeGP cofactor, but also unfolded protein, contaminants from the enzyme preparation, methanol, 2-mercaptoethanol, ammonia, and water (Lyon *et al.*, 2004b). The protein can be removed by ultrafiltration, but ultrafiltration of extract containing 60% methanol takes a very long time (>6 h for 15 ml; Pilak *et al.*, 2006); during this time, much of the cofactor can decompose. Therefore, the protein is instead removed by salt precipitation with $CaCl_2$ (5 m*M* final concentration), which induces the aggregation of protein, and the precipitated

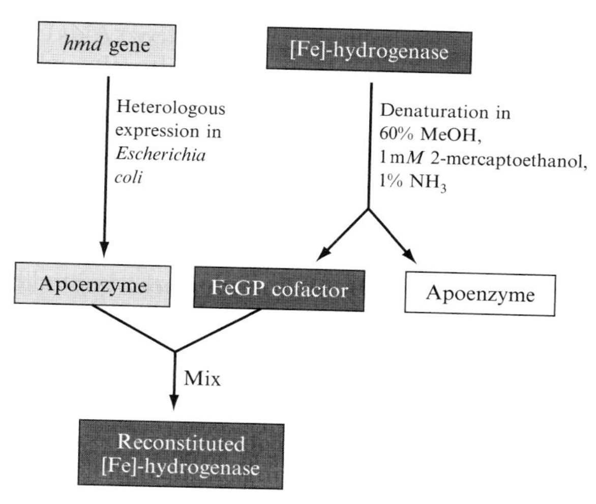

Figure 7.4 Procedure for [Fe]-hydrogenase reconstitution. Inactive [Fe]-hydrogenase apoenzyme is heterologously produced in *Escherichia coli* and purified. Active FeGP cofactor is extracted from native [Fe]-hydrogenase by denaturation in 60% methanol, 1 m*M* 2-mercaptoethanol, and 1% NH_3. Finally, reconstituted holoenzyme is formed from the apoenzyme and the FeGP cofactor mixture.

proteins are removed by centrifugation at 13,000×*g* for 20 min at 4 °C. After removal of most of the methanol from the supernatant by evaporation at 4 °C (see below), any remaining proteins in the supernatant can be quickly removed using Amicon Ultra centrifugal filter units (10-kDa cutoff; Millipore) for 10 min. Instead of using $CaCl_2$, the denatured proteins can also be precipitated by adding NaCl (0.2 m*M* final concentration) in 50 m*M* Tris/HCl, pH 8.0. The low-molecular-weight contaminants cannot be removed by salt precipitation and ultrafiltration. The FeGP cofactor preparation obtained is therefore only applicable for the reconstitution of [Fe]-hydrogenase for crystallization but not for other purposes. The cofactor extracts must be stored at −80 °C.

4.2. Purification of FeGP cofactor

To obtain highly purified FeGP cofactor, the $CaCl_2$ precipitation solution mentioned above is centrifuged, and the supernatant is diluted with the same volume of 0.01% ammonia/1 m*M* 2-mercaptoethanol (NH_3-mercaptoethanol) and then applied to a Q Sepharose High Performance (HiTrap, 5 ml) column equilibrated with NH_3-mercaptoethanol. The unfolded apoenzyme elutes in the flow-through fractions; the FeGP cofactor is then eluted by a linear-increasing gradient of NaCl from 0 to 1 *M* in 100 ml at a flow rate of 5 ml min^{-1}; 5 ml fractions are collected.

The FeGP cofactor elutes at around 0.46 *M* NaCl. Decomposed FeGP cofactor (guanylyl pyridinol) and other contaminants are removed by this method (Shima *et al.*, 2004). These FeGP cofactor fractions are pooled and applied to a Sephadex G-25 (HiPrep 26/10) column equilibrated with NH_3-mercaptoethanol. The FeGP cofactor binds to the Sephadex G25 resin in the presence of high concentrations of NaCl. Therefore, NaCl is eluted first; the FeGP cofactor is eluted from the resin when the NaCl concentration decreases to a conductivity of 10 mS cm^{-1}. Ammonia and 2-mercaptoethanol in the purified fraction can be removed by evaporation at 4 °C. Using this method, we have purified 60,000 U of cofactor (see below) from an extract containing 100,000 U cofactor from 140 mg [Fe]-hydrogenase.

4.3. Concentration of FeGP cofactor by evaporation

The FeGP cofactor decomposes when heated. Therefore, any evaporation has to be done quickly at lower temperature. We concentrated the FeGP cofactor sample by evacuating through an electric cold trap. The sample vial is placed in a water bath at room temperature in which the temperature of the sample is maintained lower than room temperature by continuous evaporation of water from the sample. The cooling medium in the water bath should be water rather than ice water. If ice water is used, the evaporation time increases, which results in decomposition of the cofactor. Lyophilization is not recommended because it requires longer periods for complete drying, and the cofactor decomposes during this time.

4.4. Characterization of purified FeGP cofactor

The purified FeGP cofactor is characterized by the cofactor activity assay (see below), UV–Vis spectroscopy, and chemical determination of the iron and phosphate contents. The UV–Vis spectrum of the FeGP cofactor is similar to that of [Fe]-hydrogenase in the region of 300–500 nm (Fig. 7.3; Lyon *et al.*, 2004b). Upon inactivation of the cofactor, the peak at 360 nm decreases and becomes less significant, and the shoulder at 300 nm increases. Thus, the quality of the FeGP cofactor preparation can be judged by its UV–Vis spectrum. Because the quality of the UV–Vis spectrum of the FeGP cofactor depends on the intactness of the structure, the concentration of the cofactor estimated using its extinction coefficient ($\varepsilon_{360} \sim 2$ mM^{-1} cm^{-1} and $\varepsilon_{300} \sim 5.5$ mM^{-1} cm^{-1}) is not highly accurate. An alternative means of determining the concentration of the FeGP cofactor preparation is to completely decompose an aliquot of the active FeGP cofactor to guanylyl pyridinol with high-intensity light (e.g., Cold Light Sources KL 2500 LC, Schott) and then to estimate the concentration of guanylyl pyridinol using its extinction coefficient ($\varepsilon_{300} \sim 9.0$; Shima *et al.*, 2004).

The amount of FeGP cofactor can also be determined by the chemical analysis of the phosphate group in the guanylyl pyridinol part or an analysis of the iron (Lyon *et al.*, 2004b).

5. Preparation of Reconstituted [Fe]-Hydrogenase

Many *hmd* genes encoding [Fe]-hydrogenase have been found in sequenced genomes (e.g., see Fig. 7.5). We have heterologously overexpressed some of the *hmd* genes in *E. coli* (Table 7.1). The holoenzyme can functionally be reconstituted from the apoenzyme and the FeGP cofactor at a scale for crystallographic analysis. Because the FeGP cofactor cannot yet be chemically synthesized, the cofactor for the reconstitution experiments has to be isolated from [Fe]-hydrogenase purified from methanogenic archaea. Heterologous synthesis of FeGP is also not yet possible; the *hcg* genes, which are clustered with the *hmd* gene in some methanogenic archaea, are proposed to be responsible for the biosynthesis of FeGP cofactor (Fig. 7.6).

5.1. Heterologous overproduction of apoenzyme in *E. coli*

The *hmd* genes from *M. jannaschii* and other methanogens can be overexpressed in *E. coli* BL21(DE3) by using T7lac promoter vectors with kanamycin resistance. When a vector encoding ampicillin resistance is used, carbenicillin should be used instead of ampicillin as a selective antibiotic in the growth medium. The proteins produced in *E. coli* cells cultivated in Luria–Bertani (LB) medium or in M9 minimal medium tend to aggregate and to produce inclusion bodies. Soluble [Fe]-hydrogenase proteins from *M. jannaschii* and some methanogens are successfully produced in the recombinant *E. coli* cells cultivated in trypton–phosphate (TP) medium (Pilak *et al.*, 2006). TP medium was developed by Moore *et al.* (1993) for producing soluble proteins in *E. coli*. Unfortunately, even when TP medium is used, many [Fe]-hydrogenase apoenzymes are produced in *E. coli* as inclusion bodies (Table 7.2). [Fe]-hydrogenase from *M. marburgensis* is heterologously produced only when the expression vector contains a synthetic *hmd* gene, whose codon usage and encoded mRNA structure are optimized for *E. coli* production. Still, the apoenzyme is mainly produced as inclusion bodies (Vogt *et al.*, unpublished results).

The *hmd* gene from *M. jannaschii* has been cloned into the vector pET24b (Buurman *et al.*, 2000). *E. coli* BL21(DE3) harboring the expression vector carrying *hmd* is precultivated at 37 °C overnight in 100 ml LB medium containing 30 μg ml^{-1} kanamycin in a 200-ml flask with shaking at 150 rpm. The whole preculture is inoculated into 2 l of prewarmed TP

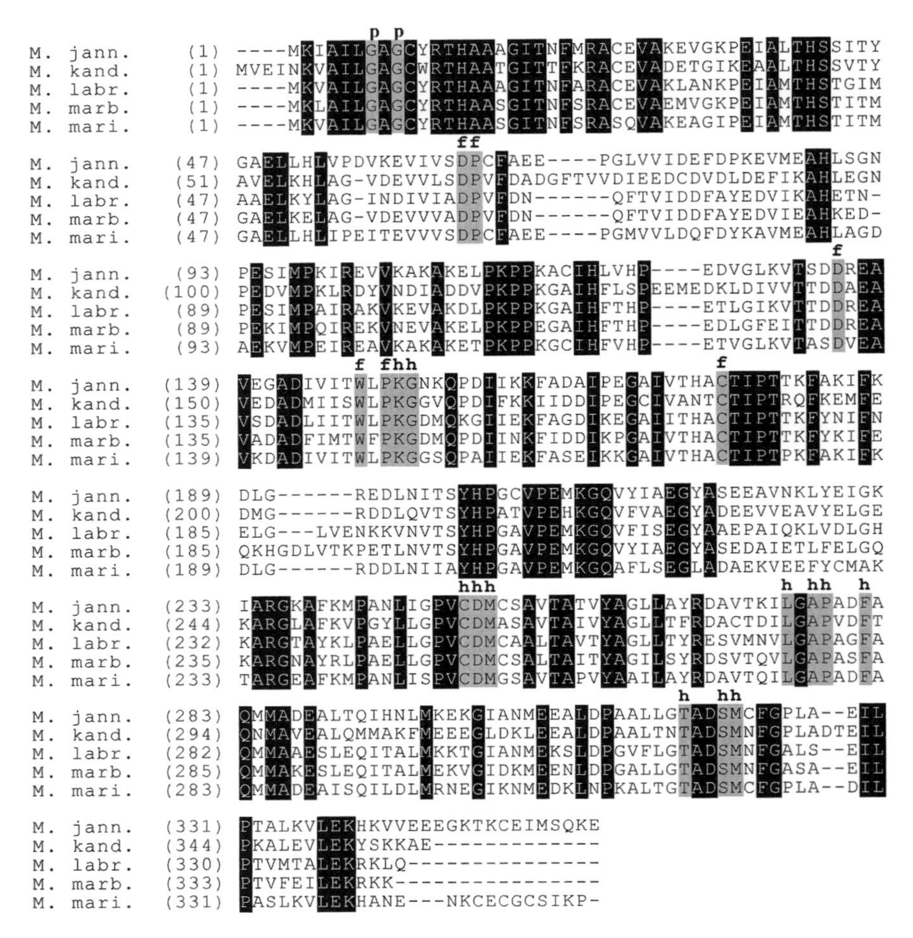

Figure 7.5 Alignment of some primary structures of [Fe]-hydrogenases. Conserved amino acid residues are shaded black, and those in the active site are shaded gray. (p) The monophosphate-binding motif GXG. (f) The residues contacting the FeGP cofactor. (h) The residues of the tetrahydromethanopterin-binding site. M. jann., *Methanocaldococcus jannaschii*; M. kand., *Methanopyrus kandleri*; M. labr., *Methanocorpusculum labreanum*; M. marb., *Methanothermobacter marburgensis*; M. mari., *Methanococcus maripaludis*. Sequences were aligned using Vector NTI (Invitrogen).

medium containing 30 μg ml^{-1} kanamycin in a 5-l flask; the culture is stirred at 650 rpm with a magnetic stirrer. When the ΔOD_{578} of the culture reaches 1.0, 1 m*M* IPTG (final concentration) is added to induce gene expression. After 3 h induction, the culture is cooled in ice water, and the cells are harvested under air by centrifugation at 3000×*g* for 20 min at 4 °C. The cells are frozen and stored at −80 °C.

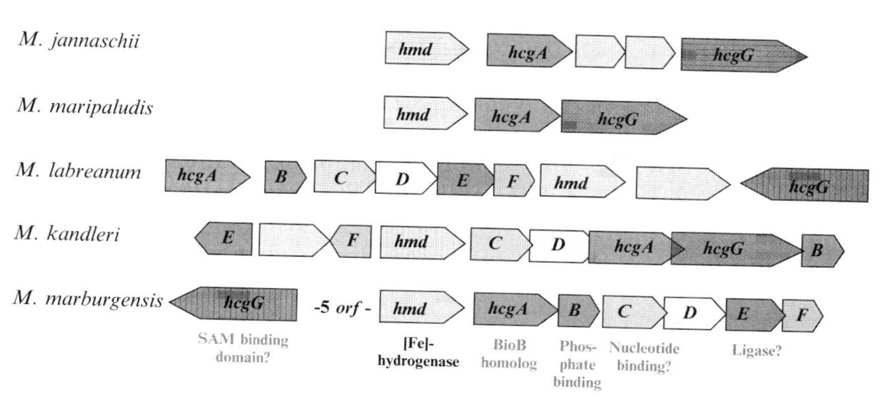

Figure 7.6 Genes neighboring the *hmd* gene for [Fe]-hydrogenase in *Methanococcus jannaschii*, *Methanococcus maripaludis*, *Methanocorpusculum labreanum*, *Methanopyrus kandleri*, and *Methanothermobacter marburgensis* are shown. *hcg* are putative genes for biosynthesis of FeGP cofactor found near some *hmd* genes (Thauer *et al.*, 2010). The genomes of *M. jannaschii* and *M. maripaludis* also contain *hcg* B–F apart from the *hmd* genes.

[Fe]-hydrogenase apoenzyme is purified under oxic conditions at 4 °C. The cells are suspended in 50 m*M* Mops/KOH pH containing 1 m*M* dithiothreitol, and the cell suspension is disrupted at 4 °C by sonication in ice water with a 50% cycle and 160 W (five times 4 min, with 4-min pauses). The supernatant is fractionated by ultracentrifugation at 130,000×*g* for 40 min at 4 °C. The *E. coli* proteins in the supernatant are precipitated by heating at 70 °C for 15 min, and the protein aggregates are removed by centrifugation at 13,000×*g* for 20 min at 4 °C. Ammonium sulfate powder is added to a final concentration of 2 *M*. After removing the precipitated protein by centrifugation at 13,000×*g* for 20 min at 4 °C, the supernatant is applied to a Phenyl Sepharose (26/20) column equilibrated with 2 *M* ammonium sulfate in 50 m*M* Mops/KOH buffer, pH 7.0 with 1 m*M* dithiothreitol. After washing the column with the equilibration buffer, the [Fe]-hydrogenase apoenzyme is eluted by a linear-decreasing gradient of ammonium sulfate from 2 to 0 *M* in 200 ml. The apoenzyme is eluted at 0.6 *M* ammonium sulfate. The fractions containing the apoenzyme are pooled and concentrated to 50 mg ml^{-1} and then desalted on a HiPrep G25 desalting column equilibrated with 50 m*M* Mops/KOH, pH 7.0 containing 1 m*M* dithiothreitol. The purified apoenzyme is frozen and stored at −80 °C.

5.2. Reconstitution of the holoenzyme from the FeGP cofactor and apoenzyme

The [Fe]-hydrogenase holoenzyme is reconstituted under strictly anoxic conditions (Shima *et al.*, 2008). The purified apoenzyme and FeGP cofactor are mixed to give final concentrations of 0.125 and 0.175 m*M*, respectively, in 2 ml of 50 m*M* Tris/HCl, pH 8.5. Reconstitution takes place immediately at 8 °C. The reconstituted enzyme is separated from the excess FeGP cofactor on a HiPrep G25 desalting column equilibrated with 10 m*M* Mops/KOH, pH 7.0. The reconstituted enzyme typically has an activity of 300–400 U mg^{-1} at 40 °C using 20 μ*M* methylene-H_4MPT as substrate in the standard assay (see below). This level of activity is equivalent to the full activity of the enzyme purified from *M. marbugensis*.

6. Assay of the Enzyme Activity

[Fe]-hydrogenase enzyme activity is routinely assayed photometrically (Zirngibl *et al.*, 1992) at 40 °C in 0.7 ml buffer in 1-ml quartz cuvettes (1 cm light pass) under a N_2 atmosphere and closed with rubber stoppers. The reconstitution ability of the isolated cofactor and the apoenzyme is determined using similar conditions after reconstitution of the holoenzyme in the assay cuvettes. H_4MPT and methenyl-H_4MPT^+ are purified from the cell extract of *M. marburgesis*, as described previously (Shima and Thauer, 2001). Methylene-H_4MPT is generated by spontaneous reaction from H_4MPT (2 μmol) and formaldehyde (4 μmol) in 0.5 ml of 120 m*M* potassium phosphate, pH 6.0. Formaldehyde is removed by evaporation at 4 °C, and methylene-H_4MPT is subsequently dissolved in anoxic water. Removal of formaldehyde is necessary because it somewhat inactivates the cofactor.

6.1. Assay of the holoenzyme

The oxidation of methylene-H_4MPT by the [Fe]-hydrogenase holoenzyme is assayed under a 100% N_2 atmosphere. The gas phase is exchanged with 100% N_2 gas by evacuation and refilling with N_2 three times. After release of the N_2 overpressure by punching the rubber stopper with a thin needle, 0.68 ml anoxic 120 m*M* potassium phosphate, pH 6.0 with 1 m*M* EDTA is introduced into the cuvette, and the overpressure is again released. The assay cuvette is incubated at 40 °C for at least 5 min and then transferred to the photometer at 40 °C. The substrate methylene-H_4MPT is added to a final concentration of 20 μ*M*, and the reaction is then started by the addition of 5–10 μl of the enzyme solution (~3 U ml^{-1}). The conversion of methylene-H_4MPT to methenyl-H_4MPT^+ is measured via the increase in absorbance at 335 nm. The activity is calculated using $\varepsilon_{335} = 21.6\ mM^{-1}\ cm^{-1}$. One unit enzyme activity refers to 1 μmol methylene-H_4MPT converted to methenyl-H_4MPT^+ per minute at 40 °C. At 60 °C, the specific activity is 2.5 times higher. At 40 μ*M* methylene-H_4MPT, the specific activity is 1.5 times higher.

The reverse reaction, hydrogenation of methenyl-H_4MPT^+, is assayed at 40 °C in 1 ml quartz cuvettes, essentially as described above for the determination of the forward reaction activity. However, the gas phase of the cuvette is 100% H_2, and the assay buffer is 120 m*M* potassium phosphate, pH 7.5, 1 m*M* EDTA (Zirngibl *et al.*, 1992). Methenyl-H_4MPT^+ is injected into the assay cuvette to a final concentration of 20 μ*M*. The reverse reaction is measured as the decrease of methenyl-H_4MPT^+ concentration by following the decrease in absorbance at 335 nm. One unit of enzyme activity refers to 1 μmol methenyl-H_4MPT^+ converted to methylene-H_4MPT per minute at 40 °C.

6.2. Assay of the cofactor

The FeGP cofactor activity is assayed by measuring the activity of [Fe]-hydrogenase reconstituted from a limited amount of FeGP cofactor and an excess amount of the apoenzyme. Ten microliters of 1 mg ml^{-1} apoenzyme from *M. jannaschii* is added to the 0.7-ml assay solution at 40 °C, and then 10 μl of 0.25 μ*M* (3 U ml^{-1}) FeGP cofactor solution is added. The final concentration of the apoenzyme and FeGP cofactor in the assay is 370 and 3.7 n*M*, respectively. The reaction is started by the addition of substrates to give a final concentration of 20 μ*M*.

6.3. Assay of the apoenzyme

The apoenzyme activity is assayed by measuring the activity of [Fe]-hydrogenase reconstituted from an excess amount of FeGP cofactor and a limited amount of the apoenzyme in the assay cuvette. Ten microliters of 0.01 mg ml^{-1} apoenzyme from *M. jannaschii* is added to the assay solution at 40 °C, and then 10 μl of 0.1 m*M* (1200 U ml^{-1}) FeGP cofactor solution is added. The final concentration of apoenzyme and FeGP cofactor in the assay is 3.7 and 1400 n*M*, respectively. The reaction is started by the addition of substrates to give a final concentration of 20 μ*M*.

ACKNOWLEDGMENTS

This work was supported by the Max Planck Society and the PRESTO program of the Japan Science and Technology Agency (JST). The authors were financed by a grant to R. K. Thauer from the Max Planck Society, who is also thanked for helpful discussions and critical reading of the chapter.

REFERENCES

Afting, C., Hochheimer, A., and Thauer, R. K. (1998). Function of H_2-forming methylenetetrahydromethanopterin dehydrogenase from *Methanobacterium thermoautotrophicum* in coenzyme F_{420} reduction with H_2. *Arch. Microbiol.* **169,** 206–210.

Anderson, I., Ulrich, L. E., Lupa, B., Susanti, D., Porat, I., Hooper, S. D., Lykidis, A., Sieprawska-Lupa, M., Dharmarajan, L., Goltsman, E., Lapidus, A., Saunders, E., *et al.* (2009). Genomic characterization of Methanomicrobiales reveals three classes of methanogens. *PLoS ONE* **4,** e5797.

Bradford, M. M. (1976). Rapid and sensitive method for quantitation of microgram quantities of protein utilizing principle of protein-dye binding. *Anal. Biochem.* **72,** 248–254.

Buurman, G., Shima, S., and Thauer, R. K. (2000). The metal-free hydrogenase from methanogenic archaea: Evidence for a bound cofactor. *FEBS Lett.* **485,** 200–204.

Hartmann, G. C., Klein, A. R., Linder, M., and Thauer, R. K. (1996). Purification, properties and primary structure of H2-forming N5,N10-methylenetetrahydromethanopterin dehydrogenase from *Methanococcus thermolithotrophicus*. *Arch. Microbiol.* **165,** 187–193.

Hiromoto, T., Ataka, K., Pilak, O., Vogt, S., Salmone-Stagni, M., Meyer-Klaucke, W., Warkentin, E., Thauer, R. K., Shima, S., and Ermler, U. (2009a). The crystal structure of C176A mutated [Fe]-hydrogenase suggests an acyl-iron ligation in the active site iron complex. *FEBS Lett.* **583,** 585–590.

Hiromoto, T., Warkentin, E., Moll, J., Ermler, U., and Shima, S. (2009b). The crystal structure of an [Fe]-hydrogenase-substrate complex reveals the framework for H_2 activation. *Angew. Chem. Int. Ed.* **48,** 6457–6460.

Korbas, M., Vogt, S., Meyer-Klaucke, W., Bill, E., Lyon, E. J., Thauer, R. K., and Shima, S. (2006). The iron-sulfur cluster-free hydrogenase (Hmd) is a metalloenzyme with a novel iron binding motif. *J. Biol. Chem.* **281,** 30804–30813.

Lyon, E. J., Shima, S., Boecher, R., Thauer, R. K., Grevels, F. W., Bill, E., Roseboom, W., and Albracht, S. P. J. (2004a). Carbon monoxide as an intrinsic ligand to iron in the active site of the iron-sulfur-cluster-free hydrogenase H_2-forming methylenetetrahydromethanopterin dehydrogenase as revealed by infrared spectroscopy. *J. Am. Chem. Soc.* **126,** 14239–14248.

Lyon, E. J., Shima, S., Buurman, G., Chowdhuri, S., Batschauer, A., Steinbach, K., and Thauer, R. K. (2004b). UV-A/blue-light inactivation of the 'metal-free' hydrogenase (Hmd) from methanogenic archaea: The enzyme contains functional iron after all. *Eur. J. Biochem.* **271,** 195–204.

Ma, K., Zirngibl, C., Linder, D., Stetter, K. O., and Thauer, R. K. (1991). N^5,N^{10}-Methylenetetrahydromethanopterin dehydrogenase (H_2-forming) from the extreme thermophile *Methanopyrus kandleri*. *Arch. Microbiol.* **156,** 43–48.

Moore, J. T., Uppal, A., Maley, F., and Maley, G. F. (1993). Overcoming inclusion body formation in a high-level expression system. *Protein Expr. Purif.* **4,** 160–163.

Pilak, O., Mamat, B., Vogt, S., Hagemeier, C. H., Thauer, R. K., Shima, S., Vonrhein, C., Warkentin, E., and Ermler, U. (2006). The crystal structure of the apoenzyme of the iron-sulfur cluster-free hydrogenase. *J. Mol. Biol.* **358,** 798–809.

Salomone-Stagni, M., Stellato, F., Whaley, C. M., Vogt, S., Morante, S., Shima, S., Rauchfuss, T. B., and Meyer-Klaucke, W. (2010). The iron-site structure of [Fe]-hydrogenase and model systems: An X-ray absorption near edge spectroscopy study. *Dalton Trans.* **39,** 3057–3064.

Schönheit, P., Moll, J., and Thauer, R. K. (1980). Growth parameters (K_s, μ_{max}, Y_s) of *Methanobacterium thermoautotrophicum*. *Arch. Microbiol.* **127,** 59–65.

Schwörer, B., and Thauer, R. K. (1991). Activities of formylmethanofuran dehydrogenase, methylenetetrahydromethanopterin dehydrogenase, methylenetetrahydromethanopterin reductase, and heterodisulfide reductase in methanogenic bacteria. *Arch. Microbiol.* **155,** 459–465.

Shima, S., and Thauer, R. K. (2001). Tetrahydromethanopterin-specific enzymes from *Methanopyrus kandleri*. *Methods Enzymol.* **331,** 317–353.

Shima, S., and Thauer, R. K. (2007). A third type of hydrogenase catalyzing H_2 activation. *Chem. Rec.* **7,** 37–46.

Shima, S., Lyon, E. J., Sordel-Klippert, M. S., Kauss, M., Kahnt, J., Thauer, R. K., Steinbach, K., Xie, X. L., Verdier, L., and Griesinger, C. (2004). The cofactor of the iron-sulfur cluster free hydrogenase Hmd: Structure of the light-inactivation product. *Angew. Chem. Int. Ed.* **43,** 2547–2551.

Shima, S., Lyon, E. J., Thauer, R. K., Mienert, B., and Bill, E. (2005). Mössbauer studies of the iron-sulfur cluster-free hydrogenase: The electronic state of the mononuclear Fe active site. *J. Am. Chem. Soc.* **127,** 10430–10435.

Shima, S., Pilak, O., Vogt, S., Schick, M., Stagni, M. S., Meyer-Klaucke, W., Warkentin, E., Thauer, R. K., and Ermler, U. (2008). The crystal structure of [Fe]-hydrogenase reveals the geometry of the active site. *Science* **321,** 572–575.

Shima, S., Thauer, R. K., and Ermler, U. (2009). Carbon monoxide as intrinsic ligand to iron in the active site of [Fe]-hydrogenase. *In* "Metal Ions in Life Sciences, Vol. 6," (A. Sigel, H. Sigel, and R. K. O. Sigel, eds.), pp. 219–240. The Royal Society of Chemistry, Cambridge.

Thauer, R. K. (1998). Biochemistry of methanogenesis: A tribute to Marjory Stephenson. *Microbiology* **144,** 2377–2406.

Thauer, R. K., Klein, A. R., and Hartmann, G. C. (1996). Reactions with molecular hydrogen in microorganisms: Evidence for a purely organic hydrogenation catalyst. *Chem. Rev.* **96,** 3031–3042.

Thauer, R. K., Kaster, A. K., Goenrich, M., Schick, M., Hiromoto, T., and Shima, S. (2010). Hydrogenases from methanogenic archaea, nickel, a novel cofactor, and H_2 storage. *Annu. Rev. Biochem.* **79,** 507–536.

Vogt, S., Lyon, E. J., Shima, S., and Thauer, R. K. (2008). The exchange activities of [Fe] hydrogenase (iron-sulfur-cluster-free hydrogenase) from methanogenic archaea in comparison with the exchange activities of [FeFe] and [NiFe] hydrogenases. *J. Biol. Inorg. Chem.* **13,** 97–106.

Zirngibl, C., Hedderich, R., and Thauer, R. K. (1990). N^5,N^{10}-Methylenetetrahydromethanopterin dehydrogenase from *Methanobacterium thermoautotrophicum* has hydrogenase activity. *FEBS Lett.* **261,** 112–116.

Zirngibl, C., van Dongen, W., Schwörer, B., von Bünau, R., Richter, M., Klein, A., and Thauer, R. K. (1992). H_2-forming methylenetetrahydromethanopterin dehydrogenase, a novel type of hydrogenase without iron-sulfur clusters in methanogenic archaea. *Eur. J. Biochem.* **208,** 511–520.

CHAPTER EIGHT

Assay of Methylotrophic Methyltransferases from Methanogenic Archaea

Donald J. Ferguson Jr.,* David G. Longstaff,† and Joseph A. Krzycki‡

Contents

Abstract

The family Methanosarcinaceae has an expanded repertoire of growth substrates relative to most other methanogenic archaea. Various methylamines, methylated thiols, and methanol can serve as precursors to both methane and carbon dioxide. These compounds are mobilized into metabolism by methyltransferases that use the growth substrate to methylate a cognate corrinoid protein, which in turn is used as a substrate by a second methyltransferase to methylate Coenzyme M (CoM), forming methyl-SCoM, the precursor to both

* Department of Microbiology, University of Miami, Oxford, Ohio, USA
† Department of Microbiology, Ohio State University, Columbus, Ohio, USA
‡ Department of Microbiology and the OSU Biochemistry Program, Ohio State University, Columbus, Ohio, USA

Methods in Enzymology, Volume 494
ISSN 0076-6879, DOI: 10.1016/B978-0-12-385112-3.00008-1

methane and carbon dioxide. Orthologs of the methyltransferases, as well as the small corrinoid proteins, are found in many archaeal and bacterial genomes. Some of these are homologs of the methylamine methyltransferases predicted to require pyrrolysine, an atypical genetically encoded amino acid, for synthesis. As a resource for the study of these sizable families of proteins, we describe here techniques our laboratories have used for the study of methanogen corrinoid-dependent methyltransferases, focusing especially on isolation and assay techniques useful for various activities of components of the methylamine- and methylthiol-dependent CoM methyltransferase systems.

1. Introduction

Organisms such as *Methanosarcina barkeri* and *Methanosarcina acetivorans* have expanded the ancestral pathway of methanogenesis from carbon dioxide to include a greater variety of substrates. The majority of these compounds are those bearing methyl groups not bonded to another carbon, including methanol, monomethylamine (MMA), dimethylamine (DMA), trimethylamine (TMA), dimethylsulfide (DMS), methanethiol (MSH), and methylmercaptopropionate (MMPA). Such substrates are disproportionated to carbon dioxide and methane in a process generally known as methylotrophic methanogenesis. Both carbon dioxide and methane are formed following intermediate methylation of Coenzyme M (CoM) by analogous short pathways comprised of two methyltransferases and a cognate corrinoid protein (Krzycki, 2004). Study of these methyltransferases has led to greater understanding of the evolution of the biochemical diversity of biological methane production, and, in the case of the methylamine methyltransferases, led directly to the discovery of the 22nd genetically encoded amino acid, pyrrolysine (Krzycki, 2004). A number of homologs of these proteins are encoded in genomes of many bacteria and archaea (Rother and Krzycki, 2010). The physiological roles of these proteins are often unknown. Here, we describe techniques that may be useful for the study of such systems taking as our examples the enzymes for CoM methylation with methylotrophic substrates.

The corrinoid proteins have the fold found in the cobalamin-binding domain of methionine synthase (Drennan *et al.*, 1994) and are methylated in the Co(I) state by the first methyltransferase specific for the methylotrophic growth substrate. The resultant methyl-Co(III) protein is used by a second methyltransferase to methylate CoM. Substrates such as MMA, DMA, TMA, and methanol use the "corrinoid protein and two methyltransferases" approach (Burke and Krzycki, 1997; Ferguson *et al.*, 1997, 2000; Sauer and Thauer, 1999; van der Meijden *et al.*, 1983). Variations on this theme are seen in the use of methylthiols such as DMS or MMPA by

Methanosarcina spp. in which one methyltransferase serves to both methylate the corrinoid protein or domain with a methylthiol and then demethylate the corrinoid protein with CoM (Oelgeschlager and Rother, 2009; Tallant *et al.*, 2001).

An older naming system designated the initial methyltransferase and the cognate corrinoid protein together as MT1, and the CoM-dependent methyltransferases as MT2 (van der Meijden *et al.*, 1983). This nomenclature reflected the proclivity of the methanol methyltransferase (the first described of these systems) to form complexes with its cognate corrinoid protein. However, while tight complexes are observed with the TMA methyltransferase MttB (Ferguson and Krzycki, 1997) and methanol methyltransferase MtaB (van der Meijden *et al.*, 1983), the DMA and MMA methyltransferases (MtbB and MtmB, respectively) can both be isolated without their cognate corrinoid protein (Burke and Krzycki, 1997; Ferguson *et al.*, 2000). We have also found that the conditions of isolation sometimes favor formation of complexes that readily dissociate under other conditions, a trait that can be exploited during isolation of MtmB and its cognate corrinoid protein, MtmC.

All of the isolated methyltransferases can retain activity after lengthy exposure to air. However, it should be noted that the majority of their assays must take place under strictly anaerobic conditions, primarily so that the extremely oxygen-sensitive Co(I) state of the corrinoid protein may be maintained. The Co(I) state is easily oxidized to Co(II) and hydroxy-Co(III) forms, which remove the protein from the catalytic cycle. *In vitro*, Co(I) corrinoid proteins are generated with Ti(III) citrate, often in conjunction with methyl viologen. Alternatively, Co(I) may be generated by addition of Ti(III) citrate and an ATP-dependent redox active protein isolated from *M. barkeri* extracts called RamA (Ferguson *et al.*, 2009), which can be added as either as pure or an enriched protein fraction. The necessary anoxic conditions can be generated using evacuation and flushing with nitrogen or hydrogen of vials and solutions, along with the use of an anaerobic chamber. Full description of these methods is beyond the scope of this article, and the reader is directed to consult a more comprehensive review of the subject (Sowers, 1995). However, some anoxic techniques are described as they arise in the protocols here.

2. Preparation of Anoxic Cell Extracts of *M. barkeri*

Anoxic extracts are often made as a prelude to many of the following protocols. Even highly oxygen-sensitive enzymes can be assayed in or isolated from anoxically harvested frozen cells that were stored aerobically at −80 °C if first they were made again anaerobic before thawing. We describe here a method for the preparation of an anaerobic extract of *M. barkeri*, our organism of choice for most methylotrophic enzymes. Frozen *M. barkeri* tends to aggregate once introduced into solution, which can compromise lysis in a French pressure cell. Many organisms will not require similar rough treatment, but we find it results in higher lysis of previously frozen *M. barkeri*.

1. Frozen *M. barkeri* cells are suspended in liquid nitrogen and added to a metal Waring blender. Additional liquid nitrogen can be added as necessary to keep the cells frozen and blanketed under nitrogen until the cell paste is blended into a smooth slurry in liquid nitrogen.
2. The cell slurry is transferred to a sidearm vacuum flask with a butyl rubber stopper set loosely until most of the liquid nitrogen sublimes. The stopper is set firmly in place, and the cell paste is subjected to vacuum for several minutes followed by gassing with hydrogen, repeated for four more cycles. Take care to operate the gassing line at just several pounds over atmospheric pressure and secure the stopper when applying gas. Avoid drying the thawing cell paste.
3. When thawed, add anaerobic 50 m*M* MOPS buffer, pH 7.0, at a ratio of 1–3 ml buffer per gram of cell paste. To be made anaerobic, large volumes of buffer should be stirred rapidly with a magnetic stir bar under vacuum until the liquid is thoroughly degassed, then subjected to several more cycles of evacuation and flushing with nitrogen.
4. The cell suspension is loaded into a nitrogen-flushed French pressure cell whose outlet is fitted with rubber tubing with an 18-gauge needle which is then inserted into a hydrogen-flushed rubber-stoppered polycarbonate ultracentrifuge tube that can later be closed with an o-ringed cap for centrifugation. Finish assembling the French pressure cell. The pressure cell can also be loaded in an anaerobic chamber (Coy Laboratory Products, Grass Lakes, MI).
5. The cells are lysed at 20,000 lb/in^2 in the French press, and the lysate collected in the stoppered centrifuge tube. A vent needle is introduced into the tube stopper as collection proceeds. Alternatively, a hydrogen line can be used to gentle flush the collection tube during lysis.
6. The stopper is exchanged for the centrifuge bottle lid inside the anaerobic chamber. Make sure the o-ring seals well. The lysate is then centrifuged at 150,000×*g* for 30 min in an ultracentrifuge.
7. The supernatant can be frozen at −80 °C in a stopped vial. For subsequent use in anaerobic assays, it should be thawed under a gentle stream of hydrogen. Nitrogen can replace hydrogen throughout the protocol, as it displaces oxygen more efficiently, but hydrogen helps prevent oxygen-mediated disruption of metal centers, such as iron–sulfur clusters, as the extract is kept reduced by endogenous

hydrogenases. However, once made anaerobic, an extract should not be exposed to oxygen, as this can result in oxygen radical damage to proteins. If desired, 1–2 m*M* dithiothreitol can be added to extracts as a reductant. Dithionite should be avoided, as it can lead to formation of stable sulfito derivatives of the corrinoid cofactor, thereby inactivating the corrinoid proteins.

3. Isolation and Resolution of MtmBC Complexes

Escherichia coli requires five genes for pyrrolysine biosynthesis and insertion into recombinant MtmB, which is disappointingly insoluble (Longstaff *et al.*, 2007). Therefore, we describe an aerobic isolation protocol for native MtmB as a complex with cognate corrinoid protein MtmC. When reduced, the two proteins separate.

1. The following protocol assumes some familiarity with commonly used protein purification techniques. Aerobically made extract (50 ml) of MMA-grown *M. barkeri* (20–30 mg protein/ml) is applied to a DE-52 column (5 × 10 cm) equilibrated with 50 m*M* Tris–HCl, pH 8.0, and then eluted with a 2 L gradient of 50–500 m*M* NaCl in 50 m*M* Tris, pH 8.0. MtmB, MtmC, and MtbA copurify from the column around 280 m*M* NaCl in 200 ml. The MMA:CoM activity may be detected using the Ellman's reagent for assay of the free thiol of CoM described below. The active fractions are reddish-brown due to the presence of hydroxy-Co(III)-MtmC and coeluting iron–sulfur proteins.
2. The pool of active fractions is loaded onto a Q-Sepharose column (15 × 5 cm) equilibrated with 50 m*M* MOPS, pH 6.5, and eluted with a 1.1 L gradient of 100–500 m*M* NaCl in the same buffer. MtmBC elutes from the Q-Sepharose column in approximately 10% of gradient volume centered at 380 m*M* NaCl. At this stage, fractions assayed for MMA:CoM methyltransferase activity may require addition of MtbA, the amine-specific methylcorrinoid:CoM methylase (Ferguson *et al.*, 1996), as MtmBC and MtbA may separate. MtbA containing fractions may be identified in the column fractions using the methylcob(III)alamin:CoM assay described below. Alternatively, recombinant MtbA can be employed (Gencic *et al.*, 2001).
3. The active MtmBC fractions must be concentrated using ultrafiltration with a YM-30 (Millipore, Inc., Billerica, MA) or equivalent membrane to approximately 10–20 ml. This sample is then applied to a 400 ml (2.5 × 80 cm) Sephacryl S-100 column equilibrated with 50 m*M* NaCl

in 50 m*M* MOPS, pH 7.0. MtmBC elutes after 200 ml in pink to red colored fractions in a total of 20 ml.

4. MtmB and MtmC are copurified to near homogeneity with a Mono Q 10/10 column equilibrated with 50 m*M* MOPS, pH 6.5, and then eluted using a 160-ml linear gradient from 100 m*M* NaCl to 500 m*M* NaCl in 50 m*M* MOPS, pH 7.0. This is a medium-pressure chromatography column, and the manufacturer's instructions for operation of an appropriate liquid chromatography system should be followed. Elution of the corrinoid protein can be monitored at 530 nm and by the red color of the hydroxy-Co(III) form of MtmC.
5. Concentrate the sample by ultrafiltration using a YM-30 membrane. The MtmBC fraction is then made anaerobic by flush/evacuation. Care must be taken to avoid excessive foaming of the proteins under vacuum; this can inactivate the proteins. Many gases for operating the manifold are contaminated with sufficient oxygen to cause problems. The nitrogen gas supplied to the manifold should be equipped with an oxygen-scrubbing column to remove residual oxygen from the inlet gas. We employ a No-Air glass catalyst column that is thermal jacketed (Ace Glass, Inc., Vineland, NJ) which can be heated for activation of BASF catalyst R3-11G or R3-15 (Research Catalysts, Inc., The Woodlands, TX) under a stream of hydrogen in nitrogen as per manufacturer's instructions. Once activated, it can be used at room temperature.
6. Prepare a stock solution of Ti(III) citrate by an amendment of the techniques of these workers (Seefeldt and Ensign, 1994). Never add Ti(III) citrate directly to aerobic protein solutions, as oxidative damage will result. Typically, we prepare a 90–100 m*M* Ti(III) citrate solution by addition of 2.23 ml of 500 m*M* trisodium citrate and 2.25 ml of saturated Tris–NaOH to 0.5 ml of a commercial preparation of 10% Ti (III) chloride in 20–30% HCl (Sigma-Aldrich, St. Louis, MO) inside the anaerobic chamber. The Ti(III) citrate stock solution should have a final pH of 8.0–8.8 when checked with pH paper. Due to the variable acid concentration in different commercial preparations, the formula may need to be adjusted to include less or more base. Alternatively, solid Ti (III) chloride can be converted to Ti(III) citrate (Seefeldt and Ensign, 1994).
7. The final purification step is performed inside an anaerobic chamber. The concentrated MtmBC from the preceding column will be reduced and loaded onto a Sephacryl-S100 gel filtration column (80 × 2.5 cm) that has been pre-equilibrated with anaerobic 50 m*M* NaCl in 50 m*M* MOPS, pH 7.0. Buffers and column material can be made anoxic outside the chamber. Stirring bars should be avoided in degassing the slurry of Sephacryl-S100 used to make the column, as they will damage the beads. Instead, manually agitate the flask when under vacuum. The column may then be poured in the anaerobic chamber.

8. Once the column is prepared, make 20 ml of anaerobic 5 m*M* sodium dithionite in equilibration buffer and load it onto the column. Elute with equilibration buffer to reduce the matrix and continue elution until two column volumes of anoxic equilibration buffer have passed through the column. At this point, add 5 m*M* Ti(III) citrate to the anaerobic concentrated MtmBC complex in the chamber. Allow to incubate for half an hour and then load onto the column and elute with buffer. MtmB and MtmC will elute after 200 and 260 ml, respectively. This procedure will yield pure MtmC in the orange Co(II) form. If small amounts of MtmC are still bound to MtmB, a second reduction and purification may be required to fully separate the proteins.

4. Ellman's Assay of Methylotrophic Substrate: CoM Methyl Transfer

4.1. Assay for a component protein of a methyltransferase system using diluted extracts

The rate of methylotrophic substrate conversion to methyl-CoM is most easily measured by the substrate-dependent loss of the free thiol of CoM using Ellman's reagent, 5,5-dithio-bis(2-nitrobenzoic acid) or DTNB (Ellman, 1958). The reaction forms 5-thio-2-nitrobenozoic acid (TNB) and the mixed dithiol of TNB and CoM. The former has an estimated extinction coefficient of 13.6 $mM^{-1}cm^{-1}$ at 412 nm, but see a recent discussion of alternative estimates (Owusu-Apenten, 2005). Here, we describe using this assay to identify and isolate a single component protein of an unresolved methyltransferase system by monitoring the component's stimulation of methylotrophic substrate:CoM methyltransferase activity in diluted extract. This approach was used to isolate the first proteins of the TMA and MMA methyl transfer systems (Burke and Krzycki, 1995; Ferguson and Krzycki, 1997) The protocol is provided as an aide to the identification of novel methyltransferases from other systems but can be easily adapted to assay of methylamine:CoM or methanol:CoM methyltransferase activities directly in extracts.

1. All assays are carried out in anaerobic 2 ml glass vials sealed with a tight-fitting Wheaton sleeve type red rubber stopper (Wheaton Scientific, Millville, NJ). The vials can be attached to a flush/evacuation manifold fitted with 22 guage needles for five cycles of evacuation and flushing with hydrogen. For assay of methyltransferase systems in extracts of methanogens, the addition of hydrogen provides reducing equivalents to the RamA protein or homologs that reduce the cognate corrinoid protein of the desired methyltransferase system in an MgATP-dependent reaction (Ferguson *et al.*, 2009).

2. Prepare an anaerobic solution that contains a mix of reagents that will be added to the assay vials. Stocks of each solution can be made aerobically and combined to give a final concentration of 62.5 m*M* ATP, 125 m*M* $MgCl_2$, 50 m*M* CoM, and 20 m*M* bromoethanesulfonate (BES) in 50 m*M* MOPS buffer, pH 7.0. This assay mix is then made anaerobic by flush/evacuation. Smaller volumes of solution can be flushed/evacuated without a stirring bar. For the first several applications of the vacuum, the sides of the vials can be gently bumped with a rubber stopper to assist degassing of the solution. Five to six cycles of evacuation over a period of half an hour is generally sufficient to make a solution ready for anaerobic assay. Care should be taken that the solution does not boil up into the manifold, which can lead to contamination between stock solutions for various assays. Note, most ATP solutions in water are highly acidic and will need to be adjusted to near neutrality. BES is added to prevent reduction of methyl-CoM to methane.
3. Prepare 1 ml of anaerobic 400 m*M* methylated substrate (TMA, DMA, MMA, or methanol). This will be used to initiate the reaction.
4. Thaw protein fractions to be tested in the stimulation assay, as well as extract for use in the assay. These should be prepared anoxically, and if stored frozen in stoppered vials, a vent needle should be inserted into the stopper before thawing, and the vial placed on the gassing manifold for thawing under a gentle stream of nitrogen or hydrogen.
5. Prepare a 1 ml solution of 90–100 m*M* Ti(III) citrate solution in the anaerobic chamber.
6. Prepare a 150-ml stoppered vial with anaerobic 50 m*M* MOPS, pH 7.0, containing 1 m*M* Ti(III) citrate. This will be used to flush syringes prior to additions to the assay vials.
7. Prepare 0.5 m*M* DTNB in 150 m*M* Tris–HCl pH 8.0; then dispense 250 μl aliquots into the wells of a microtiter plate to which time points of the assay will be added. Add reagent to sufficient wells for samples, standards, and a blank for zeroing the plate reader. Ellman's reagent will slowly hydrolyze over time, leading to yellow TNB. The solution should be made fresh, and solutions of a similar age used for all samples and standards in an assay series.
8. Prepare 8 and 2 m*M* standards of CoM. Add 3 μl of each to different plate wells containing Ellman's reagent to determine the average ratio of CoM concentration to A_{412} in the assay wells. The reaction of the Ellman's reagent and CoM is rapid and usually complete in less than 5 min.
9. The volume of methylotrophic substrate grown extract added to the reaction is variable depending on the activity of the extract. Therefore, perform a determination of the proper dilution by using increasing amounts of the extract in the assay below, but without any column fractions being tested for a particular component of the methyltransferase system under assay. A dilution of extract yielding a relatively low

rate of CoM methylation (< 5% of total CoM) over a 30-min period is sufficient. Note, if your aim is to assay methyltransferase activity solely in extracts, up to 100 μl of the extract can be added in place of the MOPS buffer and the column fraction in step 11.

10. A gas-tight Hamilton microliter syringe (Hamilton, Inc., Reno, NV) is used for injecting the components of the assay. In order to eliminate oxygen from the syringe, first flush the syringe by removing and discarding several syringe volumes of gas from the headspace of the bottle containing Ti(III) citrate in MOPS, then remove and discard several syringe volumes from the vial containing only anaerobic MOPS.
11. Using the flushed syringe, begin additions to the anaerobic reaction vials while on ice. Add 20 μl of the reaction mix, the predetermined volume of cell-free extract, 50 μl of a column fraction suspected to contain a component of the methyltransferase system under study, along with a sufficient volume of 50 m*M* MOPS, pH 7.0, to the vial to bring the final volume to 120 μl. Finally, add 5 μl of the methylated donor substrate solution to each vial. It is advisable to use a series of syringes in order to prevent contamination of stock solution, especially when testing dependence on methylated substrate, ATP, extract, or added column fractions.
12. We have found that, when on ice-methyltransferase activity is negligible over a short time period, and often initiate the reactions by transferring the vial from ice to a shaking water bath at 37 °C. As each vial is transferred to the bath, sample the vial contents for a T_0 determination of CoM by transferring 3 μl of the reaction mixture to a plate well with Ellman's reagent. Vials can be started in a staggered fashion so that a reaction is initiated every 30 s. This will allow similarly staggered sampling of the vials beginning at the next time point so that all reactions will be sampled at essentially identical times into the reactions. The vials should be placed in water deep enough to submerge nearly the entire vial. This prevents condensation at the top of the vial and a corresponding increased CoM concentration in the reaction mixture with time in controls lacking enzyme.
13. Sampling should be done in a similar way for each vial. First the syringe should be rinsed by withdrawing and discarding anaerobic buffer; following this, the syringe should be flushed with the gas phase above the anaerobic solution of Ti(III) citrate–MOPS. Approximately 5 μl of the gas phase should be removed, and then the syringe immediately used to remove 3 μl from the liquid phase of the sample vial. As the syringe is withdrawn from the liquid, sample 2 μl of the headspace of the assay vial, and withdraw the syringe. Displace 8 μl of the syringe volume into the Ellman's reagent in the plate. This procedure guards

against oxidation of the reaction vial, and accurately samples the vial CoM concentration without danger of contaminating the sample with the contents of other vials.

14. Measure the absorbance at or near 412 nm by use of a microplate reader. Zero the instrument using a blank well in which reagent is dispensed, but which receives no further additions. Calculate the thiol concentration based on the standards tested above. Plot the millimolar concentration of CoM in each assay vial versus time. As some variation in sampling is expected, duplicate or triplicate vials for each reaction condition is prudent. Assay vials containing a column fraction with a component of a tripartite substrate:CoM methyl transfer system are stimulated above those containing only the slow activity seen with diluted extract alone.
15. Essential controls include reactions without the methylated substrate, extract, or column fraction. For uncharacterized methyltransferase systems, it is important to monitor both the loss of the methylated substrate and production of the predicted product in order to establish stoichiometry to CoM methylation as deduced by loss of the thiol of CoM.
16. Novices to this assay may find it helpful to test if they can accurately measure the same concentration of CoM in a reaction vial lacking enzyme when sequential samples are taken over time. Practice can solve most problems, except those caused by a syringe with a blocked needle or a leak in the syringe housing. The scatter observed around the average concentration of CoM in a vial where CoM is not consumed gives the lower limit of detection for the rate of the methyltransferase reaction in a given assay volume over a given time period.
17. Very seldom it may be found that CoM concentration appears to decrease in a vial lacking enzyme. This is most likely due to CoM oxidation, caused by gross contamination of some reagents or gases with oxygen.

4.2. Assay of CoM during methylation by purified methylamine methyltransferases

In this assay, purified or near-homogenous proteins are used and corrinoid proteins are activated with Ti(III) citrate and methyl viologen, rather than relying on cellular reductive activation systems fueled by MgATP and hydrogen. The negative Ti(III) citrate complex is unable to directly donate electrons to the methylamine corrinoid proteins for activation to Co(I) state, and requires methyl viologen protein as redox mediator (Burke and Krzycki, 1997; Ferguson and Krzycki, 1997; Ferguson *et al.*, 2000).

1. Prepare anaerobic stock solutions of 100 m*M* CoM, 100 m*M* of the desired methylamine substrate, 12.5 m*M* methyl viologen, all in 50 m*M* MOPS, pH 7.0. Additionally prepare 90–100 m*M* Ti(III) citrate as described above. The methyl viologen should be prereduced to the blue monovalent cationic form by adding a slight molar excess of Ti(III) citrate to the solution.
2. Although the methylamine methyltransferases and corrinoid proteins are oxygen stable, they must be rendered anoxic for assay by gentle evacuation and flushing to degas the solution and headspace. Once anaerobic, they may be stored in stoppered vials. Such vials draw in oxygen as they freeze, and these must be thawed under a gentle stream of nitrogen.
3. Typical MMA:CoM or DMA:CoM reactions in our labs contain 20 μg of MtbA, 20 μg of purified corrinoid protein (MtmC or MtbC), and 40 μg of purified MtmB or MtbB. TMA:CoM reactions are typically carried out using 40 μg of MtbA and 80 μg of MttBC.
4. Prepare vials flushed with hydrogen following step 1 of the protocol described above. Nitrogen displaces oxygen better than hydrogen; however, hydrogen may help stabilize reduction of protons by the highly reducing solution.
5. As above, prepare 0.5 m*M* DTNB in 150 m*M* Tris–HCl, pH 8.0. Dispense 90 μl aliquots into the wells of a microtiter plate. Sufficient wells should be filled to accommodate standards, blank, and anticipated samples.
6. Make all additions with an anoxic Hamilton gas-tight syringe. Add 5 μl of the prereduced methyl viologen solution to each vial on ice. Further reduce the methyl viologen to the neutral species by adding 2–3 μl of 90–100 m*M* Ti(III) citrate. The neutral species is colorless, as is the divalent oxidized species. Thus, it is important to titrate to the blue monovalent cation which is then further titrated with Ti(III) citrate to the colorless species. Difficulty in doing this may be due to the presence of oxygen in the vial. If so, the source must be identified, as oxygen will not only prevent the proper reduction of the corrinoid protein to the active form, but also produce oxygen radicals that can damage the proteins. Typical sources included improperly degassed solutions, failed catalyst columns, leaks in the hoses of the manifold, and occasionally, gross oxygen contamination of a commercial gas tank.
7. Using a gas-tight anoxic Hamilton syringe, add 2.5 μl of CoM solution, 2.5 μl of methyl donor solution, appropriate volumes of purified enzymes, and sufficient volume of 1 m*M* Ti(III) citrate in 50 m*M* MOPS, pH 7.0, to each vial to bring the volume to 125 μl.
8. Initiate the staggered reactions by moving the vials from ice and nearly submerging in a water bath at 37 °C. Agitation is not necessary.

9. Remove samples (10 μl) at specific time points and mix in wells of the microtiter plate with the DTNB solution. Follow the sampling procedure described in the protocol above, but modify the volume of assay liquid sampled and delivered to the plate wells accordingly.
10. Measure the absorbance in each well at 412 nm by using a microtiter plate reader, and construct plots of millimolar CoM in the reactions versus time using the standard curve. The CoM standards used should begin at 2 m*M* and include several lower concentrations. Typical results are seen in Fig. 8.1.

5. Assay of Methylthiol:CoM Methyltransferases

DMS- or MMPA-dependent CoM methyltransferase assays have a complication. Methylation of CoM with either DMA or MMPA leads to no net change in thiol concentration, as MSH and 3-mercaptopropionate (MPA) are, respectively, made. To circumvent this problem, reaction products can be derivatized with monobromobimane before separation by reverse phase high-performance liquid chromatography (HPLC) (Tallant and Krzycki, 1997). The bimane derivatives can be readily detected most sensitively by fluorescence (Fahey *et al.*, 1981), but UV absorbance allowed

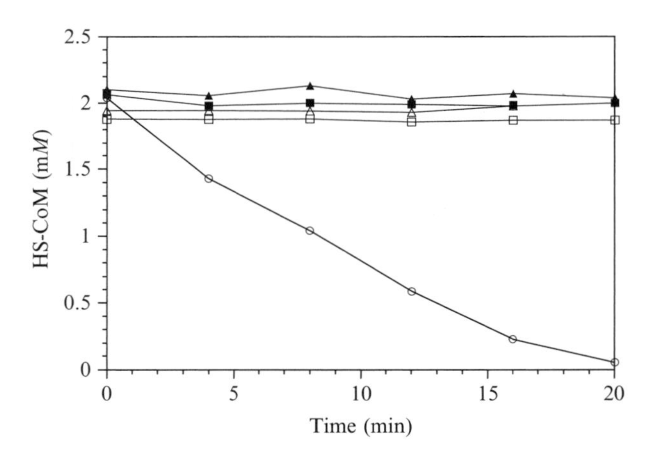

Figure 8.1 TMA:CoM methyl transfer catalyzed by purified MttBC and MtbA and assayed by loss of the free thiol of CoM using DTNB. The complete reaction (open circles) contained MttBC, MtbA, Ti(III) citrate, methyl viologen, CoM and TMA at the concentrations indicated in the text, while other reactions lacked MttBC (open squares), MtbA (open triangles), Ti(III) citrate (closed squares), or methyl viologen (closed triangle).

us to detect below 0.5 nmol CoM. This assay is appropriate for extracts or purified proteins. We used Ti(III) citrate/methyl viologen to activate the methylthiol:CoM methyltransferase in extracts of *M. barkeri*, but the reader is encouraged to also try MgATP and hydrogen, as the endogenous activation proteins will probably prove more robust than chemical reduction. The DMS:CoM methyltransferase activity is highest in extracts of *M. barkeri* grown on acetate.

1. Please consult the procedures above for preparation of anoxic extracts.
2. Prepare a set of 13.5 ml assay vials fitted with sleeved Wheaton red rubber stoppers. Flush/evacuate with hydrogen for five cycles, allowing several minutes for lowest vacuum pressure before applying gas.
3. Prepare 1 ml anaerobic stock solutions of 100 m*M* methyl viologen and 200 m*M* CoM, each in 50 m*M* MOPS, pH 7.0. Prepare an anaerobic stock of 90–100 m*M* Ti(III) citrate. For rinsing sample syringes, prepare two 50 ml volumes of 50 m*M* MOPS, pH 7.0, and 1 m*M* Ti(III) citrate in the same buffer in anaerobic 150-ml bottles fitted with Wheaton stoppers.
4. This procedure must be carried out in a fume hood. Several milliliters of DMS liquid can be degassed by slowly bubbling nitrogen through the solution in a vial, which is then stoppered. Once anoxic, insert a vent needle and fill the rest of the vial with anaerobic water. The vial is agitated, and left to equilibrate, resulting in an approximately 2% (w/v) DMS aqueous solution (~300 m*M*).
5. Place the assay vials on ice and inject 10 μl CoM, 20 μl Ti(III) citrate, 10 μl methyl viologen, and 50 μl MOPS buffer into each vial using a gas-tight syringe that was first made anoxic by flushing with Ti(III) citrate in MOPS. The methyl viologen should reduce to blue and then to the colorless neutral species. If necessary, add 10 μl more Ti(III). If excessive reduction (beyond a 2:1 molar excess of Ti(III) citrate to methyl viologen) is required, oxygen contamination should be suspected.
6. Add 5 μl of the saturated DMS to the vial; this results in a final assay concentration of approximately 10 m*M* DMS in solution, with the remainder in the gas phase.
7. Add 0.2 ml of extract to the vials. The assay is initiated by removal from the ice to the shaking 37 °C water bath. Samples will be removed at time zero, and at 1–3 min intervals as appropriate for the expected rate of reaction.
8. See methods above for anaerobic sampling during assay. For this assay, samples of 10 μl are withdrawn at designated times and added to aerobic 190 μ*M* sodium phosphate buffer, pH 8.9, containing 1 m*M* monobromobimane. Both monobromobimane and bimane derivatives should be protected from light by wrapping in foil. Following incubation at room temperature for at least 30 min, samples are frozen for later analysis.

9. Following thawing, samples are placed in a microcentrifuge and spun at maximum speed (10,000×*g*) for 5 min to remove any precipitated protein or salts.
10. The HPLC is required for this procedure, and as details of operation vary considerably, the analysis will only be described in general.
11. Attach a 25 × 0.46 cm Microsorb C18 column (Varian, Inc., Santa Clara, CA) or equivalent to the HPLC. Set the UV detector for 245 nm.
12. Equilibrate the column with 15% (v/v) methanol in 50 m*M* sodium acetate buffer, pH 3.5. Program a gradient from the 15–99% methanol in the same buffer over 8 min, with continued elution at the higher methanol concentration for 5 min, followed by reequilibration to 15% methanol in acetate buffer.
13. Prepare a series of standards of different concentrations of CoM and 3-MPA, if testing MMPA:CoM activity, in 50 m*M* MOPS, pH 7.0. The concentrations should vary from 1 to 0.1 m*M*. Remove 10 μl samples and add to 190 μl potassium phosphate buffer, pH 7.0, containing 1 m*M* monobromobimane. Allow to react for 30 min. Inject 100 μl of the sample into the HPLC, and initiate the gradient program. The bimane derivative of CoM elutes at 8–9 min, while that of MPA elutes at 11–12 min. The area of the peak observed with several different concentrations of a standard constitutes the standard curve. Injection of dummy samples lacking CoM or 3-MPA should confirm identity of the bimane derivative peaks.
14. Once the standards are complete, analyze samples by injection of 100 μl onto the column and repeating the gradient procedure.
15. Unfortunately the bimane derivative of MSH elutes at 12–13 min, along with unreacted monobromobimane. This might be overcome using a monobromobimane trapping agent such as thiol–agarose (Fahey *et al.*, 1981).

6. Assay of Methylotrophic Substrate: Corrinoid Methyltransferase Activity

6.1. Spectrophotometric assay of methylamine:cob(I)alamin methyltransferases

The DMA and methanol methyltransferase carry out methylamine-dependent methylation of free cob(I)alamin (Ferguson *et al.*, 2000; Sauer and Thauer, 1999). This reaction allows for the direct measurement of the methyltransferase independent of its cognate corrinoid protein, although rates of methyl group transfer are much slower than to cognate corrinoid proteins, typically, 10–20 nmol min^{-1} mg^{-1}. The technique is broadly applicable to many cobalamin-dependent methyltransferases and has been used with both purified proteins (Ferguson *et al.*, 2000; Sauer and Thauer, 1999) and cell extracts (Kreft and Schink, 1993). Cob(I)alamin is produced by reduction of hydroxycob(III)alamin with Ti(III) citrate. Formation of methylcob(III)alamin from cob(I)alamin is monitored at 540 nm, the isosbestic point of cob(I)alamin and cob(II)alamin, which absorb less strongly than methylcob(III)alamin at this wavelength. This eliminates interference by the possible formation of cob(II)alamin by oxidation of cob(I)alamin during assay (Kreft and Schink, 1993).

1. The reactions are performed using a 2 mm cuvette that can be readily stoppered, such as a Starna Type 21 "stopper top" cuvette (Starna Cells, Inc., Atascadero, CA). Soft red natural rubber size 000 stoppers can be used, depending on length, as can some sleeved Wheaton serum vial stoppers. Quartz cuvettes are preferable for this assay to visualize complete spectra.
2. The reaction can be followed in any spectrophotometer equipped with kinetic programs; however, a photodiode array spectrophotometer allows spectra to be collected for comparison to the known spectra of cob(I)alamin, methylcob(I)alamin, and cob(II)alamin. See Kreft and Schink (1993) for good examples of the spectra of these different forms of cobalamin.
3. Prepare anaerobic stock solutions of 90–100 m*M* Ti(III) citrate and 50 m*M* MOPS, pH 6.5. As above, prepare an anaerobic vial of 1 m*M* Ti(III) citrate in MOPS for making syringes anaerobic. Prepare also in 50 m*M* MOPS, pH 6.5, solutions of 5 m*M* hydroxycob(III)alamin as well as a solution of 2 *M* methyl donor substrate.
4. Attach the stoppered cuvettes via 22-gauge needles to the manifold and flush/evacuate with hydrogen for five cycles.
5. When the cuvette is anaerobic, switch the manifold to gassing at a low pressure and rate, leaving the cuvette connected to the manifold. This allows filling the cuvette without a buildup of pressure. Flush a gas-tight Hamilton syringe several times with anaerobic buffer prior to injecting each of these anaerobic additions to the cuvette on the manifold: 97 μl of Ti(III) citrate, 150 μl of methyl donor solution, and 50 m*M* MOPS, pH 6.5. The amount of MOPS buffer should be varied depending on the protein volume to be added to the cuvette, with the final volume being 0.5 ml. Finally, add 100–250 μg of methyltransferase. The high amount of protein is due to the slow rate of reaction.
6. Remove the cuvette from the gassing manifold and place into the spectrophotometer and record the spectrum for use as a blank.
7. Inject anoxically 100 μl of the hydroxycob(III)alamin solution. The reduction to cob(I)alamin is fairly rapid and is complete in under a minute, whereas the formation of methylcob(I)alamin by the methylamine methyltransferases can continue linearly for over an hour.

8. Follow the change in absorbance at 540 nm. The first few minutes may be distorted due to the formation of hydroxycob(III)alamin as the protein is injected, which like methylcob(III)alamin absorbs more than the reduced forms. This is quickly reduced, after which a linear rise in A_{540} is seen due to the formation of methylcob(III)alamin. The differential of the extinction coefficient of methylcob(III)alamin to that of cob (II)alamin/cob(I)alamin is 4.4 mM^{-1} cm^{-1} (Kreft and Schink, 1993). As always, enzyme and the substrate dependence of the activity should be established.

6.2. Monitoring cognate corrinoid protein methylation by HPLC

Essentially, the protocol for the reactions below is similar to that described above for the Ellman's assay of CoM during methylation with purified proteins, except that CoM and MtbA are not added to the assays, and analysis requires extraction of the corrinoid cofactor and measurement by reverse phase HPLC (Ferguson *et al.*, 2000).

1. All assays are carried out in stoppered 2 ml glass vials with an atmosphere of H_2.
2. Prepare anaerobic stock solutions of 50 m*M* MOPS, pH 7.0. Prepare also anaerobic stock solutions of 2.0 *M* methyl donor substrate in MOPS and 12.5 m*M* methyl viologen in MOPS buffer, as well as 90–100 m*M* Ti(III) citrate. The methyl viologen solution is pre-reduced to the blue monovalent cationic form of methyl viologen by adding a slight molar excess of Ti(III) citrate to the methyl viologen.
3. These reactions can be carried out using 50 μg (10 μ*M*) of either MtmB or MtbB and 35 μg of either MtmC or MtbC (12 or 15 μ*M* respectively). The reaction could also be carried out using MttBC containing 12–15 μ*M* MttC in the reaction and TMA as the methyl donor.
4. Using a gas-tight Hamilton syringe flushed with anaerobic buffer, add 6 μl of reduced methyl viologen and 4 μl of Ti(III) citrate to an anaerobic reaction vial; the methyl viologen should reduce from the blue cation to the colorless neutral species. Remember, fully oxidized and fully reduced methyl viologen are both colorless.
5. On ice, inject into each assay vial 50 μl of methyl donor substrate, the appropriate volumes of methyltransferase and corrinoid protein solutions, and a sufficient volume of 50 m*M* MOPS, pH 7.0, to bring the total reaction volume to 100 μl.
6. Initiate the reaction by transferring the vials to a shaking 37 °C water bath and incubating under dim red light for 20 min. For all further manipulation, the corrinoid protein or cofactor must be handled under dim red light, as this minimizes photolysis of the methyl–cobalt bond.

7. Terminate the reaction by removing the vial from the water bath and combining the sample with an equal volume of aerobic 95% ethanol. Mix the sample several times by inversion and transfer the solution to a polypropylene microfuge tube.
8. Transfer the tube to an 80 °C heat block for 10 min.
9. Place the tube in an ethanol–dry ice bath for 10 min. This step will accelerate precipitation of the protein, leaving the soluble cofactor.
10. Centrifuge the microfuge tube at 10,000×*g* for 10 min.
11. Remove the supernatant quantitatively and concentrate to dryness under vacuum. This is most conveniently done in a Savant Speed-Vac (Thermo Scientific, Inc., Waltham, MA).
12. The residue from the dried supernatant is dissolved in 110 μl of deionized water. The solution should have a faint pink color when seen briefly under room light.
13. Equilibrate a 25 cm × 4.6 mm Microsorb-MV columns or equivalent column with 20% methanol in 25 m*M* sodium acetate, pH 6.0. The column effluent should be monitored at 260 nm. Program the HPLC controller for a 60-min linear gradient from 20–99% methanol buffered with 25 m*M* sodium acetate, pH 6.0.
14. The extracted corrinoid migrates with a retention time on reverse HPLC very near that of authentic cobalamin (Burke and Krzycki, 1997), although the most probable form of the corrinoid cofactor in *M. barkeri* differs slightly from cobalamin (Pol *et al.*, 1982). Therefore, prepare 5–20 μ*M* solutions of methylcob(III)alamin and hydroxycob (III)alamin. Inject samples (100 μl) and elute with the gradient program. The peaks attributed to methylcob(III)alamin and hydroxycob (III)alamin should increase linearly with increasing analyte concentration. The retention times are approximately 12–13 min for the hydroxy form of the cofactor and 22 min for methylated cofactor. Recovery of the corrinoid cofactor averages 60% and has been similar for both methylated and hydroxyl forms of the corrinoid protein. Therefore, the rate of methylation in a time course can be calculated from the ratio of the two cofactors and the known concentration of the corrinoid protein.

7. Spectral Assay of Methylcobalamin:CoM Methyltransferases

As mentioned above, methylation of cob(I)alamin can be monitored by following the increased absorbance at 540 nm due to formation of methylcob(III)alamin. The reverse reaction, CoM-dependent demethylation of methylcob(III)alamin to form cob(I)alamin (and methyl-CoM), can also be monitored by monitoring the decrease in A_{540}, an isosbestic point for cob(I)alamin and cob(II)alamin.

1. This assay uses 1 cm cuvettes fitted with rubber stoppers. Place cuvettes on the manifold and evacuate/flush to create a nitrogen atmosphere. Meanwhile, prepare an assay solution of 0.5 m*M* methylcob(III)alamin in 50 m*M* phosphate buffer, pH 7.2, and make anaerobic by five cycles of flush/evacuation. For all manipulations of methylcob(III)alamin, work primarily under red light and shield solutions from bright light to prevent photolysis. Photolysis in the presence of oxygen will yield hydroxycob(III)alamin in the starting solutions, which will complicate analysis if it is not reduced to cob(II)alamin. Excess CoM can alleviate the problem, but if low and stable CoM concentrations are needed for kinetic analysis, 2 m*M* dithiothreitol can be added as a precaution to the assay mixture.
2. With the cuvettes anaerobic but connected to the manifold, inject 1 ml of the assay mixture.
3. Blank the spectrophotometer against a matching cuvette containing only phosphate buffer.
4. The reaction may be initiated by anoxic injection of the methylcorrinoid:CoM methyltransferase solution (either extracts or purified protein) and then 5 μl of 1 *M* CoM. Monitor enzyme and/or CoM-dependent decrease in absorbance at 540 nm. The methyl-CoM consumed can be calculated using the differential extinction coefficient of 4.4 mM^{-1} cm^{-1}.
5. A useful alternative to the above assay is the dicyano-derivatization assay (Grahame, 1989).
6. A variation on the A_{540} assay is useful in purification of methylcorrinoid:CoM methyltransferases. The methylcob(III)alamin reaction is monitored at 620 nm and under air. Cob(I)alamin is then oxidized to both cob(II)alamin and hydroxylcob(III)alamin, both of which absorb differently, but both less strongly, than methylcob(III)alamin at 620 nm. True methyltransferase-specific activities therefore cannot be determined with this assay, but the decrease in absorbance is roughly proportional to the amount of enzyme, and useful when initially scanning aerobic or anaerobic column fractions for methylcobalamin:CoM methyltransferase activity.

ACKNOWLEDGMENTS

The authors would like to acknowledge their coworkers who help pioneer many of these techniques in application to the study of methylamine and methylthiol:CoM methyltransferase reactions, in particular, Steve Burke, Tsuneo Ferguson, Thomas Tallant, Ligi Paul, and Jitesh Soares. Research in the laboratory of JAK is supported by grants from the Department of Energy (DE-FG0202-91ER200042) and the National Institute of Health (GM070663).

REFERENCES

Burke, S. A., and Krzycki, J. A. (1995). Involvement of the "A" isozyme of methyltransferase II and the 29-kilodalton corrinoid protein in methanogenesis from monomethylamine. *J. Bacteriol.* **177,** 4410–4416.

Burke, S. A., and Krzycki, J. A. (1997). Reconstitution of monomethylamine:Coenzyme M methyl transfer with a corrinoid protein and two methyltransferases purified from *Methanosarcina barkeri*. *J. Biol. Chem.* **272,** 16570–16577.

Drennan, C. L., Matthews, R. G., and Ludwig, M. L. (1994). Cobalamin-dependent methionine synthase: The structure of a methylcobalamin-binding fragment and implications for other B12-dependent enzymes. *Curr. Opin. Struct. Biol.* **4,** 919–929.

Ellman, G. L. (1958). A colorimetric method for determining low concentrations of mercaptans. *Arch. Biochem. Biophys.* **74,** 443–450.

Fahey, R. C., Newton, G. L., Dorian, R., and Kosower, E. M. (1981). Analysis of biological thiols: Quantitative determination of thiols at the picomole level based upon derivatization with monobromobimanes and separation by cation-exchange chromatography. *Anal. Biochem.* **111,** 357–365.

Ferguson, D. J. Jr., and Krzycki, J. A. (1997). Reconstitution of trimethylamine-dependent coenzyme M methylation with the trimethylamine corrinoid protein and the isozymes of methyltransferase II from *Methanosarcina barkeri*. *J. Bacteriol.* **179,** 846–852.

Ferguson, D. J. Jr., Krzycki, J. A., and Grahame, D. A. (1996). Specific roles of methylcobamide: Coenzyme M methyltransferase isozymes in metabolism of methanol and methylamines in *Methanosarcina barkeri*. *J. Biol. Chem.* **271,** 5189–5194.

Ferguson, D. J. Jr., Gorlatova, N., Grahame, D. A., and Krzycki, J. A. (2000). Reconstitution of dimethylamine: Coenzyme M methyl transfer with a discrete corrinoid protein and two methyltransferases purified from *Methanosarcina barkeri*. *J. Biol. Chem.* **275,** 29053–29060.

Ferguson, T., Soares, J. A., Lienard, T., Gottschalk, G., and Krzycki, J. A. (2009). RamA, a protein required for reductive activation of corrinoid-dependent methylamine methyltransferase reactions in methanogenic archaea. *J. Biol. Chem.* **284,** 2285–2295.

Gencic, S., LeClerc, G. M., Gorlatova, N., Peariso, K., Penner-Hahn, J. E., and Grahame, D. A. (2001). Zinc-thiolate intermediate in catalysis of methyl group transfer in *Methanosarcina barkeri*. *Biochemistry* **40,** 13068–13078.

Grahame, D. A. (1989). Different isozymes of methylcobalamin: 2-mercaptoethanesulfonate methyltransferase predominate in methanol- versus acetate-grown *Methanosarcina barkeri*. *J. Biol. Chem.* **264,** 12890–12894.

Kreft, J. U., and Schink, B. (1993). Demethylation and degradation of phenylmethylethers by the sulfide-methylating homoacetogenic bacterium strain TMBS 4. *Eur. J. Biochem.* **226,** 945–951.

Krzycki, J. A. (2004). Function of genetically encoded pyrrolysine in corrinoid-dependent methylamine methyltransferases. *Curr. Opin. Chem. Biol.* **8,** 484–491.

Longstaff, D. G., Larue, R. C., Faust, J. E., Mahapatra, A., Zhang, L., Green-Church, K. B., and Krzycki, J. A. (2007). A natural genetic code expansion cassette enables transmissible biosynthesis and genetic encoding of pyrrolysine. *Proc. Natl. Acad. Sci. USA* **104,** 1021–1026.

Oelgeschlager, E., and Rother, M. (2009). In vivo role of three fused corrinoid/methyl transfer proteins in *Methanosarcina acetivorans*. *Mol. Microbiol.* **72,** 1260–1272.

Owusu-Apenten, R. (2005). Colorimetric analysis of protein sulfhydryl groups in milk: Applications and processing effects. *Crit. Rev. Food Sci. Nutr.* **45,** 1–23.

Pol, A., van der Drift, C., and Vogels, G. D. (1982). Corrinoids from *Methanosarcina barkeri*: Structure of the α-ligand. *Biochem. Biophys. Res. Commun.* **108,** 731–737.

Rother, M., and Krzycki, J. A. (2010). Selenocysteine, pyrrolysine, and the unique energy metabolism of methanogenic archaea. *Archaea* doi: 10.1155/2010/453642.
Sauer, K., and Thauer, R. K. (1999). Methanol:Coenzyme M methyltransferase from *Methanosarcina barkeri*-substitution of the corrinoid harbouring subunit MtaC by free cob(1)alamin. *Eur. J. Biochem.* **261,** 674–681.
Seefeldt, L. C., and Ensign, S. A. (1994). A continuous spectrophotometric activity assay for nitrogenase using the reductant titanium (III) citrate. *Anal. Biochem.* **221,** 379–386.
Sowers, K. R. (1995). Techiques for anaerobic biochemistry. *In* "Archaea: A Laboratory Manual," (K. R. Sowers and H. J. Schreier, eds.), pp. 141–160. Cold Spring Harbour Laboratory Press, Plainview, NY.
Tallant, T. C., and Krzycki, J. A. (1997). Methylthiol:coenzyme M methyltransferase from *Methanosarcina barkeri*, an enzyme of methanogenesis from dimethylsulfide and methylmercaptopropionate. *J. Bacteriol.* **179,** 6902–6911.
Tallant, T. C., Paul, L., and Krzycki, J. A. (2001). The MtsA subunit of the methylthiol: coenzyme M methyltransferase of *Methanosarcina barkeri* catalyses both half-reactions of corrinoid-dependent dimethylsulfide:Coenzyme M methyl transfer. *J. Biol. Chem.* **276,** 4485–4493.
van der Meijden, P., Heythuysen, H. J., Pouwels, A., Houwen, F., van der Drift, C., and Vogels, G. D. (1983). Methyltransferases involved in methanol conversion by *Methanosarcina barkeri*. *Arch. Microbiol.* **134,** 238–242.

CHAPTER NINE

Methyl-Coenzyme M Reductase from *Methanothermobacter marburgensis*

Evert C. Duin, Divya Prakash, *and* Charlene Brungess

Contents

Abstract

Methyl-coenzyme M reductase catalyzes the reversible synthesis of methane from methyl-coenzyme M in methanogenic and ANME-1 and ANME-2 Archaea. The purification procedure for methyl-coenzyme M reductase from *Methanothermobacter marburgensis* is described. The procedure is an accumulation of almost 30 years of research on MCR starting with the first purification described by Ellefson and Wolfe (Ellefson, W.L., and Wolfe, R.S. (1981). Component C of the methylreductase system of *Methanobacterium*. *J. Biol. Chem.* **256,** 4259–4262). To provide a context for this procedure, some background information is provided, including a description of whole cell experiments that

Department of Chemistry and Biochemistry, Auburn University, Alabama, USA

Methods in Enzymology, Volume 494
ISSN 0076-6879, DOI: 10.1016/B978-0-12-385112-3.00009-3

provided much of our knowledge of the behavior and properties of methyl-coenzyme M reductase.

1. Introduction

Methyl-coenzyme M reductase (MCR) is the enzyme that is responsible for catalyzing the methane producing step in the process of methanogenesis (Duin, 2007; Ragsdale, 2003; Thauer, 1998). In the so-called hydrogenothropic pathway (Fig. 9.1), the substrate CO_2 is bound to, and subsequently transferred between, a set of carrier molecules. At each of the designated steps, two reducing equivalents from H_2 are added until the C_1 unit has become a methyl group bound as a thioether to the carrier coenzyme M (HS-CoM), forming methyl-coenzyme M (CH_3–S–CoM), which is one of the two substrates for MCR (Fig. 9.2). CH_3–S–CoM is converted into CH_4 with the help of the cosubstrate coenzyme B (HS-CoB), which forms a disulfide bond with the produced coenzyme M forming the heterodisulfide, CoM–S–S–CoB.

Given the importance of methane as an energy source for microbes and for man, MCR is an important and interesting enzyme to study. Furthermore, MCR catalyzes the first step in anaerobic methane oxidation. Enzymologists and bioinorganic chemists have been intrigued by MCR and by the unique nickel-containing tetrapyrrole, cofactor 430 (F_{430}—Fig. 9.2) that is at the center of the reaction mechanism. The enzyme and its cofactor have been studied extensively using a wide range of spectroscopic techniques but no conclusive evidence has been obtained concerning the mechanism by which MCR catalyzes the formation of methane, and several hypothetical reaction mechanisms are currently being tested (Duin and Mckee, 2008; Grabarse *et al.*, 2001; Horng *et al.*, 2001; Pelmenschikov and Siegbahn, 2003; Pelmenschikov *et al.*, 2002; Shima and Thauer, 2005).

Because the formation of methane by methanogenic archaea is thermodynamically favorable, it came as a surprise that some microbes accomplish the anaerobic conversion of methane into CO_2 to sustain growth, albeit very slowly, in microbial consortia of ANME archaea and sulfate reducing bacteria (Boetius *et al.*, 2000; Shima and Thauer, 2005). It is clear that both processes, the anaerobic oxidation of methane and the reduction of sulfate, have to be somehow coupled for methanogenesis to go in reverse, but it is not clear how this is achieved and what compound(s) is (are) exchanged between the two consortia partners to couple the two processes. Most of the steps in the hydrogenothropic pathway are considered reversible, except for the reaction catalyzed by MCR ($\Delta G^{o'} = -30 \pm 10$ kJ mol^{-1}). MCR, however, is present in the consortia ANME archaea. These consortia were

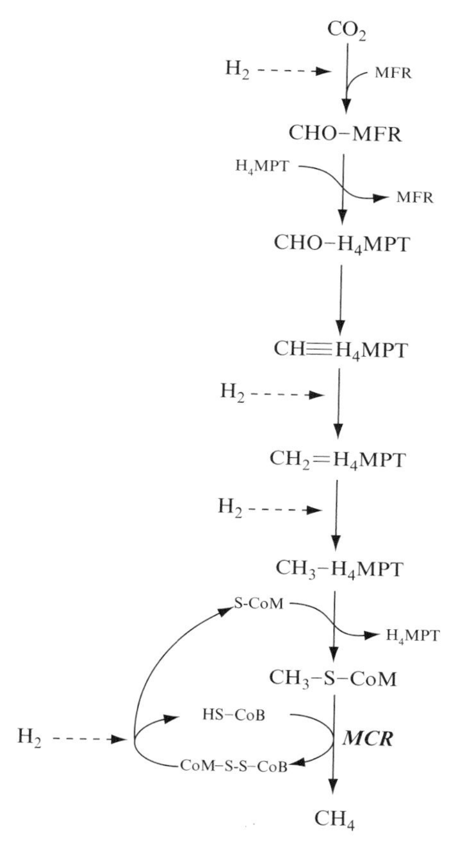

Figure 9.1 Hydrogenothropic pathway used by *Methanothermobacter marburgensis*. CHO–MFR, *N*-formylmethanofuran; CHO–H_4MPT, N^5-formyltetrahydromethanopterin; CH≡H_4MPT^+, N^5,N^{10}-methenyltetrahydromethanopterin; CH_2=H4MPT, N^5, N^{10}-methylenetetrahydromethanopterin; CH_3–H_4MPT^+, N^5-methyltetrahydromethanopterin; CH_3–S–CoM, methyl-coenzyme M; CoB, coenzyme B; MCR: methyl-coenzyme M reductase.

isolated from biomats that grow on top of methane hydrates at the bottom of the black sea. The high methane concentration in this environment would make the reversible reaction a possibility. In a recent paper, it was shown that under certain conditions MCR can indeed catalyze the conversion of methane into methyl-coenzyme M (Scheller *et al.*, 2010). This was not shown for MCR obtained from an ANME archaeon, but for the enzyme from *Methanothermobacter marburgensis*, an enzyme that is normally only involved in methane production. Here, we will present the purification procedure for the *M. marburgensis* enzyme.

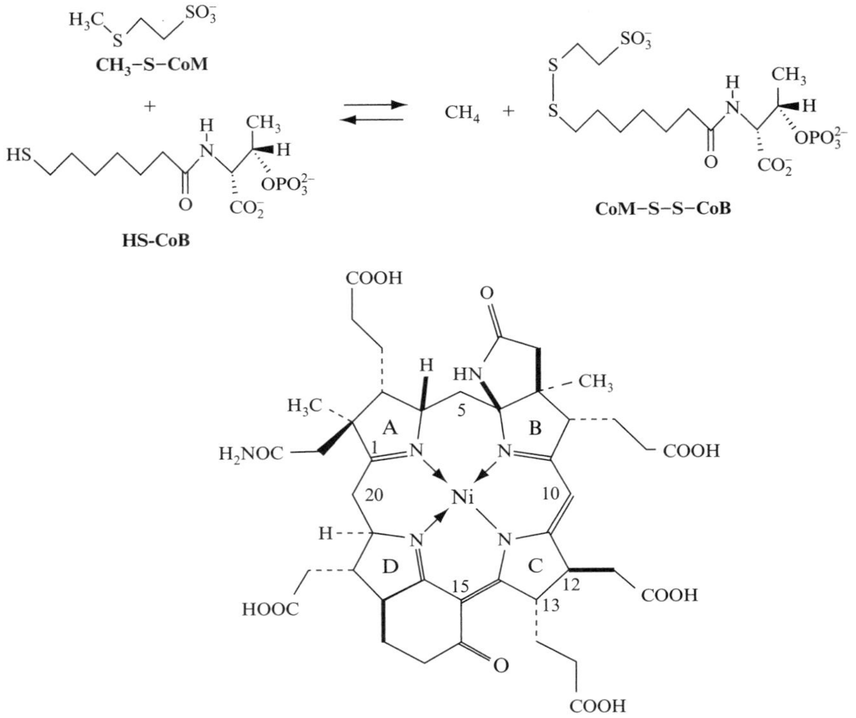

Figure 9.2 Reaction catalyzed by methyl-coenzyme M reductase (MCR). CH_3–S–CoM, methyl-coenzyme M (2-(methylthio)ethanesulfonate); HS-CoB, coenzyme B (*N*-7-mercaptoheptanoyl-*O*-phospho-L-threonine); CoM–S–S–CoB, heterodisulfide of coenzyme M (2-mercaptoethanesulfonate) and coenzyme B; F_{430}, factor 430.

2. Equipment

M. marburgensis cells and the purified MCR enzyme are highly oxygen sensitive. This requires that growth of cultures, harvesting of cells, protein purification, and sample handling all have to be performed under conditions that completely exclude air. Several essential pieces of equipment will be needed to do this. First of all, a fermenter is needed that is suitable for anaerobic growth. In our laboratory, we use the New Brunswick BioFlo Benchtop System that can be fitted with vessels of different sizes. Heating of the fermenter is done with a heating blanket that is wrapped around the glass vessel. One of the main reasons to choose this system was that the blanket can be easily removed and the whole vessel can be placed in an ice bath which is an essential step in the preparation of active MCR (see below). For harvesting the cells, a flow-through centrifuge is used. We use a Sorvall Centrifuge Stratos (Thermo Scientific) fitted with a Titanium Continuous Flow Rotor. With this setup, 10 L of cell suspension can be harvested within 1–1.5 h. Breaking the cells and all the subsequent steps are then performed in an anaerobic chamber. At Auburn, a vinyl anaerobic chamber (Coy Laboratories) is used. Any anaerobic box will work, but the Coy Type B chamber provides sufficient space for all the necessary equipment. In addition, the catalyst that removes the oxygen from the chamber atmosphere is not sensitive towards higher humidity levels, which is important due to the use of large amounts of aqueous solutions during the purification protocol. Any chamber or box has to be large enough to house a sonifier (Branson Digital Sonifier) for breaking the cells, several stirring plates, a small centrifuge for sample concentration (Hettich Mikro 120), and an FPLC for protein purification. In this case, an Äkta FPLC (GE Healthcare) is used. There will also be a need to centrifuge protein samples at speeds of ~35000 rpm during the purification procedure; we use aBeckman XL70 Ultracentrifuge and a Beckman Ti45 rotor, which can hold six 50-ml centrifuge tubes. The OptiSeal tubes (Beckman) are air tight and can be taken out of the chamber without exposing their contents to air.

For characterization of the enzyme and assessing the amount of active enzyme that is present, access to a UV–vis spectrophotometer is necessary. Samples can be removed from the chamber in stoppered cuvettes for these measurements; alternatively, one can purchase a system that allows measurements inside the chamber. We use two systems: an Ocean Optics USB 2000 miniature fiber optic spectrometer and an Agilent 8453 UV–vis spectrophotometer that is connected via an Agilent Fiber Optic Coupler (Custom Sample Systems, Inc.) to a Qpod temperature-controlled sample compartment (Quantum Northwest) in a second anaerobic box (Vacuum Atmospheres Company). For kinetic measurements, a gas chromatograph is needed that can detect and quantify methane production by integration of the peaks and comparing to the integrated intensity of a series of methane standards. Although the absorption spectra are useful, the spectra for some of the different states of MCR overlap and, in some cases, are identical. Therefore, there is the need for a technique like electron paramagnetic resonance (EPR) spectroscopy that can discern the different enzyme states more easily. A basic continuous-wave X-band spectrometer (Bruker, EMX model) is sufficient for these types of measurements. More information on this technique can be found in previous issues of this series (Beinert, 1978; Beinert *et al.*, 1978; Fee, 1978; Palmer, 1967) and a recent book by Wilfred Hagen on Biomolecular EPR Spectroscopy (Hagen, 2009).

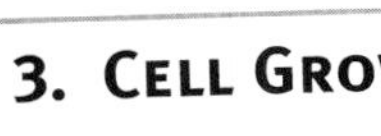

3. Cell Growth: Basic Procedure

Several culture collections are available where you can obtain the culture of a methanogen. We use *M. marburgensis*, which can be ordered from The Deutsche Sammlung von Mikroorganismen und Zellkulturen GmbH (DSMZ—www.dsmz.de/), catalog #: DSM 2133; The American Type Culture Collection (ATCC—www.atcc.org/), catalog #: BAA-927; and the NITE Biological Resource Center (NBRC—www.nbrc.nite.go.jp/e/index.html, catalog #: NBRC 100331. We generally culture 1–2 L of cells, which can be kept alive in stoppered bottles at 4 °C for 2–3 months after which a new 2-L culture has to be prepared. These bottles should be pressurized (0.5 atm) to prevent oxygen from entering. Use bottles with a protective coating (Kimax plastic safety-coated media bottles) in case a bottle cracks. When large volumes of culture are required, for example, for purifying MCR, the 2 L cultures (with an OD_{568} of 6–8) are used as inoculum. It is possible to propagate the cultures on plates using methods that are described in Chapters 1 and 2. This, however, is not essential.

If you are growing *M. marburgensis*, there is no need to sterilize the fermenter and the medium. The growth conditions are unique for *Methanothermobacter* species and of these *M. marburgensis* has the highest growth rate and grow to the highest cell concentrations. Small quantities of thermophilic *Clostridia*, however, can be present, which grow on *M. marburgensis* lysis products. Thus, when the goal is to isolate DNA from *M. marburgensis*, penicillin G should be added to the medium to kill the *Clostridia*. Penicillin G transiently slows down the growth of *M. marburgensis* probably due to clostridial lysis products (R. Thauer, personal communication).

The medium for the cell culture contains (for 10 L)

- 90 g KH_2PO_4
- 27 g NH_4Cl
- 33 g Na_2CO_3
- 15 mL trace element solution
- 0.5 mL 0.2% resazurine (not more, sticks to MCR!)

Procedure to make the element solution

- 30.0 g nitrilotriacetic acid
- Add water to a volume of 500 mL
- Set pH at 6.7 with 10 *M* NaOH—this will dissolve the nitrilotriacetic acid
- 40 g $MgCl_2 \cdot 6H_2O$
- 10 g $FeCl_2 \cdot 4H_2O$

- 0.2 g $CoCl_2 \cdot 6H_2O$
- 0.2 g $NaMoO_4 \cdot 2H_2O$
- 1.2 g $NiCl_2 \cdot 6H_2O$
- Add water to a total volume of 1 L

Gasses and gas mixtures

- Mixture of 80% H_2/20% CO_2
- H_2S
- Mixture of 80% N_2/20% CO_2
- 100% H_2

For growth, the cell culture has to be gassed with the 80% H_2/20% CO_2 mixture. A small amount of H_2S (~1%) is also needed. To not waste any gas and to prevent air from coming into the fermenter, it is recommended to install a changeover panel (Airgas). This system will automatically change to a new full gas bottle when the one in use is running low and allows the replacement of the old bottle without interrupting the gas flow to the fermenter.

In our setup, the H_2/CO_2 gas pressure is set at 1 atm overpressure since the fermenter can only handle up to 1.5–2 atm overpressure. The gas flow is regulated with a valve and a flow meter and set to 1200 mL min^{-1} for a 10-L culture and 600 mL min^{-1} for a 2-L culture. The H_2S gas, from a 1-L lecture bottle, is mixed into the gas stream through a T-connection. The H_2S flow is regulated with a valve and a small bubble counter and set to about one bubble per 10 s. Initially, the flow of H_2S can be higher to help with removing oxygen from the fermenter, but a black precipitate will form very quickly if the H_2S gas flow is kept too high for too long.

The next couple of steps will describe the procedure when using the BioFlo system (Fig. 9.3) but can be adjusted to any other system. After the medium is put in the fermenter, it is sealed and placed inside a container (in our case an aluminum cooking pot) in a hood. The fermenter is not completely gas tight but will stay anaerobe when the inside pressure is higher than the outside pressure. This means, however, that hydrogen gas can leak out and sufficient ventilation will be required. Connect the stirring motor, temperature sensor, heating blanket, and gas lines. An additional blanket can be wrapped around the exposed part of the fermenter to contain the heat produced by the heating blanket. This will greatly speed up the heating process since the required growth temperature of 65 °C is the maximum temperature for this system. The extra blanket is particularly useful in the hood due to the high air flow which can slow down the heating process considerably. The stirring motor is set to 1000 rpm for a 10-L culture and 400 rpm for a 2-L culture. A higher speed will cause cell lyses in the 2 L cultures.

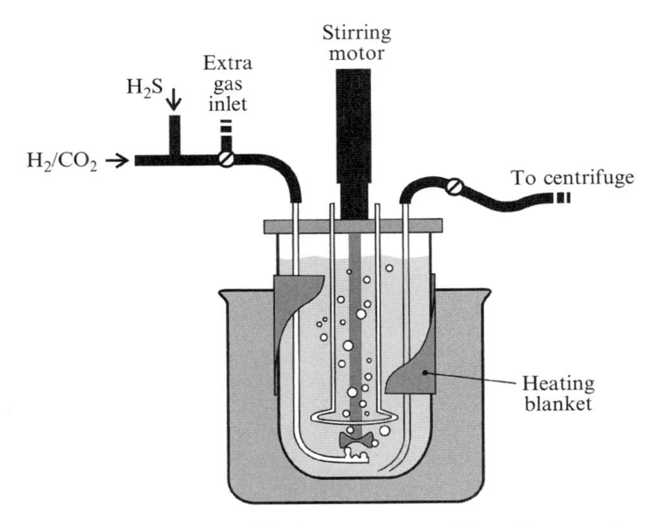

Figure 9.3 Schematic overview of the fermenter assembly and connections during cell growth.

Wait for 1.5 h to achieve anaerobicity before inoculation (or until 65 °C is reached). The resazurine in the medium will change color, from dark blue to red to colorless, as an indication that anaerobicity is reached. In general, this happens within the first 10–20 min but it is recommended to wait a little longer. Inoculate with ±200 mL cells for a 10-L culture or ±50 mL for a 2-L culture. Use the earlier prepared 2-L stock. The cells can be drawn from the 2 L bottle with a disposable syringe, and injected into the fermenter through a septum or rubber stopper present on the fermenter top plate. Grow the cells for ±13 h at 65 °C. The OD_{568} of the cells should reach 4 or higher for a 10-L culture or 6–8 for a 2-L culture.

Dependent on your needs, the cell solution can be transferred anaerobically into a bottle and stored at 4 °C, or the cells can be harvested using the flow-through rotor (Fig. 9.4). The system comes with a pump but it is better to use the pressure of the fermenter to push the cell solution into the rotor since this will prevent air leaking in. To make the rotor space anaerobic, it is first flushed with nitrogen gas. Subsequently, 2 L anaerobic 50 m*M* phosphate buffer is pumped through the rotor. After this, the cell culture can be pumped through the system. Note that the essential part in the flow-through rotor, connecting the moving and nonmoving parts, is a single ball bearing. This ball bearing is supposed to last for 50 h, however, our experience indicates that it is better to replace it after every run. Having a ball bearing lock up in mid-run has been one of my most horrifying moments in the lab, not even mentioning the pain in my wallet to get the centrifuge and the rotor repaired.

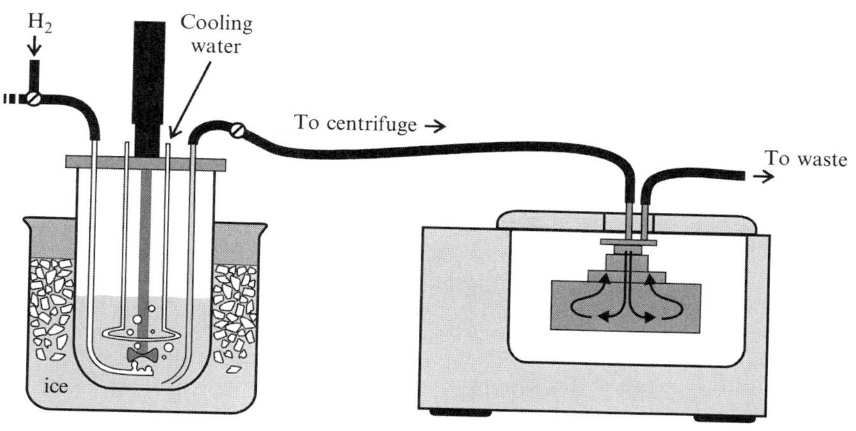

Figure 9.4 Schematic overview of the fermenter assembly and connections during cell harvest.

4. Whole-Cell Experiments

A problem with the early research on MCR was that the protein turned out to be very unstable and that upon purification pretty much all activity was lost. Based on the MCR content of cells and methane formation in cell suspensions, it was estimated that only 1–4% of the purified protein was active (Albracht *et al.*, 1988). The biggest drop in activity is observed when the cells are broken. A set of whole-cell experiments, however, indicated how MCR with a much higher activity could be obtained (Albracht *et al.*, 1986, 1988; Rospert *et al.*, 1991). Figure 9.5 shows one of these experiments. We describe these types of experiments here since they show the quick activation of MCR that can be achieved in whole cells. This process is not well understood and cannot be performed *in vitro*.

For these types of experiments, either fresh cell cultures can be used or cultures that have been stored at 4 °C. Due to the high content in the cell of enzymes involved in the hydrogenothropic pathway, *M. marburgensis* cells are ideal to study directly in EPR spectroscopy. Of all the protein present in the cell, 10% is MCR (Ellefson and Wolfe, 1981). Still the signal intensity in cell cultures with an OD_{568} of 6–8 is not high enough. The cells have to be concentrated by at least a factor 10 by centrifuging them and resuspending them in a smaller volume of medium. About 10 mL of this concentrated cell suspension was placed in a 100-mL bottle which was closed off with a rubber stopper. The cell suspension was incubated at different temperatures and using different gasses in the gas phase. The gas was applied as a

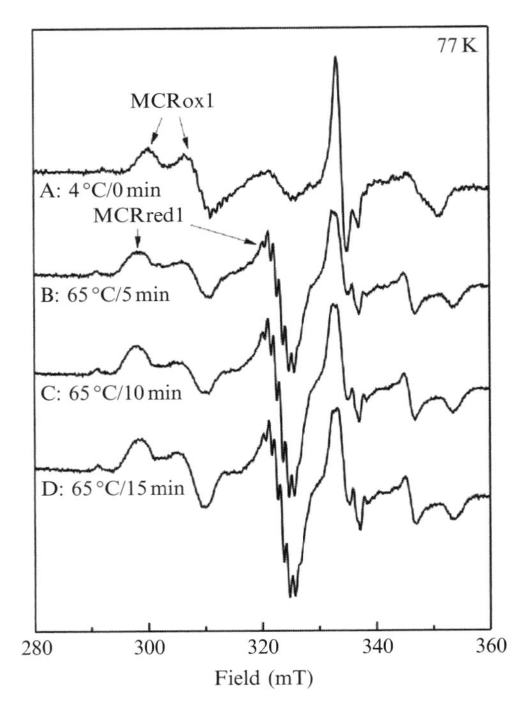

Figure 9.5 EPR spectra obtained for whole cells incubated with a gas mixture of 80% H_2 and 20% CO_2. (A) Cell suspension kept at 4 °C, (B) Cell suspension incubated for 5 min at 65 °C, (C) for 10 at 65 °C, (D) for 15 min at 65 °C. EPR conditions: microwave frequency, 9.385 GHz; microwave power incident to the cavity, 2.01 mW; temperature, 77 K; modulation frequency, 100 kHz; modulation amplitude, 1.0 mT.

continuous gas flow to remove any methane formed in the process and any air that might leak in. In this experiment, the activation was followed with EPR spectroscopy. If that technique is not available, the methane formation can be used as an indicator. In that case, the gas phase should be stationary and gas samples can be withdrawn with a gas-tight syringe at certain times and injected into a gas chromatograph. After a set amount of time, an aliquot of cell suspension is removed with a gas-tight syringe and inserted into an EPR tube that previously was made anaerobe and sealed with a piece of closed off rubber tubing. In between transfers, the syringe can be cleaned with anaerobic buffer. By inserting the syringe needle into a stoppered bottle that contains the gas mixture from the chamber, the syringe can be kept oxygen free until it is used. If there are problems with maintaining anaerobicity, however, dithiothreitol can be added to the cell suspension.

Figure 9.5 shows the effect of the incubation with the growth gas mixture of 80% H_2/20% CO_2. Trace A (Fig. 9.5) shows the EPR spectra of the cells kept at 4 °C. Both the MCRox1 and MCRred1 signals can be recognized

(For an overview of the different EPR-active forms of MCR see Fig. 9.6). Traces B, C, and D (Fig. 9.5) show the change in EPR signals when the temperature is changed to 65 °C. A large increase in the MCRred1 form is detected. The signal reaches maximum intensity within 10 min of incubation time. The MCRox1 signal becomes less intense and seems to change somewhat. This is due to the formation of other signals including MCRred2 and MCRox2 (see below). When the cells are kept at 4 °C, no change in the EPR spectrum is observed (not shown). When 100% H_2 is used in the gas phase, most of the MCRox1 signal is lost and the development of two new species can be detected, MCRred1 and MCRred2 (Fig. 9.7, panel A). When the mixture of 80% N_2/20% CO_2 is used, the MCRox1 signal becomes much more intense (Fig. 9.7, panel B). A full description of this work can be found in the literature (Albracht *et al.*, 1988).

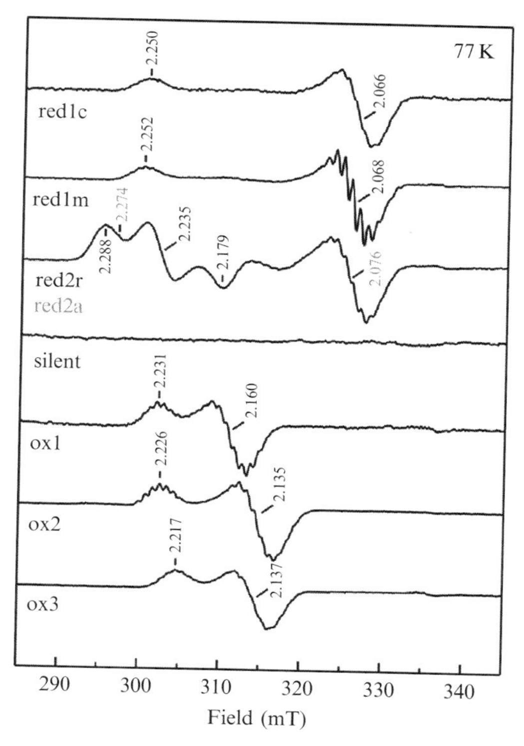

Figure 9.6 Overview of the different EPR-active species in methyl-coenzyme M reductase. Note that the indicated *g*-values are the apparent *g*-values. None of the species displays a completely axial EPR spectrum. Check the literature for a complete set of *g*-values (Finazzo *et al.*, 2003; Kern *et al.*, 2007; Mahlert *et al.*, 2002a). For EPR conditions see Fig. 9.5.

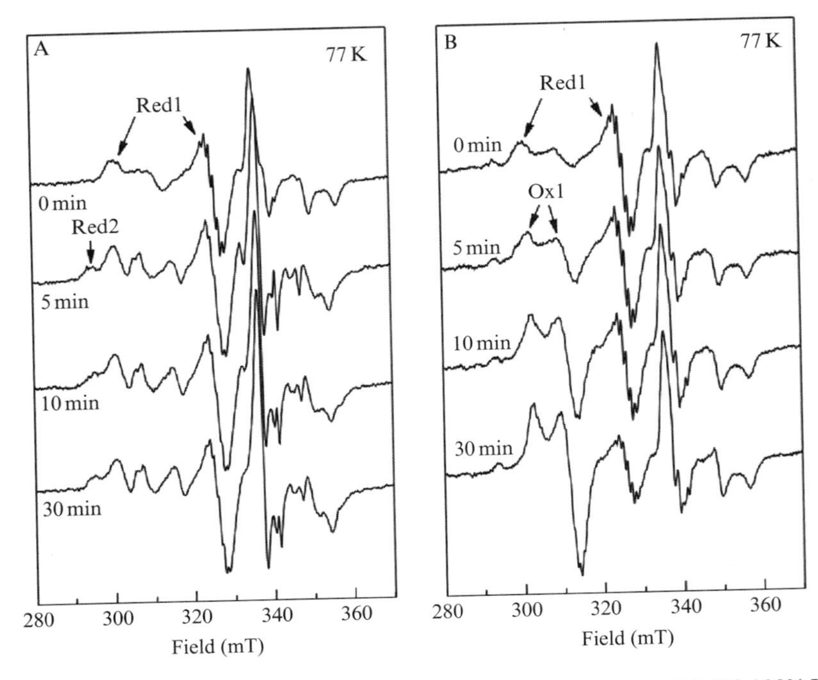

Figure 9.7 Incubation of the cell suspension in the 10-L fermenter with (A) 100% H_2, (B) 80% N_2/20% CO_2. See the text for a complete description of the preparation of the samples. For EPR conditions see Fig. 9.5. (Reproduced from reference (Duin, 2007) with permission from Landes Bioscience and Springer Science+Business Media.)

The selection of the gas mixtures has to do with the hydrogenothropic pathway, where H_2, through hydrogenases, is the reductant and the substrate CO_2 can be considered an oxidant. Therefore, the signals that were induced by gassing with H_2 were dubbed "red" and the number indicated the order in which they were discovered. In the same way, the signals, which were detected upon incubation with the N_2/CO_2 mixture, were called "ox" signals (Albracht *et al.*, 1988). These names are still appropriate since in the MCRred1 form, the Ni in F_{430} is in the 1+ oxidation state. It is also the active form of the enzyme. The increase in intensity of the MCRred1 signal in whole cells is accompanied by an increase in CH_4 production (Albracht *et al.*, 1988). MCRox1 is an inactive form with the Ni in the 3+ oxidation state. MCRox1, however, can be activated (see below). Note in Fig. 9.5, that there is an overall increase in EPR intensity which means that part of the MCR molecules have the Ni present in an EPR-silent form (2+ oxidation state). This form is not active and was dubbed MCRsilent.

The most important application of these studies is that if the cell culture is first treated with 100% H_2 before harvesting, most of MCR will be present in the MCRred1 (and MCRred2) forms (Fig. 9.7, panel A) and when treated with 80% N_2/20% CO_2, most of the MCR will be present as MCRox1 (Fig. 9.7, panel B). These forms will be maintained after lysing the cells as long as the cell suspension is kept cold. If the cells are harvested without pretreatment, most of the MCR will be converted to the MCRsilent form. For the data obtained in Fig. 9.7, 100 mL samples were taken from a 10-L cell culture at the indicated times. The suspension was centrifuged (10 min, 20,000 rpm, 4 °C, RCF = 31293) and the pellet was resuspended in 2 mL of buffer. The resulting suspension was used to make the EPR samples. During all these steps, the cell suspensions were kept on ice to prevent changes in the EPR signals.

5. Cell Growth: Induction of Different Forms of MCR

5.1. MCRred1

When an OD_{568} of 4–5 is reached in the 10-L culture, the MCRred1 and MCRred2 forms can be induced by changing the gas phase to 100% H_2 for 20–25 min. Although the maximal intensity of the MCRred1 and MCRred2 EPR signals is reached in 5 min of incubation (Fig. 9.7, panel A), a longer incubation is needed to keep the MCR reduced during the purification. Furthermore, the cells will have to be cooled down quickly below 20 °C to maintain the MCR forms. Below this temperature, all enzyme activity in the cells appears to be halted, in particular the activity of the hydrogenases. When the H_2 partial pressure drops during the purification procedure, these enzymes will start removing electrons by producing H_2, ultimately converting MCR back to the inactive silent form. This will be prevented by cooling down the cells. In addition, the extra 15 min of incubation time makes sure that all enzymes and cofactors in the pathway are fully reduced providing a buffer against the reoxidation of MCR. To cool the fermenter, the heat blanket is removed from the vessel and the aluminum pot is filled with ice water. The fermenter contains a coil (Fig. 9.4) that is used to guide additional cooling water through the fermenter. With this setup, the cell culture will be cooled down from 65 to 20 °C in about 10 min.

Use the pressure from the H_2 gas tank to transfer the cells into the flow-through rotor (Fig. 9.4). Make sure that the waste water is guided into the sink in the hood since it contains a lot of hydrogen. Exposure of the waste water to oxygen will immediately kill all cells that are left behind so you will not get into trouble with waste management at your institution. Alternatively, collect the waste and sterilize it with bleach or Lysol. Minimize the flow so that it takes about 60–75 min to empty the fermenter. This will give a nice solid pellet. After the fermenter has been emptied, bring the rotor into the anaerobic tent and collect the pellet. There should be around 40–60 g (wet weight).

5.2. MCRox1

When an OD_{568} of 4–5 is reached in the 10-L culture, the MCRox1 form can be induced by changing the gas phase to 80% N_2/20% CO_2 for 25–30 min. Note from Fig. 9.7, panel B, that about 30 min of incubation time is needed to get the highest amount of MCRox1. There is still some MCRred1 left, but this state will not disappear upon longer incubation. Both red1 and ox1 signals will actually lose intensity (not shown). The remaining red1 will disappear during the purification procedure. Keep the gas flow fairly low. It is important to remove any traces of oxygen from the gas stream with an in-line purifier system. The rest of the cooling down and harvesting procedures are identical to those described above for MCRred1. However, during the harvest phase, the fermenter is pressurized from the top with N_2 gas and no gas is bubbled through the suspension anymore.

6. Enzyme Purification

6.1. MCRred1

It is very important that as soon as the cells are lysed, the steps before the 100% ammonium sulfate precipitation step are performed swiftly and without any breaks. Any delay in the procedure will result in loss of active MCR. Resuspend the pellet in about 100 mL 10 m*M* Tris–HCl (pH 7.6)/ 10 m*M* coenzyme M buffer. Sonicate the cells in a sonication beaker on ice for three 3.5 min cycles (pulse 0.5 s on/0.5 s off, flat top, 100% power). The tip of the sonifier should be about 1–1.5 cm in the solution. Centrifuge the cells (20 min, 35,000 rpm, 4 °C, RCF = 95834). Bring the centrifuge tubes back into the chamber and collect the supernatant.

The next two steps are ammonium sulfate precipitation steps. Measure the volume of the supernatant and add 2.33-fold this amount of a saturated $(NH_4)SO_4$/10 m*M* coenzyme M solution to create a 70% saturated solution. The $(NH_4)SO_4$ solution has to be added slowly while stirring the protein solution on ice. Centrifuge the solution (20 min, 35,000 rpm, 4 °C, RCF = 95834) and decant the supernatant. Keep the supernatant on ice and slowly add solid $(NH_4)SO_4$ (±25 g) to obtain a 100% saturated solution, that is, continue adding salt until it does not dissolve anymore. Make sure you do not add more salt than needed because that would cause problems with the next purification step, ion-exchange chromatography. Centrifuge the solution (20 min, 35,000 rpm, 4 °C, RCF = 95834) and resuspend the pellet, which contains the MCR, in 100 mL of buffer containing 10 m*M* Tris–HCl (pH 7.6) and 10 m*M* coenzyme M buffer. If you use a smaller volume, the salt concentration will be too high for the next purification step. This 100% precipitation step mainly serves to remove the smaller soluble cofactors that are present in the cytosol or *M. marburgensis*. It also serves to quickly get the salt concentration down without adding a large volume of buffer or using time-consuming desalting steps.

The solution containing the ammonium sulfate fraction is loaded onto a Q-Sepharose XK 26/20 column (GE Healthcare). Buffer A contains 50 m*M* Tris–HCl buffer (pH 7.6) and 10 m*M* coenzyme M. Buffer B contains 50 m*M* Tris–HCl buffer (pH 7.6), 10 m*M* coenzyme M, and 1 M NaCl. The enzyme is eluted using a step gradient (Fig. 9.8), starting with 0% buffer B and increasing the concentration to 18%, 20%, 22%, 24%, and 26%, followed with wash steps of 30% and 100% buffer B. Collect the tubes that contain MCR. These tubes can be easily recognized since they have a bright yellow color. There is an additional yellow peak, however, just before the main peak that is due to the MCR isoenzyme II. Both enzymes can be collected separately, but most work described in the literature has been done with isoenzyme I. The isoenzyme II is far more unstable than the isoenzyme I, even with the protection described here. The enzyme solution is concentrated to about 2–3 mL using an Amicon Stirred Cell (Millipore) with a filter with a molecular-mass cutoff of 100 kDa.

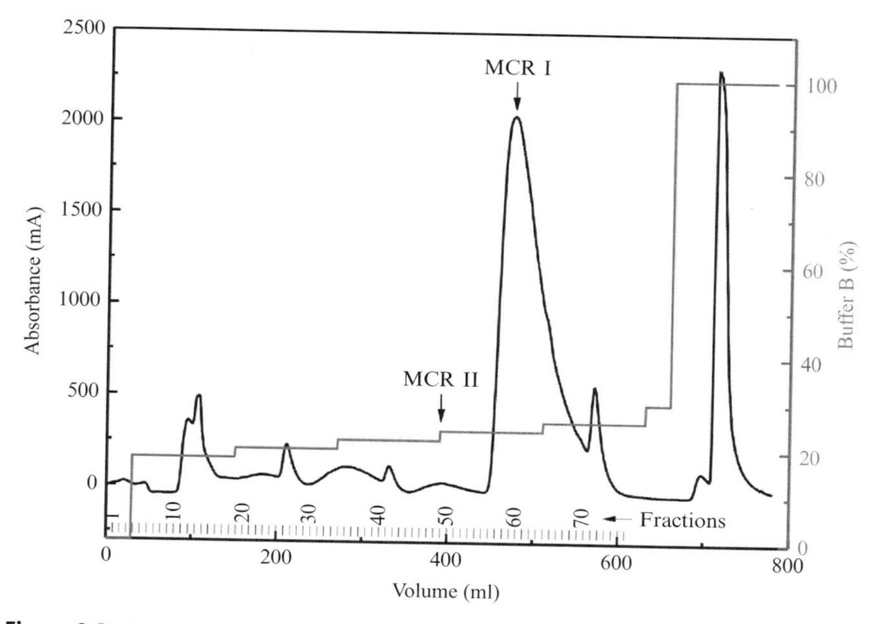

Figure 9.8 Typical purification profile for the Q-sepharose column. The black line is the absorbance at 280 nm. The gray line represents the percentage concentration of buffer B used. Note that the percentage Buffer is what is programmed and is added to the top of the column, while the absorption is measured at the column exit. The "delay" is about 30 mL. The methyl-coenzyme M peaks are indicated in the figure.

Since the buffer solution contains coenzyme M, this form of the enzyme is referred to as MCRred1c ("c" for *c*oenzyme M). Generally, 80–90% of the MCR is in the red1c form with the remainder being MCRsilent. At this stage, the protein is pure enough for kinetic and/or spectroscopic studies. For crystallization studies, an additional purification step is required using a size-exclusion column (16/60 superdex 200, GE healthcare). The buffer used contains 50 m*M* Tris–HCl (pH 7.6), 10 m*M* coenzyme M, and 0.1 *M* NaCl.

Table 9.1 shows the progress of a purification that started with 55 g of wet cells from a 10-L cell culture with an OD_{568} of 4.6. A quick way to determine the quality of the MCR preparation is to measure the absorption spectrum. The spectrum for MCRred1c is pretty unique, with the other MCR forms, being in the Ni(II) state having different spectra. Figure 9.9, solid line, shows the absorption spectrum for the enzyme as isolated. This preparation contains about 90% MCRred1c. The main peak at 368 nm is due to MCRred1. The shoulders at 420 and 445 nm are due to MCRsilent. Figure 9.9, dashed line, shows the same sample after exposure to air. This procedure will convert all enzymes into the MCRsilent form. An overview of the absorption spectra for the different forms of MCR can be found in the literature (Duin *et al.*, 2004).

The spectrum of MCRsilent can be used to determine the protein concentration by measuring the absorbance of oxidized enzyme (MCRsilent) at 420 nm using an extinction coefficient (ε) of 44,000 $M\ cm^{-1}$, assuming a molecular mass of 280,000 Da. Note that the absorbance is due to the nickel-containing F_{430} and that MCR contains two of these per molecule, so the extinction per cofactor is 22,000 $M\ cm^{-1}$. The determination will only work when the protein preparation does not contain factor 420 (F_{420}). This compound can be present when a Q-sepharose column is used that has seen several purifications or when the column is overloaded. The protein concentration can also be determined by using the method of

Table 9.1 Purification table

	Activity in units (μmol min^{-1})	Activity (%)	Total protein (mg)	Specific activity[a] (units mg^{-1})
After sonication	7970	100	3789	2.1
Cell extract	7030	88	3981	1.8
70% AS (supernatant after centrifugation)	4830	61	521	9.3
Resuspended 100% pellet	4040	51	452	8.9
Q-sepharose	5140	65	230	22.3

[a] Activity measured with coenzyme M present.

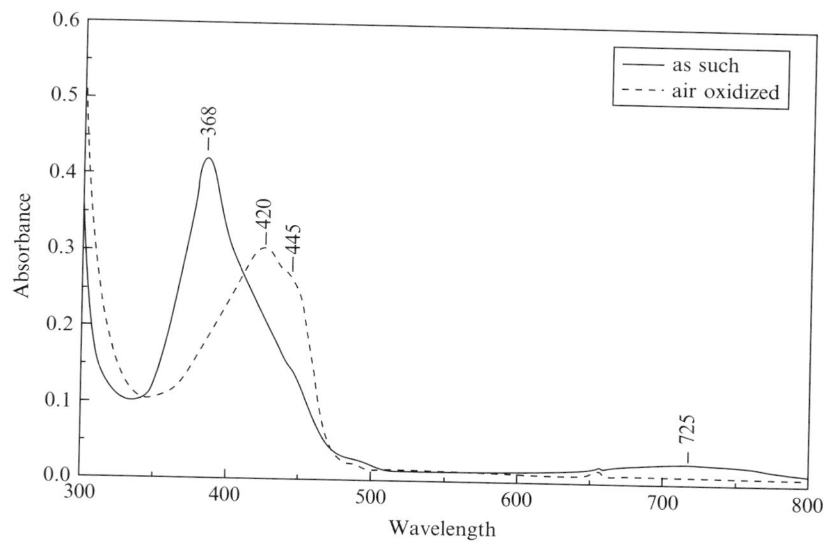

Figure 9.9 Electronic absorption spectra for methyl-coenzyme M reductase (MCR). Solid line: MCR as isolated. The enzyme is mainly (90%) in the MCRred1 form. Dashed line: After exposure to air. The enzyme is now completely MCRsilent (100%).

Bradford (1976) with bovine serum albumin (Serva) as standard. Both methods yield almost the same results.

After purification, the protein solution should be concentrated to about 2 mL and stored in the anaerobic chamber at room temperature. The solution, when pure, is stable for several months. When the protein is stored outside the chamber at 4 °C or frozen, the enzyme is less stable. The main reason is that there is a higher change of oxygen leaking into the storage container.

6.2. MCRox1

The procedure is very similar but there are a couple of important differences: First of all, the enzyme pellets from the 100% ammonium sulfate precipitation step are resuspended in 100 mL 10 m*M* Tris–HCl (pH 7.6), 10 m*M* methyl-coenzyme M buffer. The rest of the buffers *do not* contain coenzyme M or methyl-coenzyme M. After purification, only 60–70% of the MCR will be in the MCRox1 form with the remainder being MCRsilent.

6.3. Activation of MCR: Conversion of MCRox1 into MCRred1m

All forms of MCR are unstable and will eventually become MCRsilent. Currently, it is not possible to convert the silent form into active enzyme *in vitro*. As shown in Fig. 9.5, the *in vivo* process is relatively fast, but besides the fact that it is an ATP-dependent process, only little is known about the other components involved (Kuhner *et al.*, 1993; Rouvière and Wolfe, 1989). MCRred1c is very stable, as mentioned, but this is only true for pure protein. When a dirty Q-sepharose column is used or when too much protein is loaded on the column, some impurities could be present causing the enzyme to lose activity within a couple of days. This is one of the reasons that some groups initially purified the more oxygen-stable MCRox1 form and reactivated this to the MCRred1 form (Becker and Ragsdale, 1998; Horng *et al.*, 2001). The purification, however, still has to be performed under exclusion of air. MCRox1 can be easily activated by incubation with the reductant titanium(III) citrate (Becker and Ragsdale, 1998; Goubeaud *et al.*, 1997; Mahlert *et al.*, 2002b). MCR purified in the ox1 state is concentrated 20-fold by ultrafiltration in a Centricon YM-30 centrifugal Filter Device (Millipore). To 30 μl of the concentrated protein solution, 500 μl 150 m*M* TapsNaOH (pH 9) is added, containing 10 m*M* methyl-coenzyme M and 10 m*M* titanium(III) citrate. The midpoint potential of titanium(III) citrate is pH dependent and the reduction of MCR can only be achieved at pH 9 or higher. The solution was incubated at 60 °C for 30 min to activate the enzyme and then cooled to 4 °C. By adding 1 *M* Tris–HCl (pH 7.2), the pH is adjusted to pH 7.6. The sample can now be assayed for activity and the spectral properties can be determined. Note that about 80% of MCRox1 present can be converted into MCRred1m, the other 20% is lost as MCRsilent. That means that following the MCRox1 purification and activation route, preparations can be obtained with only 50% of the enzyme in the active MCRred1m form (see below for the meaning of "m"). This is sufficient to perform kinetic studies but will cause problems for some of the spectroscopic techniques that detect all forms present, like electronic absorption and X-ray absorption spectroscopies.

6.4. Activity assay

Methane formation from methyl-coenzyme M and coenzyme B is determined at 65 °C in 8-ml serum vials containing 95% N_2/5% H_2 as gas phase (chamber atmosphere) and 0.4 mL of 500 m*M* Mops (pH 7.2), 10 m*M* methyl-coenzyme M, 1 m*M* coenzyme B, 0.3 m*M* aquocobalamin, 30 m*M* titanium(III) citrate, and purified MCR. The cobalamin will reduce the formed heterodisulfide back to the single thiols, coenzyme M and coenzyme B. This is important since heterodisulfide inhibits the reaction. The samples can be premixed including MCR if they are kept on ice. The reaction is started by increasing the vial's temperature from 4 to 65 °C. Gas samples can be withdrawn every 2–3 min and analyzed by gas chromatography for methane.

The specific activity is 70–100 U (μmol min^{-1}) per mg protein calculated for 1 spin per mol F_{430}. (In this case, a spin of 1 means that all nickel in MCR is present as MCRred1. The spin is related to the metal content, not to the protein content since enzyme without cofactor should not display any activity).

6.5. Activity assay for methane oxidation

To measure the anaerobic oxidation of MCR, Scheller *et al.* devised an isotope-labeling experiment in which the formation of $^{13}CH_3$–S–CoM from $^{13}CH_4$ and CoM–S–S–CoB in the presence of $^{12}CH_3$–S–CoM was measured (Scheller *et al.*, 2010). The experiment is based on three equilibriums:

$$^{13}CH_4 + CoM-S-S-CoB \rightleftharpoons {}^{13}CH_3-S-CoM + HS-CoB, \quad \Delta G^o = +30\,kJ\,mol^{-1} \tag{9.1}$$

$$^{12}CH_3-S-CoM + HS-CoB \rightleftharpoons {}^{12}CH_4 + CoM-S-S-CoB, \quad \Delta G^o = +30\,kJ\,mol^{-1} \tag{9.2}$$

$$^{13}CH_4 + {}^{12}CH_3-S-CoM \rightleftharpoons {}^{12}CH_4 + {}^{13}CH_3-S-CoM, \quad \Delta G^o = 0\,kJ\,mol^{-1} \tag{9.3}$$

The presence of $^{12}CH_3$–S–CoM is needed to keep the concentration of HS-CoB low, via exergonic reaction (2), allowing reaction (1) to proceed. The overall reaction (3) is the apparent exchange of labeled carbon between methane and methyl-coenzyme M.

MCR was incubated at 60 °C with 2 m*M* coenzyme B and 4 m*M* $^{12}CH_3$–S–CoM in closed vials containing 1 bar $^{13}CH_4$ in the headspace. The limiting amounts of coenzyme B stop the net methane formation at 50% conversion. After variable additional equilibration times at 60 °C, the reaction was quenched by denaturation of the enzyme at 100 °C and the $^{13}C/^{12}C$ ratio in the S-methyl group of the remaining methyl-coenzyme M was analyzed by ^1H NMR. A specific activity of 11.4 nmol min^{-1} per mg MCR was found using this procedure.

7. Different MCR Forms

The addition of either the substrate methyl-coenzyme M or the analog coenzyme M will stabilize MCR. Both compounds, however, will affect the behavior of MCR in different ways and therefore different forms of MCRred1 are considered. MCRred1c is the form in the presence of ***c***oenzyme M, MCRred1m is the form in the presence of ***m***ethyl-coenzyme M. A clear difference in these forms can be detected in their respective EPR signals (Fig. 9.6). The MCRred1m EPR spectrum shows a more resolved hyperfine-splitting pattern due to the four nitrogen ligand from F_{430} to the nickel. Removal of any of these compounds results in the very unstable MCRred1a form ("a" for ***a***bsence; Mahlert *et al.*, 2002b). Figure 9.10 shows how the different forms of MCR are related and how they can be converted into each other. In most cases, it required the washing away of one compound (indicated with "–") and the addition of another one (indicated with "+"). Washing steps entail the dilution of the protein sample in the

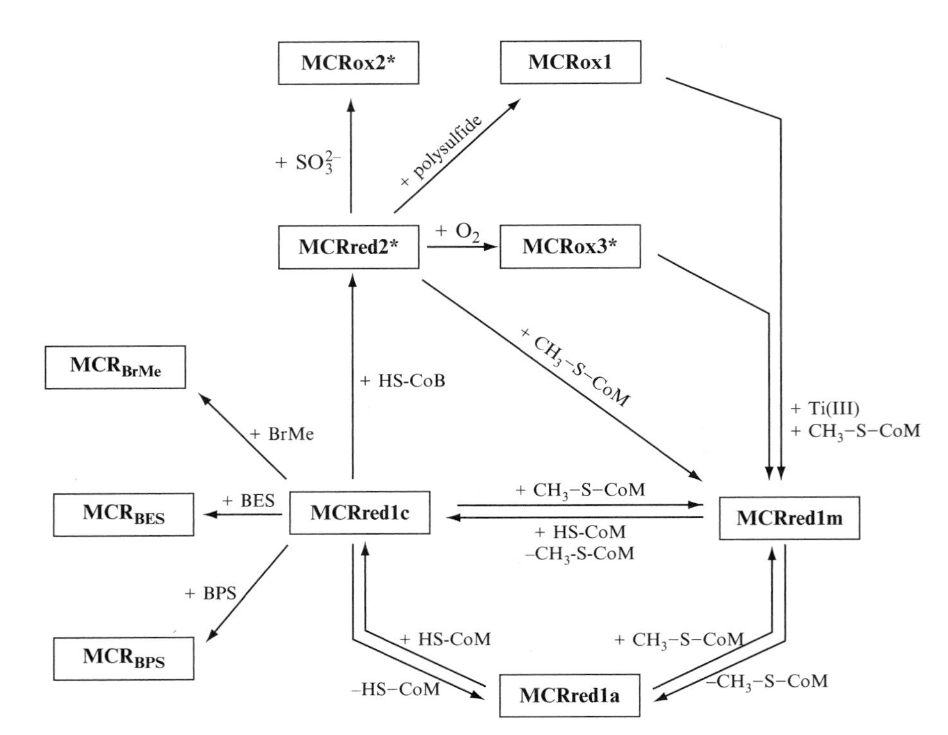

Figure 9.10 Overview of relevant forms of methyl-coenzyme M reductase (MCR) and their interrelationships. See text for details.

required buffer and the subsequent concentration of the sample. Dependent on the dilution factor, the procedure has to be repeated several times to remove most of the original buffer. Centricon centrifuge devices (Millipore) can be used for this purpose. They are also very suitable for concentrating the enzyme to higher concentrations needed for ENDOR and X-ray absorption studies.

Since coenzyme M is available commercially, it is preferred to add this compound to the buffers when purifying MCR. Methyl-coenzyme M has to be synthesized and is therefore only used when absolutely necessary, for example, in the MCRox1 purification. In this case, the methyl-coenzyme M cannot be replaced with coenzyme M. Coenzyme M is an inhibitor, albeit a weak one, and in kinetic studies methyl-coenzyme M can easily outcompete coenzyme M. Still, using the standard assay conditions (see below) a decrease in activity will be observed when coenzyme M is present. Generally, the coenzyme M should be removed before performing kinetic measurement and replaced with coenzyme M to keep the enzyme stabilized. The removal of coenzyme M is also important when the goal is to produce a particular form of MCR for spectroscopic studies.

MCRred1c (in the presence of 10 m*M* coenzyme M) can be converted into MCRred2 by the addition of 5 m*M* coenzyme B (Mahlert *et al.*, 2002a). Two forms are induced, MCRred2a ("a" for axial) and MCRred2r ("r" for rhombic), based on their differences in EPR signals (Finazzo *et al.*, 2003; Kern *et al.*, 2007). MCRred2 can be converted into several ox forms: Addition of polysulfide (0.5 m*M*) will convert it into MCRox1, addition of sulfite (10 m*M*) will convert it into the MCRox2 form, and exposure to oxygen (1 m*M*) will convert it into MCRox3 (Mahlert *et al.*, 2002a). Both MCRox1 and MCRox3 can be converted back into the active MCRred1m form (see below). Three of the forms, MCRred2, MCRox2, and MCRox3, are light sensitive. The light sensitivity of the red2 form is not very high and extensive exposure with high-intense light is needed in this case (Mahlert *et al.*, 2002b). The MCRox2 and MCRox3 forms are very light sensitive, and care has to be taken to keep EPR samples in the dark not to induce too much of the light forms since this will further complicate the EPR spectrum, which can be a problem in studies with whole cells and cell extracts (Albracht *et al.*, 1988). In Fig. 9.10, there is an additional arrow that goes from the MCRred2 box to the MCRred1m box. The addition of 10 m*M* methyl-coenzyme M to MCRred2 will start the reaction cycle, and dependent on the conditions, small CH_4 bubbles can even be detectable in the protein solution. It appears that the first step in the reaction mechanism is rate-limiting and as a result only the red1m signal can be detected in EPR. The same thing will happen when both methyl-coenzyme M and coenzyme B are added to the MCRred1 forms. As a result, other conditions have been tested to see of any reaction-intermediate-related species can be detected including the exposure to substrate analogs and inhibitors. Three that are worth mentioning are the inhibitors bromopropane sulfonate (BPS, 10 m*M*), bromoethane sulfonate (BES, 10 m*M*), and bromomethane (BrMe, 10 m*M*) (Goenrich *et al.*, 2004; Rospert *et al.*, 1992; Sarangi *et al.*, 2009). In the case of BES, the nickel-based signal disappears and a new radical signal is induced. This finding opened up the possibility that MCR might use a radical-based mechanism. This, however, seems less likely now that it is clear the reaction should be reversible. In the case of BPS and BrMe, EPR-active forms are obtained where there is a direct Ni–C bond: MCR_{BPS} and MCR_{BrMe} (Dey *et al.*, 2007; Kunz *et al.*, 2006; Yang *et al.*, 2007). It is highly debated if the Ni–CH_3 species present in MCR_{BrMe} could be an intermediate in the reaction mechanism of MCR. Absent in this scheme is MCRsilent. All forms will eventually become EPR-silent due to inherent instability of the protein. Exposure of oxygen will also turn most forms into an EPR-silent form, including MCRox1, albeit very slowly. The only exception to this is MCRred2.

8. Materials

Small amounts of solutions can be easily made anaerobic by repeated cycles of vacuum and flushing with either N_2 or Ar gas. With the larger volumes of solutions, a different approach is needed. To remove the oxygen (according to the principles of Henry's law), the buffers are boiled. This can be done with a heating plate, but the boiling will go considerably faster when a house-hold electric water boiler is used. Make sure to choose a very simple model with not too many ridges and tubes. A basic boiler with a steel "kettle" without any holes, etc. will be preferable since it is very easy to keep clean. When the buffers are brought to a boil, they are transferred into an appropriate size bottle. The bottles can be flushed with nitrogen before and while the buffer is added. Add a stirring bar to the bottle. A rubber stopper and a cap with a hole that fits over the stopper are put on the bottle. The bottle is placed on a stirring plate and hooked up to a vacuum line and is left there, under stirring, for a couple of hours. Use a pump with solvent recovery (BrandTech) that can handle the high amount of liquid being pumped off. When the bottle is cooled down, it is pressurized (0.5 atm, Ar or N_2). Now the bottle can be brought into the chamber or stored until use. Before use, the buffers can be brought into the chamber and left stirring overnight, exposed to the chamber atmosphere, to remove the last traces of oxygen. The buffers that are intended for the FPLC-based purification steps first have to be filtered before they are boiled.

This procedure will cause problems for some compounds due to heat instability. This is the reason that both methyl-coenzyme M and coenzyme

M are not added to the respective buffers until they are brought into the anaerobic chamber.

List of buffers and compound for the purification of MCRred1c

- Phosphate buffer (2 L: 14.739 g K_2P and 2.014 g KH_2)
- 10 m*M* Tris–HCl buffer (pH 7.6), 10 m*M* CoM (1 L)
- Saturated $(NH_4)SO_4$ solution + 10 m*M* CoM
- 50 m*M* Tris–HCl (pH 7.6), 10 m*M* CoM (buffer A for FPLC)
- 50 m*M* Tris–HCl (pH 7.6), 10 m*M* CoM, 2 *M* NaCl (buffer B for FPLC
- Solid $(NH_4)SO_4$

List of buffers and compound for the purification of MCRox1

- Phosphate buffer (2 L: 14.739 g K_2P and 2.014 g KH_2)
- 10 m*M* Tris–HCl buffer (pH 7.6), 10 m*M* methyl-CoM (1 L)
- Saturated $(NH_4)SO_4$ solution
- 50 m*M* Tris–HCl (pH 7.6) (buffer A for FPLC)
- 50 m*M* Tris–HCl (pH 7.6), 2 *M* NaCl (buffer B for FPLC)
- Solid $(NH_4)SO_4$

Columns

- Q-sepharose (regenerate after five to six rounds of purification)
- Superdex 200 (not needed for routine MCR preparations)

Activity assay

- 500 m*M* Mops (pH 7.2)
- 10 m*M* methyl-coenzyme M
- 1 m*M* coenzyme B
- 0.3 m*M* aquocobalamin
- 30 m*M* titanium(III) citrate (see below)

Titanium(III) citrate (~200 mM)

- Dissolve 4.4 g Na-citrate·$2H_2O$ in 45 mL water.
- Add 1.54 g $Ti(III)Cl_3$ (Fluka).
- Add solid $KHCO_3$ to neutralize the solution. Adding this will cause the formation of gas. Repeat this step until the gas formation stops. The pH should be around 7.0.

The electronic equipment in the chamber will be damaged by extended exposure to H_2S that escapes from the cell suspensions that are brought into the chamber. The H_2S can be removed from the chamber atmosphere by bubbling it through a silver sulfate solution.

- Bring 2 L of distilled water to a boil
- Add 10 g Ag_2SO_4
- Cool down to RT and add 20 mL 1 *M* sulfuric acid
- Add 2 L of glycerol

8.1. Synthesis of methyl-coenzyme M

Methyl-coenzyme M is synthesized from coenzyme M (Sigma) by methylation with methyl iodide (Fluka). Coenzyme M (3 mmol) is incubated for 12 h with 6 mmol of methyliodide in 5 mL of a 32% aqueous ammonia solution under a 100% N_2 atmosphere. Use a dark brown bottle. After incubation for 12 h at room temperature, the reaction mixture is flash evaporated to almost dryness followed by lyophilization. The now dry yellow residue is dissolved in 10 mL distilled water. Then 5-ml aliquots of the solution are loaded on a Q-Sepharose XK 26/20 column (GE Healthcare) and a linear gradient of 0–1 *M* ammonium carbonate is used for separation. It is important to remove any gas from the ammonium carbonate solution by leaving it under vacuum for a couple of hours before use. Eluate is collected and tested for methyl-coenzyme M by thin layer chromatography (TLC) on Silica Gel 60 (Merck) with butanol/acetic acid/water (2:1:1). Develop the TLC plates with 1% $KMnO_4$/5% Na_2CO_3 in water. The whole plate will turn purple with yellow band that indicates the position of the methyl-coenzyme M. The compound generally elutes from the column between 0.25 and 0.35 *M* ammonium carbonate. Fractions containing methyl-coenzyme M are combined and lyophilized at room temperature. When the sample is dry, the temperature is increased to 60 °C to remove remaining traces of methyl iodide. A dry white powder is the end product. If it has a yellow color, it was not kept long enough under vacuum at 60 °C. Ammonium carbonate is used instead of hydrochloric acid to elute methyl-coenzyme M from the Q-Sepharose column, since the nondissociated sulfonic acid form of methyl-coenzyme M and water form an azeotropic mixture.

8.2. Synthesis of Coenzyme B

8.2.1. 7,7′-Dithiodiheptanoic acid

7-Mercaptoheptanoic acid is synthesized by dissolving 17.4 g (228 mmol) of thiourea in a stirred solution of 7-bromoheptanoic acid (Matrix Scientific) (9.6 g, 46.6 mmol) in 110 mL of ethanol. This mixture is refluxed under argon at 90 °C for 17 h, cooled to room temperature, and 25 mL of a 60% aqueous solution (w/v) of sodium hydroxide is added. The mixture is refluxed under argon for an additional 2 h, and cooled to room temperature. The yellow/white solution is concentrated. A 100 mL aqueous solution of 1 *M* HCl is added followed by 20 mL concentrated HCl. The thiol is extracted three times with 100 mL dichloromethane. The dichloromethane phase is extracted three times with 150 mL of an aqueous solution of 1 *M* sodium bicarbonate. The aqueous extract is acidified with HCl to a pH of 1–2, and extracted three times with 200 mL dichloromethane. The organic phase is filtered over cotton and concentrated to 70 mL. The thiol is oxidized to a disulfide by mixing the dichloromethane phase with an aqueous solution of 10% (w/v) iodine and 20% (w/v) potassium iodide until the brown color persists. The aqueous phase is removed and the dichloromethane phase is washed three times with an aqueous solution of 1 *M* sodium thiosulfate, and three times with water. The organic phase is filtered over cotton, (dried over anhydrous magnesium sulfate), and concentrated under vacuum.

The product is crystallized twice from benzene (or toluene), and dried for 2 days under high vacuum, to give 3.3 g (46%) of white crystals (Mp. 70 °C).

8.2.2. 7,7′-dithiobis(succinimido-oxyheptanoate)

7,7′-Dithiodiheptanoic acid (388 mg, 1.2 mmol) is dissolved in 8 mL of 1,4-dioxane at room temperature and the solution is stirred while 290 mg (2.5 mmol) of *N*-hydroxysuccinimide is added. When this dissolves, 505 mg (2.4 mmol) of dicyclohexylcarbodiimide (in 3 mL 1,4-dioxane) is added drop-wise and the solution is stirred for 16 h at room temperature. The precipitated dicyclohexylurea is removed by filtration over a glass filter and the filtrate is washed two times with 5 mL 1,4-dioxane, and dried (to a clear oil) under vacuum. The product is recrystallized twice from boiling 2-propanol, and dried for 1 day under high vacuum to give 384 mg (62%) of white crystals (Mp. 105–106 °C).

8.2.3. (+)-*N*,*N*′-(7,7′-Dithiodiheptanoyl)bis(*O*-phospho-L-threonine)

A solution of *O*-phospho-L-threonine (400 mg, 2 mmol) and triethylamine (0.56 mL, 4 mmol) in 4 mL water is added under stirring to a solution of 7,7′-dithiobis(succinimido-oxyheptanoate) (362 mg, 0.7 mmol l) in 18 mL tetrahydrofuran and 4 mL acetonitrile. After stirring at room temperature under nitrogen for 36 h, the solvents are removed under vacuum at 30 °C. The resulting white residue is dissolved in 1 *M* HCl (25 mL) and washed three times with dichloromethane (3 × 8 mL). Traces of dichloromethane are removed from the aqueous phase under vacuum. The aqueous phase is applied to a 2 × 13 cm column of polystyrene XAD-2 (equilibrated with 1 *M* HCl). The column is washed with 100 mL of 1 *M* HCI, followed by

150 mL H_2O. The product is eluted by applying a methanol step gradient (80 mL H_2O/MeOH, 4:1; 120 mL H_2O/MeOH, 1:1; 80 mL H_2O/MeOH, 1:4, 100 mL MeOH). TLC analysis (solvent system: *n*-butanol/acetic acid/H_2O, 2:1:1) of the collected fractions (10 mL each) shows that CoB–S–S–CoB (Rf = 0.35) is completely separated from *O*-phosphothreonine (Rf = 0.24) and less polar products (Rf > 0.6). The combined fractions containing pure CoB–S–S–CoB are concentrated under vacuum (50 °C) to 1/10 of the original volume. Subsequently, 2 mL 2 *M* ammonia was added to obtain the ammonium salt. After lyophilization, 112 mg (20%) (+)-*N,N'*-(7,7′-dithiodiheptanoyl)bis(*O*-phospho-L-threonine), a white solid, is obtained.

8.2.4. *N*-7-Mercaptoheptanoyl-*O*-phospho-L-threonine (HS-CoB)

Seventy-five miiligram of (96 μmol) CoB–S–S–CoB (Ammonium salt) is dissolved in 2 mL anaerobic 50 m*M* potassium phosphate buffer, pH 7.0, under a nitrogen atmosphere (This procedure can be performed in the anaerobic chamber). Subsequently, 4 mL of an anaerobic KBH_4-solution (10% in 50 m*M* potassium phosphate buffer, pH 7.0) is added, and the solution is stirred for 16 h at RT.

Dependent on the available equipment, the formation of CoB–S–S–CoB and the completeness of the reaction can be followed by either HPLC or NMR spectroscopy. The problem with the HPLC procedure is that everything has to be done under exclusion of air. An RP-18 column (12.5 × 4 mm) is used with an isocratic flow of 20 m*M* potassium phosphate (pH 2) in 50% MeOH. The absorption after the column is measured at 220 nm. HS-CoB elutes first, followed at a good distance by CoB–S–S–CoB. A quicker method, however, is to use NMR spectroscopy. The CoB–S–S–CoB shows a triplet at 2.64 ppm that is shifted to 2.42 ppm for HS-CoB (measured in D_2O).

Excess $NKBH_4$ is removed by addition of 2 mL 25% HCl (aq) and the acid solution (pH 0) is loaded on an Amberlite XAD-2 column (1 × 10 cm) previously equilibrated with anaerobic 1 *M* HCl (aq). The column is washed with 50 mL 1 *M* HCl, 150 mL anaerobic water and the HS-CoB is eluted with 150 mL anaerobic water/methanol (70:30, v/v). Note that some of the HS-CoB can already come down with the water. The collected HS-CoB fractions are lyophilized, and the powder is dissolved in anaerobic 10 m*M* Tris–HCl buffer (pH 8.0). (The pH can be adjusted with a 2-*M* NaOH solution if needed). The HS-CoB concentration is determined indirectly by determining the thiol concentration with Ellman's Reagent (DTNB, 5,5′-dithiobis-(2-nitrobenzoic acid)).

In the paper by Scheller *et al.*, a method is described for the reduction of the disulfide using dithiothreitol as reductant and a precipitation method is used instead of a column to purify the HS-CoB (Scheller *et al.*, 2010).

Spectroscopic properties: ^{1}H NMR (600 MHz, D_2O = 3.700 ppm): δ 4.65 (*dqd*, *J* = 7.50, 6.48, 3.54, ^{1}H, C*H*OPO_3^{2-}); 4.21 (*dd*, *J* = 3.42, 1.92, ^{1}H, C*H*NC=O); 2.49 (*t*, *J* = 7.17, 2H, CH_2S); 2.34 (*dt*, *J* = 14.16, 7.65, 1H, CH_2C=O); 2.29 (*dt*, *J* = 14.16, 7.35, 1H, CH_2C = O); 1.58 (*quintet*, *J* = 7.47, 2H, CH_2); 1.55 (*quintet*, *J* = 7.38, ^{2}H, CH_2); 1.35 (*quintetoid*, J_{obs} = 7.5, ^{2}H, CH_2); 1.29 (*quartetoid*, J_{obs} = 7.0, 2H, CH_2); 1.25 (*d*, *J* = 6.42, 3H, CH_3). ^{13}C NMR (150 MHz, D_2O, dioxane = 67.00 ppm): δ 177.69 (C); 176.44 (C); 73.08 (CH, ^{3}JCP = 5.48 Hz); 60.01 (CH, ^{4}JCP = 7.68 Hz); 36.04 (CH_2); 33.25 (CH_2); 28.11 (CH_2); 27.54 (CH_2); 25.62 (CH_2); 24.11 (CH_2); 18.86 (CH3, ^{4}JCP = 1.10 Hz) (Scheller *et al*., 2010).

ACKNOWLEDGMENT

Pretty much everything I know about the growth of *M. marburgensis* cells and the purification of MCR, I learned and developed during my stay in the group of Rolf Thauer at the Max-Planck-Institute for Terrestrial Microbiology in Marburg, Germany. Therefore, I have to thank him here and also the other lab members that worked on MCR: Reinhard Böcher and Felix Mahlert. I also thank the group of Bernhard Jaun at the ETH in Zurich, Switzerland, and in particular Sylvan Scheller, for advising on the synthesis of coenzyme B.

REFERENCES

Albracht, S. P. J., Ankel-Fuchs, D., Van der Zwaan, J. W., Fontijn, R. D., and Thauer, R. K. (1986). A new EPR signal of nickel in *Methanobacterium thermoautotrophicum*. *Biochim. Biophys. Acta* **870,** 50–57.

Albracht, S. P. J., Ankel-Fuchs, D., Böcher, R., Ellermann, J., Moll, J., Van der Zwaan, J. W., and Thauer, R. K. (1988). Five new EPR signals assigned to nickel in methyl-coenzyme M reductase from *Methanobacterium thermoautotrophicum*, strain Marburg. *Biochim. Biophys. Acta* **955,** 86–102.

Becker, D. F., and Ragsdale, S. W. (1998). Activation of methyl-SCoM reductase to high specific activity after treatment of whole cells with sodium sulfide. *Biochemistry* **37,** 2639–2647.

Beinert, H. (1978). EPR spectroscopy of components of the mitochondrial electron-transfer system. *Methods Enzymol.* **54,** 133–150.

Beinert, H., Orme-Johnson, W. H., and Palmer, G. (1978). Special techniques for the preparation of samples for low-temperature EPR spectroscopy. *Methods Enzymol.* **54,** 111–132.

Boetius, A., Ravenschlag, K., Schubert, C. J., Rickert, D., Widdel, F., Gieseke, A., Amann, R., Jørgensen, B. B., Witte, U., and Pfannkuche, O. (2000). A marine microbial consortium apparently mediating anaerobic oxidation of methane. *Nature* **405,** 623–626.

Bradford, M. M. (1976). Rapid and sensitive method for the quantitation of microgram quantities of protein utilizing the principle of protein-dye binding. *Anal. Biochem.* **72,** 248–254.

Dey, M., Telser, J., Kunz, R. C., Lees, N. S., Ragsdale, S. W., and Hoffman, B. M. (2007). Biochemical and spectroscopic studies of the electronic structure and reactivity of a methyl-ni species formed on methyl-coenzyme M reductase. *J. Am. Chem. Soc.* **129,** 11030–11032.

Duin, E. C. (2007). Role of coenzyme F_{430} in methanogenesis. *In* "Tetrapyrroles: Their Birth, Life and Death," (M. J. Warren and A. Smith, eds.), pp. 352–374. Landes Bioscience, Georgetown.

Duin, E. C., and Mckee, M. L. (2008). A new mechanism for methane production from methyl-coenzyme M reductase as derived from density functional calculations. *J. Phys. Chem. B* **112,** 2466–2482.

Duin, E. C., Signor, L., Piskorski, R., Mahlert, F., Clay, M. D., Goenrich, M., Thauer, R. K., Jaun, B., and Johnson, M. K. (2004). Spectroscopic investigation of the nickel-containing porphinoid cofactor F_{430}. Comparison of the free cofactor in the +1, +2 and +3 oxidation states with the cofactor bound to methyl-coenzyme M reductase in the silent, red and ox forms. *J. Biol. Inorg. Chem.* **9,** 563–576.

Ellefson, W. L., and Wolfe, R. S. (1981). Component C of the methylreductase system of *Methanobacterium. J. Biol. Chem.* **256,** 4259–4262.

Fee, J. A. (1978). Transition metal electron paramagnetic resonance related to proteins. *Methods Enzymol.* **49,** 512–528.

Finazzo, C., Harmer, J., Jaun, B., Duin, E. C., Mahlert, F., Thauer, R. K., Van Doorslaer, S., and Schweiger, A. (2003). Characterization of the MCR_{red2} form of methyl-coenzyme M reductase, a pulse EPR and ENDOR study. *J. Biol. Inorg. Chem.* **8,** 586–593.

Goenrich, M., Mahlert, F., Duin, E. C., Bauer, C., Jaun, B., and Thauer, R. K. (2004). Probing the reactivity of Ni in the active site of methyl-coenzyme M reductase with substrate analogues. *J. Biol. Inorg. Chem.* **9,** 691–705.

Goubeaud, M., Schreiner, G., and Thauer, R. K. (1997). Purified methyl-coenzyme-M reductase is activated when the enzyme-bound coenzyme F430 is reduced to the nickel (I) oxidation state by titanium(III) citrate. *Eur. J. Biochem.* **243,** 110–114.

Grabarse, W., Mahlert, F., Duin, E. C., Goubeaud, M., Shima, S., Thauer, R. K., Lamzin, V., and Ermler, U. (2001). On the mechanism of biological methane formation: structural evidence for conformational changes in methyl-coenzyme M reductase upon substrate binding. *J. Mol. Biol.* **309,** 315–330.

Hagen, W. R. (2009). Biomolecular EPR Spectroscopy. CRC Press, Boca Raton.

Horng, Y.-C., Becker, D. F., and Ragsdale, S. W. (2001). Mechanistic studies of methane biogenesis by methyl-coenzyme M reductase: Evidence that coenzyme B participates in cleaving the C-S bond of methyl-coenzyme M. *Biochemistry* **40,** 12875–12885.

Kern, D. I., Goenrich, M., Jaun, B., Thauer, R. K., Harmer, J., and Hinderberger, D. (2007). Two sub-states of the red2 state of methyl-coenzyme M reductase revealed by high-field EPR spectroscopy. *J. Biol. Inorg. Chem.* **12,** 1097–1105.

Kuhner, C. H., Lindenbach, B. D., and Wolfe, R. S. (1993). Component A2 of methyl-coenzyme M reductase system from *Methanobacterium thermoautotrophicum* ΔH: Nucleotide sequence and functional expression by *Escherichia coli. J. Bacteriol.* **175,** 3195–3203.

Kunz, R. C., Horng, Y. C., and Ragsdale, S. W. (2006). Spectroscopic and kinetic studies of the reaction of bromopropanesulfonate with methyl-coenzyme M reductase. *J. Biol. Chem.* **281,** 34663–34676.

Mahlert, F., Bauer, C., Jaun, B., Thauer, R. K., and Duin, E. C. (2002a). The nickel enzyme methyl-coenzyme M reductase from methanogenic archaea: *In vitro* induction of the nickel-based MCR-ox EPR signals from MCR-red2. *J. Biol. Inorg. Chem.* **7,** 500–513.

Mahlert, F., Grabarse, W., Kahnt, J., Thauer, R. K., and Duin, E. C. (2002b). The nickel enzyme methyl-coenzyme M reductase from methanogenic archaea: In vitro interconversions among the EPR detectable MCR-red1 and MCR-red2 states. *J. Biol. Inorg. Chem.* **7,** 101–112.

Palmer, G. (1967). Electron paramagnetic resonance. *Methods Enzymol.* **10,** 594–595.

Pelmenschikov, V., and Siegbahn, P. E. M. (2003). Catalysis by methyl-coenzyme M reductase: A theoretical study for heterodisulfide product formation. *J. Biol. Inorg. Chem.* **8,** 653–662.
Pelmenschikov, V., Blomberg, M. R. A., Siegbahn, P. E. M., and Crabtree, R. H. (2002). A mechanism from quantum chemical studies for methane formation in methanogenesis. *J. Am. Chem. Soc.* **124,** 4039–4049.
Ragsdale, S. W. (2003). Biochemistry of methyl-CoM reductase and coenzyme F_{430}. *In* “Porphyrin Handbook,” (K. M. Kadish, K. M. Smith, and R. Guilard, eds.), pp. 205–228. Elsevier Science, San Diego.
Rospert, S., Böcher, R., Albracht, S. P. J., and Thauer, R. K. (1991). Methyl-coenzyme M reductase preparations with high specific activity from H_2-preincubated cells of *Methanobacterium thermoautotrophicum*. *FEBS Lett.* **291,** 371–375.
Rospert, S., Vogus, M., Berkessel, A., Albracht, S. P. J., and Thauer, R. K. (1992). Substrate-analoque-induced changes in the nickel-EPR spectrum of active methyl-coenzyme-M reductase from *Methanobacterium thermoautotrophicum*. *Eur. J. Biochem.* **210,** 101–107.
Rouvière, P. E., and Wolfe, R. S. (1989). Component A3 of the methylcoenzyme M methylreductase system of *Methanobacterium thermoautotrophicum* ΔH: Resolution into two components. *J. Bacteriol.* **171,** 4556–4562.
Sarangi, R., Dey, M., and Ragsdale, S. W. (2009). Geometric and electronic structures of the Ni-I and methyl-Ni-III intermediates of methyl-coenzyme M reductase. *Biochemistry* **48,** 3146–3156.
Scheller, S., Goenrich, M., Boecher, R., Thauer, R. K., and Jaun, B. (2010). The key nickel enzyme of methanogenesis catalyses the anaerobic oxidation of methane. *Nature* **465,** 606–609.
Shima, S., and Thauer, R. K. (2005). Methyl-coenzyme M reductase and the anaerobic oxidation of methane in methanotrophic Archaea. *Curr. Opin. Microbiol.* **8,** 643–648.
Thauer, R. K. (1998). Biochemistry of methanogenesis: A tribute to Marjory Stephenson. *Microbiology* **144,** 2377–2406.
Yang, N., Reiher, M., Wang, M., Harmer, J., and Duin, E. C. (2007). Formation of a nickel-methyl species in methyl-coenzyme M reductase, an enzyme catalyzing methane formation. *J. Am. Chem. Soc.* **129,** 11028–11029.

CHAPTER TEN

Methods for Analysis of Acetyl-CoA Synthase: Applications to Bacterial and Archaeal Systems

David A. Grahame

Contents

Abstract

The nickel- and iron-containing enzyme acetyl-CoA synthase (ACS) catalyzes *de novo* synthesis as well as overall cleavage of acetyl-CoA in acetogens, various other anaerobic bacteria, methanogens, and other archaea. The enzyme contains a unique active site metal cluster, designated the A cluster, that consists of a binuclear Ni–Ni center bridged to an $[Fe_4S_4]$ cluster. In bacteria, ACS is tightly associated with CO dehydrogenase to form the bifunctional

Department of Biochemistry and Molecular Biology, Uniformed Services University of the Health Sciences, Bethesda, Maryland, USA

Methods in Enzymology, Volume 494
ISSN 0076-6879, DOI: 10.1016/B978-0-12-385112-3.00010-X

heterotetrameric enzyme CODH/ACS, whereas in archaea, ACS is a component of the large multienzyme complex acetyl-CoA decarbonylase/synthase (ACDS), which comprises five different subunits that make up the subcomponent proteins ACS, CODH, and a corrinoid enzyme. Characteristic properties of ACS are discussed, and key methods are described for analysis of the enzyme's multiple redox-dependent activities, including overall acetyl-CoA synthesis, acetyltransferase, and an isotopic exchange reaction between the carbonyl group of acetyl-CoA and CO. Systematic measurement of these activities, applied to different ACS protein forms, provides insight into the ACS catalytic mechanism and physiological functions in both CODH/ACS and ACDS systems.

1. Introduction

Methanogens growing on one-carbon substrates such as $CO_2 + H_2$, formate, methanol, methyl amines, or methyl sulfides synthesize acetyl groups in the form of acetyl-CoA, which is then used for assimilation into cell carbon (Ferry, 2010). The biochemical mechanism of synthesis is unusual because acetyl groups are formed by direct condensation of separate C-1 species, one at the level of CH_3 and the other as CO, and because it involves one- and two-carbon organometallic intermediates generated at a unique nickel- and iron-containing metallocofactor, the A cluster (Fig. 10.1) located at the active site of the enzyme acetyl-CoA synthase (ACS). In acetogenic bacteria and most species of bacteria able to carry out complete oxidation of acetate under anaerobic conditions, ACS is part of a 310 kDa $\alpha_2\beta_2$ bifunctional enzyme known as CO dehydrogenase/acetyl-CoA

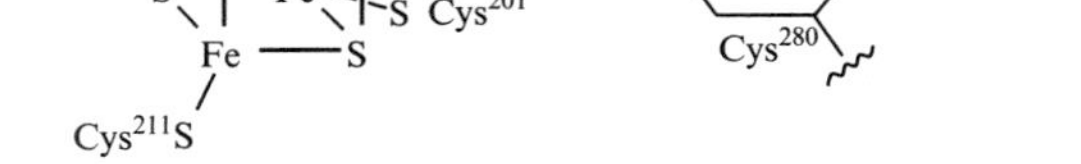

Figure 10.1 The A cluster at the active site of acetyl-CoA synthase. Amino acid numbers refer to the ACDS β subunit. Ni_p is the proximal nickel of an Ni–Ni binuclear site that is connected through a bridging thiolate to an Fe_4S_4 cluster. Ni_d represents the distal Ni atom, which exhibits square-planar geometry and shares two cysteine thiolate ligands with Ni_p, and is further coordinated by two deprotonated amide nitrogen atoms from the protein backbone. Crystallographic structures of A cluster-containing enzymes have been solved by Doukov *et al.* (2002), Darnault *et al.* (2003), and Svetlitchnyi *et al.* (2004).

synthase (CODH/ACS; Ragsdale and Pierce, 2008; Thauer, 1988; Xia *et al.*, 1996). By contrast, in methanogens and other species of Archaea, ACS is resident as an integral part, the β subunit, of the five-subunit-containing, 2000 kDa $(\alpha_2\varepsilon_2)_4\beta_8(\gamma\delta)_8$ multienzyme complex known as acetyl-CoA decarbonylase/synthase (ACDS; Gencic *et al.*, 2010). In the ACDS complex, ACS not only is closely associated with carbon monoxide dehydrogenase, an $\alpha_2\varepsilon_2$ archaeal-type CODH enzyme subcomponent that produces CO from reduction of CO_2, but also is in proximity to its methyl donating partner, the $\gamma\delta$ corrinoid protein subcomponent that secures methyl groups on its bound corrinoid cofactor by autocatalytic transfer from N^5-methyl-tetrahydromethano(sarcina)pterin (CH_3–H_4MPt or CH_3–H_4SPt). The overall reaction catalyzed by the ACDS β subunit is as follows:

$$\text{CO} + \text{CH}_3\text{-Co(III)} + \text{CoA} \rightleftarrows \text{acetyl-CoA} + \text{Co(I)} + \text{H}^+, \quad K'_{eq} \approx 1 \times 10^8 \text{atm}^{-1} \tag{10.1}$$

The ACS reaction is complex, and various exchange reactions and partial reactions are notable contributors, studies on which have given insight into how the ACDS β subunit functions in the overall condensation. The purpose of this chapter is to describe key analytical methods that have been developed to measure the overall ACS reaction, as well as those applied to analyze important exchange and partial reaction activities catalyzed by such A cluster-containing proteins.

1.1. Acetyl-CoA synthesis and cleavage

The large value of K'_{eq} for Eq. (10.1) holds that the ACS reaction is highly favorable under standard conditions, and under a variety of experimental assay conditions, this leads to extensive substrate utilization and product formation. The equilibrium constant for reaction (10.1) is calculated from data in Eq. (10.2) (the portion of the overall reaction of the ACDS complex that excludes CO oxidation and reduction of ferredoxin; Grahame and DeMoll, 1995) and the equilibrium constant determined for methyl group transfers from CH_3–H_4SPt to the ACDS $\gamma\delta$ corrinoid protein, Eq. (10.3) (Grahame and DeMoll, 1996).

$$\text{CO} + \text{CH}_3\text{-H}_4\text{SPt} + \text{CoA} \rightleftarrows \text{acetyl-CoA} + \text{H}_4\text{SPt}, \quad K_{eq} \approx 2 \times 10^9 \text{atm}^{-1} \tag{10.2}$$

$$\text{H}_4\text{SPt} + \text{CH}_3\text{-Co(III)}\gamma\delta \rightleftarrows \text{CH}_3\text{-H}_4\text{SPt} + \text{Co(I)}\gamma\delta + \text{H}^+, \quad K'_{eq} = 0.057 \tag{10.3}$$

Thus, the K'_{eq} value given for the ACS reaction (Eq. (10.1)) pertains to the native corrinoid methyl donor. However, there is indication that the equilibrium position is influenced by the reactivity of the particular Co (III)-CH_3 corrinoid substrate employed in ACS assays, as base-off methyl corrinoid substrates are significantly more reactive methyl group donors than base-on configurations (Norris and Pratt, 1996; Pratt, 1999; Stich *et al.*, 2006). For example, acetyl-CoA synthesis by the ACDS β subunit using methylcobalamin (a base-on corrinoid) results in incomplete substrate consumption, even after prolonged incubation, and proceeds in the forward direction at an initial velocity of only 1/1500th of the reaction carried out with methylcobinamide, which, like the native corrinoid protein cofactor, is base-off (Gencic and Grahame, 2003, 2008). A similar relative rate difference has been measured for bacterial ACS (Seravalli *et al.*, 2001).

Despite the highly favorable standard free energy change for acetyl-CoA formation evident from the large K_{eq} values for Eqs. (10.1) and (10.2), under cellular conditions the synthesis and cleavage of acetyl-CoA by the ACDS complex is freely reversible. One factor that contributes in large part to the decomposition of acetyl-CoA by reversal of reaction (10.2) is the oxidation of CO to CO_2, which is favorable under a wide range of cellular redox conditions, and is tightly coupled to acetyl C–C bond cleavage by specific subunit-to-subunit communications within the complex (Gencic *et al.*, 2010). Another factor that favors the reversal of reaction (10.2) is the removal of CH_3–H_4SPt and regeneration of H_4SPt, which takes place as methyl groups are continually transferred to coenzyme M and reduced to methane. A third consideration that favors acetate decomposition under physiological conditions where levels of ATP and acetate are high is the reacetylation of CoA, which increases the ratio of acetyl-CoA to CoA. Thus, in species of methanogens such as *Methanosarcina* and *Methanosaeta* growing on acetate as sole source of carbon and energy, ACDS plays a major, central role in acetate cleavage providing reducing equivalents (from CO oxidation) and methyl groups (from the CH_3 group of acetate) to fuel methanogenesis and provide for energy conservation. These methanogens contain the highest amounts of ACDS protein, estimated to be 20–25% of the buffer-soluble protein in cell extracts; as such, the ACDS complex was first isolated (Terlesky *et al.*, 1986) and later shown to catalyze the cleavage of acetyl-CoA and direct methylation of H_4SPt (Grahame, 1991) using extracts from species of *Methanosarcina* grown on acetate.

The individual steps in cleavage of acetyl-CoA by ACDS (Fig. 10.2) reduce to the following balanced overall reaction:

$$\begin{aligned}&\text{acetyl-CoA} + H_4SPt + 2Fd_{ox} + H_2O \rightleftarrows \\ &\text{CoA} + CH_3\text{-}H_4SPt + CO_2 + 2Fd_{red} + 2H^+\end{aligned} \qquad (10.4)$$

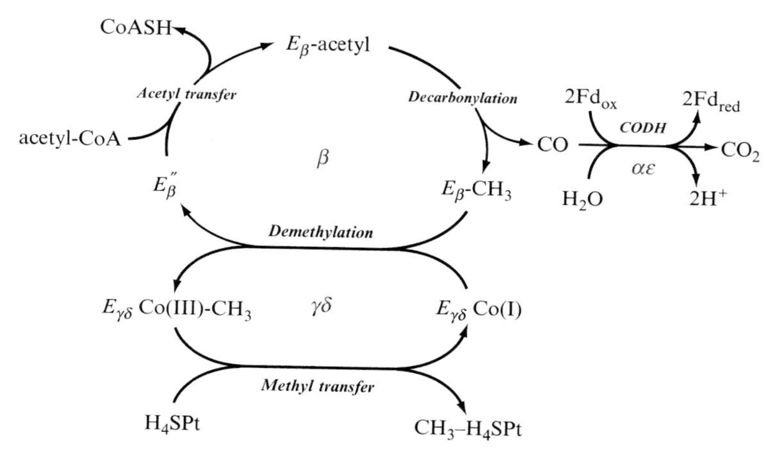

Figure 10.2 Overall process of acetyl-CoA synthesis and cleavage by the ACDS complex and the partial reactions catalyzed by its individual protein subcomponents. Reactions are drawn in the direction of net acetyl-CoA cleavage, such as occurs in methanogens growing on acetate as sole source of carbon and energy. β, ACDS acetyl-CoA synthase subunit; αε, ACDS CO dehydrogenase (CODH) subcomponent; γδ, ACDS corrinoid protein subcomponent; Fd, ferredoxin; H_4SPt, tetrahydrosarcinapterin; CH_3–H_4SPt, N^5-methyl-tetrahydrosarcinapterin.

In the first step, acetyl-CoA reacts with the β subunit with its A cluster in a reduced form to generate an acetyl-enzyme intermediate (Fig. 10.2; acetyl transfer). Two methods for assay of the acetyltransferase activity of the β subunit and other ACS enzymes are described here, and both must be carried out anaerobically and under strongly reducing conditions to obtain maximal activity. One is an isotopic exchange assay that follows the incorporation of radioactivity into acetyl-CoA due to enzyme-catalyzed acetyl transfer from acetyl-CoA to radiolabeled CoA as substrates. The other is a nonradioactive assay that measures the transfer of acetyl groups from acetyl-CoA to the CoA analog 3′-dephospho-CoA. Both assays yield information on the rate of acetylation and deacetylation of the enzyme. The formation of an acetyl-enzyme species, which is central to the acetyl transfer process, was implicated by early kinetic studies (Bhaskar *et al.*, 1998; Lu and Ragsdale, 1991) and more recently this intermediate has been chromatographically isolated and observed directly (Gencic and Grahame, 2008). The physicochemical nature of the acetyl-intermediate remains of substantial interest because its formation is proximate to the central events of acetyl C–C bond formation and scission. An organometallic A cluster Ni_p^{2+}-acetyl derivative is postulated, and extensive kinetic studies on the isolated ACDS β subunit indicate that the acetyl-intermediate is a high energy species, with a free energy of hydrolysis somewhat more negative than that of the acetyl-CoA thioester from which it is derived. Acetyl-enzyme intermediate formation by reaction of the reduced enzyme with acetyl-CoA (Eq. (10.5)) is unfavorable with K_{eq} of around 0.2.

$$E_{\beta}'' + \text{acetyl-CoA} \rightleftarrows E_{\beta}\text{-acetyl} + \text{CoA}, \quad K_{eq} \approx 0.2 \tag{10.5}$$

Nevertheless, when reactions are carried out with CO and CH_3–cob(III) inamide, the enzyme becomes quantitatively converted into the acetylated form, as indicated in reaction (10.6).

$$E_{\beta}'' + \text{CO} + \text{CH}_3\text{-Co(III)} \rightleftarrows E_{\beta}\text{-acetyl} + \text{Co(I)} + \text{H}^+, \quad K_{eq}' \approx 2 \times 10^7 \text{atm}^{-1} \tag{10.6}$$

The isolated acetyl-enzyme form can be quantified by HPLC analysis of the amount of acetyl-CoA produced upon reaction of the acetylated enzyme with excess CoA. By reversal of reaction (10.5), this results in the transfer of nearly all of the acetyl groups from the enzyme to CoA, forming acetyl-CoA in an amount essentially equivalent to that of the acetyl-enzyme. Fragmentation of the acetyl C–C bond in the acetyl-enzyme intermediate results in the eponymous decarbonylation reaction (Fig. 10.2, decarbonylation), that produces CO and a methylated enzyme form, likely to correspond to an Ni_p^{2+}–CH_3 species. The CO is oxidized to CO_2 by the $\alpha_2\varepsilon_2$ CO dehydrogenase component (CODH) with electrons being donated to ferredoxin. An intersubunit tunnel found in crystal structures of the 310 kDa CODH/ACS enzyme reveals how CO is transferred over the ~67 Å span that separates the A and C clusters in that protein (Darnault *et al.*, 2003; Doukov *et al.*, 2002). For ACDS, structural information is limited, however, a shorter distance between clusters of around 34 Å has been predicted (Volbeda *et al.*, 2009). Reaction of the β subunit Ni_p^{2+}–CH_3 species with the γδ corrinoid protein in its Co(I) form brings about demethylation of the A cluster and regenerates the reduced β subunit (Fig. 10.2, *demethylation*). Subsequent methyl group transfer from the corrinoid protein to the cellular pool of H_4SPt (Fig. 10.2, *methyl transfer*) completes the job of the ACDS complex, providing CH_3 groups for reduction to methane and the electrons from CO group oxidation required for that reduction. The partial reactions of acetyl transfer, CO oxidation, and methyl transfer are all substantially more rapid than the overall synthesis or cleavage of acetyl-CoA. Thus, steps involved in methylation/demethylation of the A cluster, and/or the decarbonylation process (C–C bond fragmentation) itself are likely to be rate limiting. Further, recent studies on different recombinant ACS enzyme forms suggest that protein conformational changes, involving intersubunit (ACDS) or interdomain (CODH/ACS) interactions, required both for access of the methylated corrinoid protein and for promoting acetyl C–C bond scission, may have a direct influence on reactions at the A cluster and determine the overall rates of acetyl-CoA synthesis and cleavage (Gencic *et al.*, 2010).

2. Assays of A Cluster Protein (ACS) Catalytic Activities

2.1. Requirement for reductive activation

ACS activity is critically dependent upon the presence of a low potential reductant. In the as-isolated enzyme, the proximal and distal Ni sites both exist in the 2+ oxidation state, and theory predicts that the distal Ni site, as long as it remains in square-planar geometry, is unlikely to undergo change of oxidation state. In general, the redox state and electronic configuration of the A cluster in its reduced and active state, designated here as E″, are not well defined. Controlled potential enzymatic assays display one-electron Nernst behavior (Bhaskar *et al.*, 1998; Gencic and Grahame, 2008; Lu and Ragsdale, 1991). In addition, one electron-reduced forms such as Ni^{1+} (Bender *et al.*, 2010; Funk *et al.*, 2004; Tan *et al.*, 2008) and Ni^{1+}–CO (e.g., Bender *et al.*, 2010; Ragsdale *et al.*, 1985; Seravalli *et al.*, 2002) have been characterized spectroscopically. Evidence also exists that two low-potential electrons are trapped by the A cluster upon formation of the acetyl intermediate, which has been interpreted to indicate that two separate one-electron steps are involved (Gencic and Grahame, 2008). However, it remains to be established whether methylation (and acetylation) of a one-electron reduced Ni_p^{1+} state takes place along with concurrent or subsequent uptake of the second electron, or whether a two-electron reduced Ni_p^{0} (Lindahl, 2004) or Ni_p^{2+}-hydride state must first be attained. Nevertheless, regardless of the formal oxidation state or number of electrons needed for initial stages of the reaction, the properties of Ni and its coordination environment at the proximal site in the A cluster lie at the heart of the decarbonylation mechanism. Nickel, surrounded by electron density which is especially high in the reduced state, serves to initiate nucleophilic attack on the acetyl thioester C–S bond, brings about decarbonylation by insertion into the C–C bond, and catalyzes remarkable rearrangements of organometallic one- and two-carbon fragments within its coordination sphere unique in chemistry and biology.

Over a range of pH values, reductive activation of the A cluster involves the concomitant uptake of approximately one proton, such that the apparent midpoint reduction potential for enzyme activation becomes increasingly more negative as the pH is raised. The result is that half maximal activity, which can be achieved for acetyl-CoA synthesis by the ACDS β subunit at around −460 mV at pH 6.6, requires much stronger reducing potential around -550 mV at pH 8.2. Therefore, to obtain activities around 90% of maximum within this pH region, the effective redox potentials in the reaction mixtures must be in the range of -520 to -610 mV. Strong reductants such as sodium dithionite are not always "innocent" in the sense that they, or their reaction products, such as SO_2 and sulfite, have the potential to engage in unwanted side reactions with other components in the assay, such as corrinoids. Difficulties may be encountered as well because of the limited number of redox mediators suitable for reactions at such low potentials. For example, methylviologen, with $E_0 \approx -458$ mV, must be over 90% reduced to obtain a potential of -520 mV, leaving slightly less than 10% of its oxidized redox buffer capacity able to resist further reduction. Further, overreduction of methylviologen, by addition of even slight excesses of dithionite or other strong reductant, triggers a process in which continued chemical decomposition of the mediator occurs, a reaction that resembles a radical chain process, and results in major loss of redox control such that measured potentials quickly begin to drift to less negative values. Different coordination complexes of Ti^{3+} such as brown Ti^{3+} citrate, blue Ti^{3+} nitrilotriacetate (Ti^{3+} NTA), and purple Ti^{3+} EDTA, are readily prepared from commercial titanous chloride; they exhibit different apparent redox potentials (which are also highly pH sensitive), as well as differences in their kinetics for reduction of mediators, and are useful for reductive activation of ACS samples under conditions that minimize noninnocent effects of the reducing system.

3. Acetyltransferase Reactions

Acetyltransferase activity is one of several important redox-dependent reactions catalyzed by functional A cluster-containing proteins, and measurement of the rate at which acetyl groups are accepted by ACS from acetyl-CoA and then transferred back to CoA, or to a CoA analog, is probably the least complicated of the various A cluster enzyme assays to set up. Acetyltransferase specific activity quantifies the ability of enzyme samples to react with acetyl-CoA to form an A cluster–acetyl intermediate and to reversibly transfer the acetyl group back to CoA — steps that are absolutely essential for overall net cleavage and synthesis of acetyl-CoA. Substrate levels of acetyl-CoA (100 μM in the standard assay, for which the ACDS β subunit exhibits $K_m \approx 45$ μM) provide the source of acetyl groups for the reaction. The initial rate of acetyl transfer from acetyl-CoA to a second substrate, either 3′-dephospho-CoA or radiolabeled CoA (radioactive method), is followed by HPLC analysis of the relative amounts of reactants and products present in aliquots removed over time. In the nonradioactive method, acetyl transfer from acetyl-CoA to

3′-dephospho-CoA converts those substrates to products CoA and *S*-acetyl-3′-dephospho-CoA. The reverse reaction also takes place, and as the reaction is followed to equilibrium the ratio of [*S*-acetyl-3′-dephospho-CoA][CoA]/[acetyl-CoA][3′-dephospho-CoA] approaches a value very close to 1.0, which simply indicates that the thioester bond in acetyl-CoA is preserved in an isoenergetic state in the product. Further, the reaction can be followed similarly starting in the reverse direction (this requires the preparation of *S*-acetyl-3′-dephospho-CoA for use as the acetyl donor substrate instead of acetyl-CoA).[1]

The standard assay is carried out under strictly anaerobic conditions inside a suitable anaerobic chamber, for example, from Coy Laboratories. All materials are kept inside the chamber for at least 24 h to allow for outgassing, and all highly gas-permeable materials, such as plastic and rubber items, are brought inside at least 3–4 days in advance. Oxygen levels in the chamber should be below 1 ppm during assay procedures and are monitored continuously using a Teledyne model 3190 trace oxygen analyzer. Reducing agent to be added to the reaction mixture is either purple Ti^{3+} EDTA or blue Ti^{3+} NTA, and optimal concentrations must be determined empirically. Further adjustments in reductant concentration may be needed, particularly if conditions are varied, such as changes in assay buffer or pH. Ti^{3+} citrate can also be used, but there is a greater tendency toward over-reduction of the enzyme which results in loss of activity. Levels of around 50–200 μ*M* Ti^{3+} NTA or 1–5 m*M* Ti^{3+} EDTA are appropriate for initial trials, and usually a two- to fourfold range of reductant concentration over which activity is constant can be found in which to work (with lower activity both above and below that range). Aquacobalamin is included in the assay and is converted to cob(I)alamin prior to adding the enzyme. Although cob(I)alamin is not required for the acetyltransferase reaction, it supports higher and generally more reproducible activity when present, an effect that is probably due to improved control of redox poise. The standard assay is carried out at 25 °C at pH 6.7 (the pH optimum for acetyltransferase activity by ACDS β subunit). A short 1-min incubation is used to reduce the enzyme in the presence of 3′-dephospho-CoA and all other assay components except for acetyl-CoA. Thereafter, the reaction is initiated by addition of acetyl-CoA, and aliquots are removed at timed intervals, mixed with stop solution, frozen, and subsequently analyzed by HPLC. The step-by-step procedure is as follows.

[1] Synthesis of *S*-acetyl-3′-dephospho-CoA is carried out by reaction of 3′-dephospho-CoA (Sigma-Aldrich) with acetic anhydride in potassium bicarbonate buffer, pH 7.5, at 0 °C according to the method of Stadtman (1957). Purification by reversed phase chromatography and subsequent lyophilization of methanol–water containing fractions provides *S*-acetyl-3′-dephospho-CoA of purity >96%, as determined by HPLC. When the ACDS β subunit is assayed in the reverse direction, using *S*-acetyl-3′-dephospho-CoA as the acetyl donor substrate and CoA as acceptor, high levels of acetyltransferase activity are observed. However, as yet, detailed kinetic analyses have been undertaken only on the reaction initiated in the forward direction.

3.1. Procedure for acetyltransferase assay

For a total assay volume of 360 μl, mix the following components in a 1.5 ml microcentrifuge tube in the order listed (see reagent list for description of specific solutions):

257 μl H_2O
18 μl 1 *M* MOPS buffer, pH 6.7
18 μl 2 *M* KCl
18 μl 1 m*M* aquacobalamin
9 μl 5.4 m*M* Ti^{3+} NTA working solution.

Incubate the mixture for 20 min at 25 °C to ensure adequate reduction of the aquacobalamin, and then add the following with mixing:

18 μl 2 m*M* 3′-dephospho-CoA, 0.4 m*M* TCEP
4 μl ACS protein (0.03 mg/ml ACDS β subunit, ~0.68 μ*M* A cluster).

After 1 min incubation, initiate the reaction by adding acetyl-CoA,

18 μl 2 m*M* acetyl-CoA (to give 100 μ*M* final concentration).

At time points of 0.5, 1, 2, 3, 4, 6, and 8 min thereafter, remove aliquots, 55 μl, from the reaction mixture and mix with 55 μl of stop solution, predispensed in labeled 12 × 75 mm glass test tubes. Mix each aliquot immediately and promptly freeze the resulting 110 μl mixture by immersing the tube in a dewar containing liquid nitrogen. Keep the tubes in a liquid N_2 freezer until ready for analysis by HPLC.

3.2. Reagent list for acetyltransferase assay

Anaerobic H_2O

Inside the anaerobic chamber, sparge 1 l of deionized water (Milli-Q apparatus, Millipore Corp.), contained in a glass bottle, with N_2 gas (<0.5 ppm O_2) at a rate of 60 l/h for 2 h, venting excess gas outside the chamber.

1 M *MOPS buffer, pH 6.7*

Make 1 *M* 3-(*N*-morpholino)propanesulfonic acid (MOPS) buffer at pH 6.7 using KOH. Sparge the solution (100 ml in a glass bottle) for 1 h inside the anaerobic chamber with N_2 gas, flow rate ca. 250 ml/min, and store wrapped in aluminum foil to protect from light.

174 mM *Ti^{3+} NTA stock solution (blue complex)*

Prepare fresh daily inside the anaerobic chamber by adding 72.5 μl of 30% (w/w) titanium (III) chloride in 2 N HCl reagent (Acros Organics) to

1.0 ml of anaerobic 0.275 *M* disodium nitrilotriacetate (NTA), 0.5 *M* Tris–HCl, pH 8.0.

*5.4 m*M *Ti^{3+} NTA working solution*

Mix 10 μl of 174 m*M* Ti^{3+} NTA stock with 310 μl of anaerobic water. Make fresh on the day of use. Titrate a freshly prepared anaerobic solution of potassium ferricyanide to obtain an exact value for the Ti^{3+} concentration.

*64 m*M *Ti^{3+} EDTA stock solution (purple complex)*

The appropriate volume of a commercial $TiCl_3$ solution in HCl is added to give a final Ti^{3+} concentration of 64 m*M* in a mixture containing 129 m*M* EDTA, 0.32 *M* Tris–HCl, pH 8.0. Commercial titanous chloride solutions that contain very high HCl concentrations may exceed the Tris buffer capacity and be unsuitable.

*10 m*M *TCEP stock solution*

Weigh 28.7 mg of *tris*(2-carboxyethyl)phosphine hydrochloride (Calbiochem), transfer into the anaerobic chamber and dissolve in 10 ml of anaerobic water. Assay with Ellman's reagent to check on reagent stability over time; our results indicate that the solution is stable inside the anaerobic chamber for months.

*2 m*M *3′-dephospho-CoA, 0.4 m*M *TCEP*

Prepare 3′-dephosphocoenzyme A (Sigma-Aldrich, $\geq$90% by HPLC, $\leq$1% CoA) as a 10 m*M* stock solution by dissolving in anaerobic water inside the anaerobic chamber. Freeze aliquots and store frozen in a liquid N_2 freezer. On the day of the assay, transport a frozen aliquot of 10 m*M* 3′-dephospho-CoA stock solution in a dewar containing liquid N_2 into the anaerobic chamber, and once inside, thaw and mix 40 μl of 10 m*M* 3′-dephospho-CoA with 160 μl of 0.5 m*M* TCEP (made from the 10 m*M* TCEP stock) to give 200 μl of 2 m*M* 3′-dephospho-CoA, 0.4 m*M* TCEP.

*2 m*M *acetyl-CoA*

Prepare an anaerobic 10 m*M* stock solution of acetyl-coenzyme A trilithium salt (Sigma-Aldrich ~95% by HPLC). Freeze aliquots and store frozen in the liquid N_2 freezer, as described for the stock 10 m*M* 3′-dephospho-CoA. Transport frozen 10 m*M* acetyl-CoA stock solution into the anaerobic chamber, and after thawing, dilute an aliquot with anaerobic water to give 2 m*M* acetyl-CoA for use on the day of assay only. Refreeze CoA stocks in liquid N_2 inside the anaerobic chamber and return to liquid N_2 freezer.

ACS protein

Procedures used in our laboratory for heterologous expression in *Escherichia coli* and purification of different forms of ACS from bacteria and archaea can be found in Gencic *et al.* (2010). For proteins obtained from *E. coli*, nickel reconstitution is required prior to assay.

Stop solution

3.2 ml of stop solution (sufficient for eight assays with seven timed aliquots each) are prepared by addition of 50 μl of 174 m*M* Ti^{3+} NTA stock to 3.15 ml of anaerobic 0.5 *M* sodium citrate, pH 4.0. On the day of the assay, stop solution is dispensed in advance into sets of 12 × 75 mm glass test tubes, 55 μl per tube, prelabeled with the reaction number and aliquot time. For freezing reaction aliquots, use a large enough dewar with sufficient liquid N_2 to accommodate all tubes produced from the number of assays planned.

3.3. Reversed-phase HPLC analyses

The reversed-phase HPLC method described here yields precise rate measurements that are highly reliable for specific detection of reactants and products needed to follow acetyltransferase and other reactions catalyzed by ACS. The advantages of high precision and specificity for the compounds being quantified are paramount and nearly always outweigh the cost of time required for individual handling and chromatography of multiple samples. In addition, HPLC analysis avoids pitfalls of coupled assays, eliminates complications of direct spectrophotometric monitoring of substrates and products at high concentrations, and provides a wide dynamic range. For example, amounts of CoA derivatives as low as around 5–10 pmol can be quantified with reasonable accuracy, which corresponds to around 0.1–0.2% conversion of acetyl-CoA to product CoA in the standard acetyltransferase reaction from which 50 μl are analyzed (i.e., from injection of 100 μl of the stopped sample). The sensitivity is such that even single turnover events, that is, amounts of acetyl-CoA equimolar with enzyme, can be quantified readily.

HPLC is performed at 25 °C on an APEX ODS 5 μm column, 250 × 4.6 mm i.d. (Jones Chromatography), using a binary gradient with acetonitrile as the organic solvent component. The Jones column packing material, which consists of spherical, 100 Å pore size, C-18 derivatized 5 μm silica particles, is relatively tolerant of samples containing protein, and the stopped reaction mixture aliquots are routinely injected directly onto the column without further sample preparation steps. A guard column is not employed, however, a precolumn inlet filter with a 0.5 μm frit is placed ahead of the column. Changes of the frit are usually successful in correcting the pressure readings when run-to-run values begin to rise, and hundreds of injections can be performed without loss of peak shape or resolution. Ultimately, when peak shape deterioration is encountered, the column surface is scraped and fresh packing material is carefully replaced to a depth of several millimeters, which substantially extends the useful lifetime of the column. Solvent flow rate is set at 0.8 ml/min, and solvents are continuously sparged with helium, which is needed to prevent oxidation of reduced compounds in the reaction mixture, including free thiol-containing reactants and products. Consistent and reproducibly high recovery of CoA is obtainable by inclusion in the mobile phase of a low concentration of 2-mercaptoethanesulfonate, a small, highly polar thiol compound with generally little tendency to interact with the stationary phase. The solvent components for gradient formation are as follows:

Solvent A

10 m*M* potassium phosphate, pH 6.5
50 μ*M* 2-mercaptoethanesulfonic acid, sodium salt

Solvent B

60% (v/v) acetonitrile
10 m*M* potassium phosphate, pH 6.5

These are prepared using an appropriately filtered 0.1 *M* phosphate buffer stock solution, filtered deionized water, and HPLC grade acetonitrile. After helium degassing is initiated, 8.2 mg of solid 2-mercaptoethanesulfonic acid, sodium salt (Sigma-Aldrich) are added per liter to complete the preparation of solvent A. The column is equilibrated in solvent A between runs before each injection, and the following linear gradient segments are initiated immediately upon injection of the sample:

Time (min)	Solvent B (%)
0	0.0
5	10.0
20	28.7
25	100.0
30	0.0

Frozen samples from reaction mixtures are thawed under a stream of N_2 gas individually, one at a time, and injected immediately to avoid exposure to air. The sample is withdrawn into an N_2-purged gas-tight syringe, whereafter, 100 μl are loaded into a suitable injection loop and promptly injected onto the column. Autoinjection procedures, although attractive as an option to increase throughput, have not been attempted because of concerns about oxidative damage and sample integrity during the waiting periods between injections.

Absorbance at 260 nm is monitored and integration of the chromatograms is carried out manually or by use of a semiautomated routine. A photodiode array detector allows for easy identification of thioester-containing species based on their relative absorbance at 230 nm (absorbance band due to the thioester bond itself), and, in practice, absorbance data at this wavelength and several others are also stored, as are complete spectra across various peaks. Other UV–visible detectable reactants and products present in other types of reactions (e.g., in acetyl-CoA synthesis assays) are followed as well, such as the corrinoid cofactors methylcobinamide and methylcobalamin which are well-resolved, eluting after the CoA derivatives. Further, when run with lower pH buffer, around pH 5, tetrahydropteridine compounds such as tetrahydrosarcinapterin (H_4SPt), N^5,N^{10}-methylene-H_4SPt ($CH_2{=}H_4SPt$), and N^5-methyl-H_4SPt (CH_3–H_4SPt) are quantifiable in reactions such as overall acetyl-CoA synthesis and cleavage catalyzed by the intact ACDS complex using H_4SPt and CH_3–H_4SPt as substrates/products. However, as the HPLC solvent pH is lowered, coenzyme A derivatives exhibit increasing peak anisotropy (i.e., tailing), and retention times shift to different extents for CoA versus 3′-dephospho-CoA derivatives. As a result, at pH 5.5 and below, certain CoA derivatives are incompletely resolved (acetyl-CoA and 3′-dephospho-CoA begin to merge). In contrast, at a higher pH of 7.0, the separation and peak symmetry are even slightly better than that at pH 6.5; however, because the column manufacturer warns that buffers higher than pH 7.5 are generally not recommended, pH 6.5 is chosen for routine operation.

3.4. Calculation of acetyltransferase initial rate

As shown in Fig. 10.3, all four CoA derivatives are fully resolved by HPLC. This provides for enhanced precision of rate measurements because rather than relying on any single substrate or product to follow the course of the reaction, changes in all substrates and products are monitored together. Determination of rates is made using the fractional composition of substrates and products, which eliminates variations in sampling and injection on the HPLC (although together these usually amount to no more than 1–2% variation in the total peak area). The integrated A 260 nm peak areas of the reactants and products in each chromatogram are individually divided by the sum of all four peak areas to give the fractional composition, y, for each CoA derivative. The fractional areas for each compound are then fit as a function of reaction time, t, to the first-order kinetic equation for appearance of products (CoA and *S*-acetyl-3′-dephospho-CoA) and the

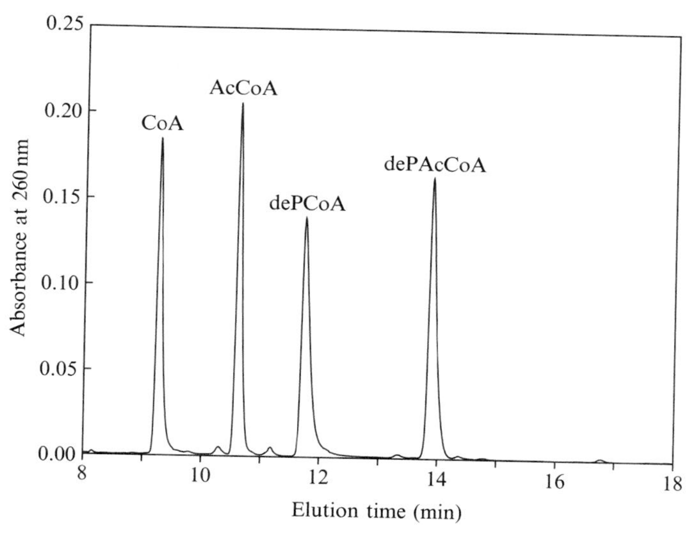

Figure 10.3 HPLC elution profile of coenzyme A derivatives. A mixture of CoA derivative standards was prepared containing 2.5 nmol of each and analyzed using the pH 6.5 HPLC method described in the text. CoA, coenzyme A; AcCoA, acetyl-coenzyme A; dePCoA, 3′-dephosphocoenzyme A; dePAcCoA, *S*-acetyl-3′-dephosphocoenzyme A.

disappearance of reactants (acetyl-CoA and 3′-dephospho-CoA) with the rate constant, k, and initial and final fractional area values (Y_0 and Y_{inf}) as adjustable parameters.For products increasing over time, the equation to fit is:

$$y = Y_{\mathrm{inf}} - (Y_{\mathrm{inf}} - Y_0)e^{-kt}$$

For reactants decreasing over time, the form is:

$$y = Y_{\mathrm{inf}} + (Y_0 - Y_{\mathrm{inf}})e^{-kt}$$

The initial fractional conversion rate for each reactant and product is calculated by multiplying the rate constant by the difference between initial and final values, that is, $k(Y_{\mathrm{inf}} - Y_0)$ and the four initial rate values, all expressed as positive numbers, are then averaged. Theoretically, all four initial rates should be identical, however, in practice, some degree of systematic variation is usually noted. The differences in absorptivity at 260 nm of the different derivatives, and the effects of differences in solvent composition on the absorptivity of the compounds (which changes slightly depending on the actual position of peak elution in the solvent gradient) are both negligible and do not explain the deviations from ideal behavior.

Reactions that contain excessive reductant are often most affected, and a small amount of selective loss or destruction of acetyl-CoA which takes place under these conditions[2] is a more likely reason for imperfect equivalence of the four initial rates.

From the average value of the initial fractional conversion rates, the rate in molar units is obtained by multiplying the fractional conversion rate by the total CoA pool concentration. As an example, for an average initial fractional conversion rate of 0.05 min^{-1} and initial concentrations of substrates acetyl-CoA and 3′-dephospho-CoA both at 100 μ*M*, with no products present under the initial conditions, the initial rate in molar units is 0.05 min^{-1}(200 μ*M*), which equals 10 μ*M*/min acetyl groups transferred (i.e., 10 μ*M*/min acetyl-CoA consumed and 10 μ*M*/min 3′-dephospho-CoA acetylated). Under conditions where fractional conversion versus time curves are nearly linear (e.g., small amounts of enzyme and early time points), rates can be reliably measured by linear least-squares fits. Standard deviations in the estimates of Y_{inf}, Y_0, and k obtained by first-order fitting under these conditions are quite large, and the usual error analyses would indicate even larger errors propagated to the calculation of the initial rate using these parameters. However, the first-order fits still afford initial rate values that are clearly within the error limits of the linear fit, and so, even in such cases, fitting to the first-order curve is useful to obtain initial slope estimates. In contrast, conditions that result in too rapid approach to equilibrium will produce too few data points in the early stages of the reaction for accurate estimation of the initial rate. From a practical standpoint, sufficiently accurate estimations of initial rates using the aliquot timing sequence indicated can be obtained as long as fractional initial rate values do not exceed about 0.2 min^{-1}, although, provided that all of the data are well fit, values as high as 0.4 min^{-1} may still give some useful information.

The radioactive procedure for acetyltransferase analysis is not presented here in detail, but it is worthwhile to mention that historically this method, also termed CoA exchange, was the first one to be applied to measure ACS-catalyzed redox-dependent transfer of acetyl groups from acetyl-CoA to CoA (by Pezacka and Wood, 1986, who used tritium-labeled CoA, and by Lu and Ragsdale, 1991 and Ramer *et al.*, 1989, using CoA-labeled with ^{32}P in the 3′ phosphate group). A similar approach has been applied with ^{14}C incorporated into the β alanine residue of CoA produced by an *in vivo* biosynthetic method (Bhaskar *et al.*, 1998). The general CoA exchange reaction conditions, pH, redox mediators, etc., are similar or equivalent to

[2] An additional HPLC peak is observed under such conditions that coelutes with adenosine 3′,5′-bisphosphate. The implication is that ACS also catalyzes a side reaction involving CoA nucleoside diphosphate hydrolysis (Nudix-like CoA pyrophosphatase activity) that is more pronounced when excessive reducing agent is added.

the nonradioactive assay, and HPLC separations done on aliquots removed over time are also involved. However, because no change takes place in the chemical concentrations of CoA and acetyl-CoA, the simple procedure of integration of their peak areas gives no information on the rate of the exchange. Instead, fractions from the HPLC are collected and counted to determine the extent of radiolabel remaining in the CoA peak and appearing in the peak of acetyl-CoA over time. The advantage of the radioactive method is that both substrates are "native" in the sense that no chemical analogs such as 3′-dephospho-CoA are used. Notwithstanding, the 3′-dephospho derivatives are readily accepted by ACS enzymes from methanogens and acetogens. The precautions involved in handling of ^{32}P and the short time frame over which ^{32}P-labeled CoA can be used after its preparation are drawbacks that can be circumvented by investing effort into biosynthetic labeling and purification to obtain ^{14}C-labeled CoA.

4. Acetyl-CoA Synthesis Reactions

A rapid continuous spectrophotometric assay can be used to follow the net synthesis of acetyl-CoA according to Eq. (10.1), which is based on the pronounced differences in the UV–visible spectra of the methylated Co(III) corrinoid substrate and its Co(I) corrinoid product (for details of the spectrophotometric assay, see Fig. 1 in Gencic and Grahame, 2008). Nevertheless, HPLC analysis of timed aliquots is often preferred because of the high specificity for detection of acetyl-CoA and to avoid spectrophotometric complications in situations where high concentrations of the corrinoid substrates or dye-containing redox buffer systems are employed. In assays of bacterial ACS, methylated corrinoid iron–sulfur protein (CH_3–CFeSP) is generally used as the source of methyl groups for acetyl-CoA synthesis (e.g., Seravalli and Ragsdale, 2008; Tan *et al.*, 2006). The methylated corrinoid protein is obtained by reduction to the Co(I) state followed either by addition of methyliodide or by reaction with N^5-methyltetrahydrofolate catalyzed in the presence of a separately purified methyltransferase. In ACDS, the corrinoid protein is tightly associated as a subcomponent protein in the complex, and conditions to reconstitute the ACDS complex are not yet fully developed. Therefore, methods to assay ACS activity of the isolated β subunit so far rely on the use of a suitable substrate mimic of the native corrinoid protein. Free corrinoid cofactor species with base-on configurations such as methylcobalamin are too unreactive to be useful for routine assays; however, the base-off compound methylcobinamide (CH_3–Cbi) is highly functional. Under saturating conditions, CH_3–Cbi provides acetyl-CoA synthesis activity as high as ~30% of the intact ACDS complex (Gencic and Grahame, 2008) and has been useful as a common methyl donor substrate to allow for side-by-side comparisons of the catalytic activities of different forms of bacterial and archaeal ACS (Gencic *et al.*, 2010).

In addition to the corrinoid methyl donor substrate and CoA, carbon monoxide is required for the ACS reaction, Eq. (10.1). Thus, specific procedures are needed to maintain control of the CO content in the reaction vessel gas phase. Experience with the use of rubber septa outside of the anaerobic chamber indicates that these are too permeable to oxygen for reproducible assays. However, inside the anaerobic chamber reaction vials/tubes sealed with rubber septa provide additional protection from oxygen, and allow for effective control over the composition of the gas phase.

A simple inexpensive system (Fig. 10.4) is used to handle CO gas inside the anaerobic chamber. A portable CO gas cylinder mounted inside the chamber serves as a convenient source of high purity CO gas (e.g., Matheson Research Purity grade CO, 99.998%, <0.5 ppm oxygen, cylinder size 6A, with attached two-stage regulator and outlet valve). The advantage of the portable cylinder is that there is no need to purge the system to remove O_2 contamination that otherwise accumulates on standing in the regulator and connections on the low pressure side of cylinders set up outside the chamber. The strategy of carrying out sealed reactions entirely inside the anaerobic chamber has the advantage of double containment, whereby exclusion of O_2 is nearly absolute; there are fewer subtle routes for O_2 entry, such as trace air leakage in vacuum lines that need to be continuously monitored and scrupulously maintained. The disadvantage is that care must be taken to prevent CO escape into the atmosphere of the anaerobic chamber so that it does not contaminate other reagents or appreciably interfere with operations of the oxygen analyzer or the catalysts used to keep the chamber anaerobic. Therefore, an exhaust system is used, which consists of a glass bubbler containing mineral oil mounted inside the chamber with its outlet connected by a long section of Tygon tubing that carries all waste CO out of the chamber and safely to a laboratory fume hood. Under static, nonflowing conditions, a slight positive pressure of the atmosphere above the reaction mixture is easily verified by observing the oil level in the bubbler, and any significant leakage of CO into the chamber, which might occur due to repeated perforation of septa, or at other connections, is readily apparent. CO-saturated water is produced as indicated in Fig. 10.4.[3] The advantage of withdrawing CO-saturated water from a septum located below the surface of the water (Fig. 10.4, septum stopper c) is that nearly indefinite facile penetration of the septum can be made without concern for gas leakage. Another important component of the gas handling system is the gas mixer. Although not necessary for certain reactions that can be carried out under 100% CO, the ability to prepare lower concentrations of CO in

[3] With the set-up as shown, CO saturation is achieved by initial 5-min bubbling at a flow rate of around 20–30 ml/min, whereafter the rate is decreased to avoid excessive gas consumption.

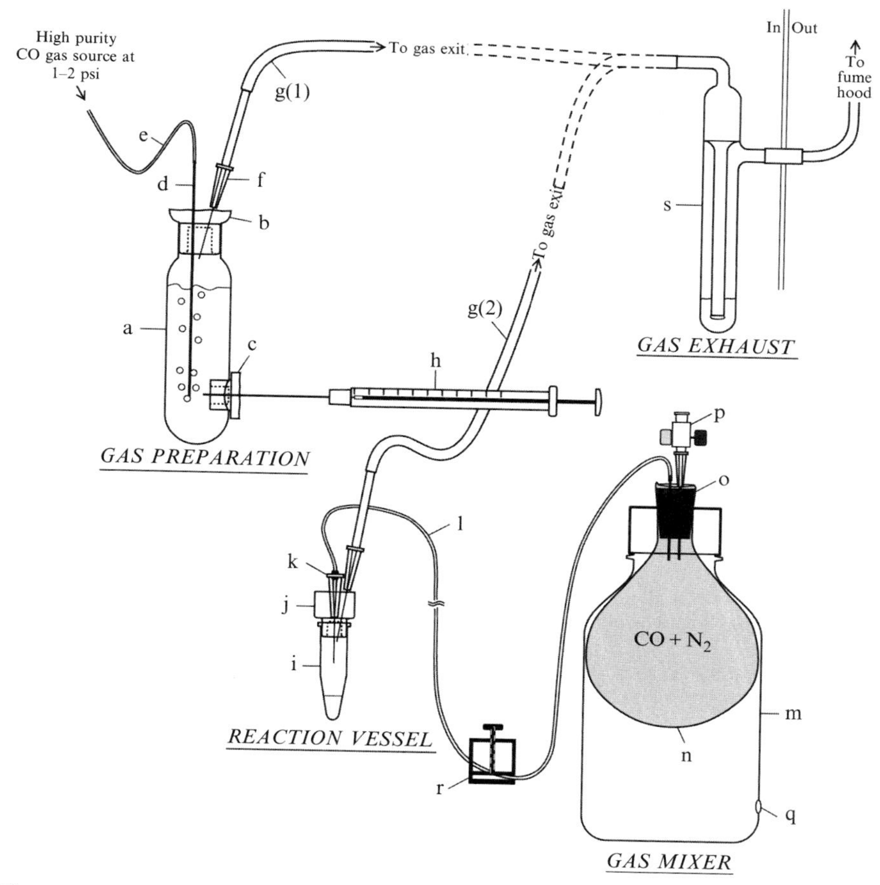

Figure 10.4 Gas handling system setup inside the anaerobic chamber for acetyl-CoA synthesis reactions. The components for gas preparation, gas exhaust, and gas mixing are shown in relation to the reaction vessel for acetyl-CoA synthesis. High purity CO gas is bubbled through water to afford a source of CO-saturated water (a), and excess gas is vented outside the anaerobic chamber through an oil bubbler, gas exhaust component (s), and is carried by a long stretch of tubing to the laboratory fume hood for safe disposal. The approximate gas flow rate through the system is judged by the rate of bubble formation in the bubbler calibrated in ml/min. Under static conditions, vertical downward displacement of the oil level in the inner glass tube of the bubbler is used to check for leaks and to verify that a slight positive pressure is maintained in the system above the reaction mixture at all times. The gas mixer uses a simple helium-quality latex balloon (n) to hold the gas and to provide pressure for gas delivery to the reaction headspace. Containment of the balloon inside the bottle (m) protects the balloon from inadvertent physical damage, provides a means for reliable leak-free attachment of an inlet valve and outlet tubing to the balloon, and affords satisfactory mechanical stability for the system overall. (a) 30 ml polycarbonate Oak Ridge style centrifuge tube (Nalgene 3183-0030) with side port, containing CO-saturated H_2O, clamped to a fixed support. (b) Sleeve stopper septum (Sigma-Aldrich Z564729,

N_2 is essential for studying ACS from bacterial sources because the pronounced inhibitory effects of CO on the reaction in such systems require work at low concentrations of CO.

The standard assay for acetyl-CoA synthesis contains 480 μ*M* dissolved CO (gas phase of 50% CO in N_2),[4] 400 μ*M* methylcobinamide, and 120 μ*M* CoA, at pH 7.2 and 25 °C, and is carried out under dim light to protect methylcobinamide from photolysis. The assay procedure is as follows.

4.1. Procedure for acetyl-CoA synthesis assay

The total assay volume is 100 μl. Mix the following in a 1.5-ml microcentrifuge tube in the order listed (see reagent list for description of specific solutions and Fig. 10.4 for setup):

bottom o.d. 15.7 mm) with top trimmed off. (c) Sleeve stopper septum (Sigma-Aldrich Z564702, bottom o.d. 14 mm) with top trimmed off, inserted into ca. 0.5 in. diameter side hole carefully cut into tube (a). (d) Stainless steel cannula, 18 gauge, 3.75 in., blunt at both ends. (e) Vinyl tubing, Tygon R-3603, 1/32 in. i.d., attached to high purity CO source (see text) by low dead-volume connections to minimize purging time. (f) Syringe needle, 25 gauge, 5/8 in., connected via a cut-off 1.0 cc plastic syringe to tubing (g) which is alternatively positioned, as needed, to provide an outlet for gas from tube (a) or from the reaction vessel (i). (g) Vinyl tubing, Tygon R-3603, carries gas to the glass bubbler (s) used in alternative positions g(1) or g(2). (h) Hamilton gas-tight syringe for transfer of CO-saturated water from tube (a) to the reaction vessel. (i) Reaction vessel, 1.5 ml microcentrifuge tube useful for small reaction volumes. (j) Rubber septum (Sigma-Aldrich Z124389, bottom o.d. 10.5 mm) with a thin, wiped-off layer of Dow-Corning high vacuum grease applied to ease insertion. (k) Syringe needle, 25 gauge, 7/8 in., attached by low dead-volume connections to the gas mixer via tubing (l). (l) 1/32 in. i.d. vinyl tubing, Tygon R-3603, used to carry gas from the gas mixer to the reaction vessel. (m) 500 ml polystyrene bottle with high density polyethylene screw-cap (Costar/Corning 8393). A smooth round hole cut in the cap accommodates a No.1 rubber stopper (o) positioned inside the stretched neck of helium-quality balloon (n). When inserted in this configuration, the force of the stopper against the hole in the bottle cap provides a tight seal on the balloon. (n) 12 in. helium-quality latex balloon. (o) No.1 rubber stopper with inserted 18-gauge needle connected to a push button valve for filling the balloon, and with a separate 18-gauge cannula inserted connected to tubing (l) for delivering gas from the mixer to the reaction vessel headspace. Needle and cannula both must be blunt-ended to prevent damage to the balloon. (p) Push button valve (Supelco Mininert style, Sigma-Aldrich 22285), remains in place at all times attached to the gas mixer. (q) Small hole drilled near the bottom of the bottle to allow for volume displacements. (r) Screw clamp, controls flow from the gas mixer. (s) 40 ml mineral oil bubbler (Ace Glass, Inc. 8761-10) clamped to a fixed support, with outlet connection exiting the anaerobic chamber and extending to a laboratory fume hood for safe disposal of carbon monoxide.

[4] Provides dissolved CO at around 480 μ*M* using the α value of 0.0214 at 25 °C (Dean, 1992).

6 μl H_2O
20 μl 0.25 *M* HEPES buffer, pH 7.2
6 μl 6.6 m*M* methylcobinamide (400 μ*M* final conc.)
2 μl 1.72 m*M* Ti^{3+} NTA
6 μl ACS protein (1.07 mg/ml ACDS β subunit, ~24 μ*M* A cluster).

Stopper the reaction tube (septum stopper j) and inject CO-saturated water (using syringe h)[5]:

50 μl CO-saturated H_2O.[6]

Insert syringe needle (k) from the CO gas mixer, and position the gas exit needle in position g(2) in the reaction vessel septum stopper, as shown in Fig. 10.4. Flush the reaction tube headspace with 50% CO in N_2 (by relaxing clamp r), and allow at least 10 ml of gas to flow gently through the reaction vessel headspace for ~1 min.

Incubate for 5 min at 25 °C under the 50% CO in N_2 atmosphere with periodic gentle agitation of the solution to effect final gas equilibration.

Initiate the reaction using a gas-tight syringe to inject CoA[7]:

10 μl 1.2 m*M* CoA, 0.3 m*M* TCEP.

At time points of 0.5, 1, 2, 3, 4, 6, and 8 min, remove aliquots of 10 μl from the reaction mixture using a gas-tight syringe[8] and mix with 100 μl of one-half concentrated stop solution, predispensed in labeled 12 × 75 mm glass test tubes. Follow procedures for aliquot mixing, freezing, storage, and HPLC analysis, as given under Sections 3.1 and 3.3.

The initial rate of acetyl-CoA formation is calculated from analysis of the 260 nm absorbance peak area of acetyl-CoA in HPLC runs as it appears over time. The amount of acetyl-CoA formed is expressed as a fraction of the total peak areas of CoA plus acetyl-CoA, and fractional conversion versus reaction time is fit to the first-order kinetic equation, as described for acetyltransferase assays (Section 3.4), with the exception that only one substrate/product pair is involved. Initial fractional conversion rate is

[5] The volume of CO-saturated water should be adjusted according to the gas phase composition. For example, here 50 μl of CO-saturated water are added in a 100 μl reaction carried out under 50% CO in N_2, but a reaction with 10% CO in N_2 requires addition of only 10 μl of CO-saturated water. In general, the purpose of adding CO-saturated water is to shorten the time needed for gas–liquid equilibration by quickly bringing dissolved levels of CO close to their required final values and then promptly establishing the correct gas phase composition thereafter.

[6] The acetyl-enzyme species is formed in the presence of CH_3–cob(III)inamide, available reducing agent, and CO.

[7] TCEP is not required for the reaction, but if omitted, reactions with little or no Ti^{3+} NTA reductant may exhibit a significant lag phase. The lag phase is overcome as levels of cob(I)inamide product increase over time, which results in reactivation of the enzyme.

[8] A 25 μl gas-tight syringe (Hamilton model 1702) can be read with sufficient accuracy and is rugged enough for the repeated septum penetrations involved.

converted to molar units as indicated for assays of acetyltransferase, multiplying by the total CoA concentration used in the reaction.

4.2. Reagents and materials for acetyl-CoA synthesis assay

50% CO, 50% N_2 gas mixture

On the day of assay, combine equal volumes of CO and N_2 gas in the gas mixer by use of a transfer syringe to inject the gas through the luer-lock valve of the mixing apparatus. Total volume of gas added should be at least 200 ml (see Section 4.3).

0.25 M *HEPES buffer, pH 7.2*

Prepare 4-(2-hydroxyethyl)piperazine-1-ethanesulfonic acid (HEPES) buffer at pH 7.2 using NaOH. Make anaerobic as described for MOPS buffer under Section 3.2.

*6.7 m*M *methylcobinamide (protect from light)*

Prepare methylcobinamide starting from methylcobalamin (Sigma-Aldrich) by use of the trifluoromethanesulfonic acid cleavage method developed by Brown and coworkers (Zou *et al.*, 1995). Carry out additional purification steps described by Gencic and Grahame (2008), as needed to obtain ca. 97% pure methylcobinamide. Dissolve the final preparation in anaerobic water and store frozen in a liquid N_2 freezer.

*1.72 m*M *Ti^{3+} NTA*

Make fresh daily by adding one part of 174 m*M* Ti^{3+} NTA stock solution to 100 parts water (see Section 3.2 for preparation of 174 m*M* Ti^{3+} NTA stock solution).

4.3. Gas mixture preparation

To prepare the gas mixture, the balloon (n) shown in Fig. 10.4 is first fully deflated by applying suction using a 60 cc syringe (not shown), leaving the push button valve (p) closed after the balloon has completely collapsed. To add CO gas to the balloon, a 22 gauge, 1-in. needle is attached to the 60 cc syringe and CO gas is withdrawn into the syringe from the headspace of tube (a) through septum (b). The flow of CO is increased at the cylinder outlet valve (not shown) to provide sufficient gas during withdrawal into the syringe. To transfer gas from the syringe to the mixer, the needle is withdrawn from septum (b) and then promptly detached from the syringe, whereafter the exposed syringe luer tip is immediately mated with the luer connection of the push button valve. The contents of the syringe are then transferred into the balloon with coordinated operation of the push button valve. Provided that the transfer is done quickly, the amount of gas leakage that takes place at the moment that the luer tip is open to the atmosphere is inconsequential. However, with very small volumes of CO it is important to eliminate such leakage altogether, and this is done by incorporating an additional valve attached to the outlet of a smaller, appropriately sized gas-transfer syringe. Transfer of N_2 gas into the balloon is done by similar manipulations of the transfer syringe and push button valve. The oxygen content of the N_2 gas added to the mixer is kept below 0.1 ppm (the detection limit of the Teledyne model 3190 oxygen analyzer) by loading the transfer syringe with standard high purity N_2 gas that has been further passed through an Oxiclear™ disposable gas purifier (Labclear, Sigma-Aldrich Z186244) located inside the anaerobic chamber. About 80–100 ml of gas must be added to the fully deflated balloon in order to develop sufficient pressure for adequate flow to the reaction vessel; a total volume in the mixing balloon of around 200 ml or more provides an ample supply to carry out numerous reactions. After all reactions are completed, the entire system is purged of residual CO by use of suction from a pump located in the fume hood.

5. The Acetyl-CoA Carbonyl:CO Exchange Reaction

Ragsdale and Wood (1985) established that CODH/ACS is the only protein required to catalyze exchange of the carbonyl group of acetyl-CoA with free CO (Eq. (10.7)). This proved that

$$\left[1\text{-}^{14}\text{C}\right]\text{acetyl-CoA} + {}^{12}\text{CO} \rightleftarrows \left[1\text{-}^{12}\text{C}\right]\text{acetyl-CoA} + {}^{14}\text{CO} \qquad (10.7)$$

CODH/ACS acts on both the acetyl C–C and C–S bonds in acetyl-CoA, and it identified CODH/ACS as the condensing enzyme that catalyzes the final steps in acetyl-CoA synthesis in acetogenic bacteria. More recently, it was found that certain ACS enzyme forms, that are nonetheless highly active in catalyzing acetyl-CoA synthesis, are essentially inactive in CO exchange. Such enzymes include the isolated archaeal ACDS β subunit, and truncated bacterial ACS lacking its N-terminal domain, a region in the bacterial protein that is characteristically absent in the archaeal enzyme. Multiple steps are required for carbonyl exchange to take place. The reduced enzyme must first react with acetyl-CoA to generate the A cluster–acetyl intermediate (Fig. 10.2; Eq. (10.5)). Next fragmentation of the acetyl C–C bond takes place, which is envisaged to produce separate CO and CH_3 species bound in the coordination sphere of Ni. From the fragmented state, CO finally is able to dissociate and exchange with free CO (Eq. (10.8)).

$$\text{E-Ni}_p{}^{2+}\text{-acetyl} \rightleftarrows \text{E-Ni}_p{}^{2+}(\text{CO})(\text{CH}_3) \rightleftarrows \text{E-Ni}_p{}^{2+}\text{-CH}_3 + \text{CO} \quad (10.8)$$

Thus, incubation of reduced bacterial ACS with [1-^{14}C]acetyl-CoA under an atmosphere containing unlabeled CO results in the rebinding of ^{12}CO and reversal of Eqs. (10.8) and (10.5), which causes a washout of the label in acetyl-CoA and escape of the label from solution to the gas phase. A simple method to measure the exchange reaction is to follow the radioactivity remaining in the solution over time as the exchange takes place (Ragsdale and Wood, 1985). This is accomplished by the transfer of timed aliquots to an acidic denaturing solution that stops the reaction, with subsequent counting of the radioactivity present in those samples. Standard assay conditions are modified slightly from the original work, and employ reducing conditions similar to those in acetyltransferase and ACS reactions described here. Standard assay conditions are 25 μ*M* [1-^{14}C]acetyl-CoA, 0.96 m*M* dissolved CO (100% CO atmosphere) at pH 6.5 and 25 °C.[9] The assay is performed as follows.

5.1. Procedure for CO exchange assay

For an assay volume of 400 μl, mix the following in a flat-bottomed 45 × 15 mm o.d. glass vial:

24 μl H_2O
80 μl 0.25 *M* MES buffer, pH 6.5
20 μl 2 *M* KCl
16 μl 3.4 m*M* Ti^{3+} NTA
20 μl ACS protein (1.5 mg/ml recombinant *Carboxydothermus hydrogenoformans* ACS, ~18 μ*M* A cluster).

Stopper the reaction tube with a septum stopper and inject CO-saturated water, 200 μl CO-saturated H_2O.

Flush the reaction headspace with 100% CO for 2 min.

Initiate the reaction by injecting [1-^{14}C]acetyl-CoA,

40 μl 0.25 m*M* [1-^{14}C]acetyl-CoA (6 μCi/μmol).

Remove aliquots, 30 μl at timed intervals, for example, after 1, 2, 4, 6, 10, 15, 20, and 30 min, and add to 300 μl of 0.1 N H_2SO_4 in 5-ml scintillation vials. After a brief wait (when no further volatile counts are lost), aqueous scintillation cocktail is added and the vials are counted.

[9] In the case of carbonyl exchange reaction catalyzed by the ACDS complex, free CO is a poor substrate owing to tight coupling between the A and C clusters. However, CO_2 exchange activity is high and can be assayed by using CO_2, made available by addition of sodium bicarbonate, along with a suitable redox buffer system that supports interconversion of free CO_2 with bound/sequestered CO by the CODH component.

5.2. Calculation of the CO exchange rate

The exchange velocity (v*) is determined by fitting the data to the equation $v* = -[A][P]/([A] + [P])\ln(1 - F)/t$ in which A and P correspond to the concentrations of substrates undergoing exchange of the label (CO and acetyl-CoA), and F is the fraction of isotopic equilibrium attained at time t, given by $F = (A_0* - A*)/(A_0* - A_\infty*)$ where $A*$ represents the label remaining in solution at time t, A_0* is the amount of label initially added as [1-^{14}C]acetyl-CoA, and $A_\infty*$ is the label remaining in A at isotopic equilibrium (Segel, 1975). With a large amount of CO in the headspace relative to the quantity of labeled acetyl-CoA available for exchange, $A_\infty*$ in principle should be essentially zero. However, if $A_\infty*$ is set to zero (e.g., as assumed by Svetlitchnyi *et al.*, 2004), then plots of $\ln(1 - F)$ versus t are non-linear, or may appear linear only at short reaction times. The non-linearity is because a portion of the radioactivity invariably remains non-volatile, as illustrated in Fig. 10.5. This effect was first investigated by Raybuck *et al.* (1988), who concluded that the remaining counts were not formed by thiolytic cleavage of acetyl-CoA to give [1-^{14}C]acetate. Our data show that neither addition of fresh enzyme nor giving supplemental Ti^{3+} NTA after 10 min to reaction mixtures identical to that shown in Fig. 10.5, was able to further reduce the final counts, ruling out loss of activity or loss of reducing potential as the cause of the incomplete washout. HPLC analyses of reaction mixtures show almost none of the residual label remains in acetyl-CoA, and only small amounts are found potentially in acetate, that is, unbound to the column. Thus, our experience is in general agreement with the findings of Raybuck *et al.* (1988). Furthermore, we observe the majority of the remaining label recovered in a single sharp peak, tentatively identified as *S*-acetyl-4′-phosphopantetheine. Thus, as found also in acetyltransferase reaction mixtures, even at the least negative potentials needed for activity, ACS catalyzes what appears to be an obligatory competing side reaction in which acetyl-CoA is destroyed by cleavage of its pyrophosphate linkage. The model in Fig. 10.5 shows how exchange with CO and the formation of the nonvolatile radioactive side-reaction product together result in complete loss of radiolabel in acetyl-CoA.

6. Concluding Remarks

In conclusion, application of the various functional assays described here to bacterial and archaeal ACS proteins in parallel provides the opportunity to gain important new insight into the catalytic properties of both systems. The remarkable direct participation of protein conformational changes in the chemical activation of the acetyl C-C bond at the A cluster

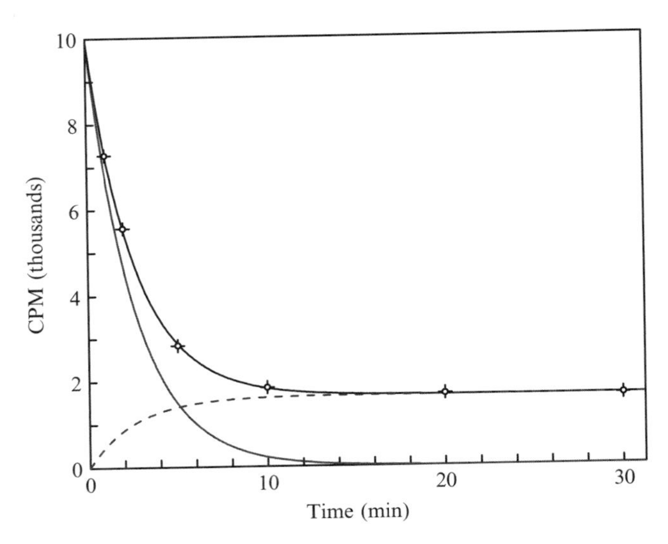

Figure 10.5 CO exchange reaction time course. The reaction was catalyzed by ACS from *Carboxydothermus hydrogenoformans* (ACS_{Ch}) under the standard assay conditions, and exhibited a turnover rate of 8.1 min^{-1}. Solid line (black), first-order decay curve fit to experimental data points; solid line (red), predicted disappearance of labeled acetyl-CoA; dashed line (blue), model for appearance of labeled side product *S*-acetyl-4′-phosphopantetheine. The CO exchange rate was 5.1 times faster than the competing reaction. (For interpretation of the references to color in this figure legend, the reader is referred to the Web version of this chapter.)

was one such finding (Gencic *et al.*, 2010). The implications for CO conservation and containment go beyond the extraordinary features of the tunnel that transfers CO between CODH and ACS active sites (Darnault *et al.*, 2003; Doukov *et al.*, 2002), with results from activity analyses indicating that the fundamental mechanism of catalysis at the A cluster itself ensures that CO is not released from acetyl-CoA in the absence of appropriate inter-subunit or inter-domain interactions. Further structural insight into differences in subunit and domain interactions in these systems along with additional enzymological studies will help to explain the overall function of ACS and to understand how it is optimized in different organisms to meet the requirements for *de novo* acetyl group synthesis and/or decomposition.

ACKNOWLEDGMENTS

This work was supported by grants from the National Science Foundation (MCB-0923766) and by the U.S. Department of Energy (DE-FG02-00ER15108).

REFERENCES

Bender, G., Stich, T. A., Yan, L., Britt, R. D., Cramer, S. P., and Ragsdale, S. W. (2010). Infrared and EPR spectroscopic characterization of a Ni(I) species formed by photolysis of a catalytically competent Ni(I)-CO intermediate in the acetyl-CoA synthase reaction. *Biochemistry* **49,** 7516–7523.

Bhaskar, B., DeMoll, E., and Grahame, D. A. (1998). Redox dependent acetyl transfer partial reaction of the acetyl-CoA decarbonylase/synthase complex: Kinetics and mechanism. *Biochemistry* **37,** 14491–14499.

Darnault, C., Volbeda, A., Kim, E. J., Legrand, P., Vernède, X., Lindahl, P. A., and Fontecilla-Camps, J. C. (2003). Ni-Zn-[Fe_4S_4] and Ni-Ni-[Fe_4S_4] clusters in closed and open α subunits of acetyl-CoA synthase/carbon monoxide dehydrogenase. *Nat. Struct. Biol.* **4,** 271–279.

Dean, J. A. (1992). Lange's Handbook of Chemistry. 14th edn. McGraw-Hill, Inc., New York.

Doukov, T. I., Iverson, T. M., Seravalli, J., Ragsdale, S. W., and Drennan, C. L. (2002). A Ni-Fe-Cu center in a bifunctional carbon monoxide dehydrogenase/acetyl-CoA synthase. *Science* **298,** 567–572.

Ferry, J. G. (2010). How to make a living by exhaling methane. *Annu. Rev. Microbiol.* **64,** 453–473.

Funk, T., Gu, W., Friedrich, S., Wang, H., Gencic, S., Grahame, D. A., and Cramer, S. P. (2004). Chemically distinct Ni sites in the A-cluster in subunit beta of the acetyl-CoA decarbonylase/synthase complex from *Methanosarcina thermophila*: Ni L-edge absorption and X-ray magnetic circular dichroism analyses. *J. Am. Chem. Soc.* **126,** 88–95.

Gencic, S., and Grahame, D. A. (2003). Nickel in subunit beta of the acetyl-CoA decarbonylase/synthase multienzyme complex in methanogens. Catalytic properties and evidence for a binuclear Ni-Ni site. *J. Biol. Chem.* **278,** 6101–6110.

Gencic, S., and Grahame, D. A. (2008). Two separate one-electron steps in the reductive activation of the A cluster in subunit beta of the ACDS complex in *Methanosarcina thermophila. Biochemistry* **47,** 5544–5555.

Gencic, S., Duin, E. C., and Grahame, D. A. (2010). Tight coupling of partial reactions in the acetyl-CoA decarbonylase/synthase (ACDS) multienzyme complex from *Methanosarcina thermophila*. Acetyl C-C bond fragmentation at the A cluster promoted by protein conformational changes. *J. Biol. Chem.* **285,** 15450–15463.

Grahame, D. A. (1991). Catalysis of acetyl-CoA cleavage and tetrahydrosarcinapterin methylation by a carbon monoxide dehydrogenase-corrinoid enzyme complex. *J. Biol. Chem.* **266,** 22227–22233.

Grahame, D. A., and DeMoll, E. (1995). Substrate and accessory protein requirements and thermodynamics of acetyl-CoA synthesis and cleavage in *Methanosarcina barkeri. Biochemistry* **34,** 4617–4624.

Grahame, D. A., and DeMoll, E. (1996). Partial reactions catalyzed by protein components of the acetyl-CoA decarbonylase synthase enzyme complex from *Methanosarcina barkeri. J. Biol. Chem.* **271,** 8352–8358.

Lindahl, P. A. (2004). Acetyl-coenzyme A synthase: The case for a Ni_p^0-based mechanism of catalysis. *J. Biol. Inorg. Chem.* **9,** 516–524.

Lu, W. P., and Ragsdale, S. W. (1991). Reductive activation of the coenzyme A/acetyl-CoA isotopic exchange reaction catalyzed by carbon monoxide dehydrogenase from *Clostridium thermoaceticum* and its inhibition by nitrous oxide and carbon monoxide. *J. Biol. Chem.* **266,** 3554–3564.

Norris, P. R., and Pratt, J. M. (1996). Methyl transfer reactions of protein-free Co corrinoids. *Biofactors* **5,** 240.

Pezacka, E., and Wood, H. G. (1986). The autotrophic pathway of acetogenic bacteria. Role of CO dehydrogenase disulfide reductase. *J. Biol. Chem.* **261,** 1609–1615.

Pratt, J. M. (1999). The roles of Co, corrin, and protein. I. Co-ligand bonding and the trans effect. *In* "Chemistry and Biochemistry of B_{12}," (R. Bannerjee, ed.), pp. 73–112. John Wiley and Sons, New York.

Ragsdale, S. W., and Pierce, E. (2008). Acetogenesis and the Wood-Ljungdahl pathway of CO_2 fixation. *Biochim. Biophys. Acta* **1784,** 1873–1898.

Ragsdale, S. W., and Wood, H. G. (1985). Acetate biosynthesis by acetogenic bacteria. Evidence that carbon monoxide dehydrogenase is the condensing enzyme that catalyzes the final steps of the synthesis. *J. Biol. Chem.* **260,** 3970–3977.

Ragsdale, S. W., Wood, H. G., and Antholine, W. E. (1985). Evidence that an iron-nickel-carbon complex is formed by reaction of CO with the CO dehydrogenase from *Clostridium thermoaceticum*. *Proc. Natl. Acad. Sci. USA* **82,** 6811–6814.

Ramer, S. E., Raybuck, S. A., Orme-Johnson, W. H., and Walsh, C. T. (1989). Kinetic characterization of the [3'-^{32}P]coenzyme A/acetyl coenzyme A exchange catalyzed by a three-subunit form of the carbon monoxide dehydrogenase/acetyl-CoA synthase from *Clostridium thermoaceticum*. *Biochemistry* **28,** 4675–4680.

Raybuck, S. A., Bastian, N. R., Orme-Johnson, W. H., and Walsh, C. T. (1988). Kinetic characterization of the carbon monoxide-acetyl-CoA (carbonyl group) exchange activity of the acetyl-CoA synthesizing CO dehydrogenase from *Clostridium thermoaceticum*. *Biochemistry* **27,** 7698–7702.

Segel, I. H. (1975). Enzyme Kinetics. John Wiley and Sons, New York.

Seravalli, J., and Ragsdale, S. W. (2008). Pulse-chase studies of the synthesis of acetyl-CoA by carbon monoxide dehydrogenase/acetyl-CoA synthase. Evidence for a random mechanism of methyl and carbonyl addition. *J. Biol. Chem.* **283,** 8384–8394.

Seravalli, J., Brown, K. L., and Ragsdale, S. W. (2001). Acetyl coenzyme A synthesis from unnatural methylated corrinoids: Requirement for "base-off" coordination at cobalt. *J. Am. Chem. Soc.* **123,** 1786–1787.

Seravalli, J., Kumar, M., and Ragsdale, S. W. (2002). Rapid kinetic studies of acetyl-CoA synthesis: Evidence supporting the catalytic intermediacy of a paramagnetic NiFeC species in the autotrophic Wood-Ljungdahl pathway. *Biochemistry* **41,** 1807–1819.

Stadtman, E. R. (1957). Preparation and assay of acyl coenzyme A and other thiol esters; use of hydroxylamine. *Methods Enzymol.* **3,** 931–941.

Stich, T. A., Seravalli, J., Venkateshrao, S., Spiro, T. G., Ragsdale, S. W., and Brunold, T. C. (2006). Spectroscopic studies of the corrinoid/iron-sulfur protein from *Moorella thermoacetica*. *J. Am. Chem. Soc.* **128,** 5010–5020.

Svetlitchnyi, V., Dobbek, H., Meyer-Klaucke, W., Meins, T., Thiele, B., Römer, P., Huber, R., and Meyer, O. (2004). A functional Ni-Ni-[4Fe-4S] cluster in the monomeric acetyl-CoA synthase from *Carboxydothermus hydrogenoformans*. *Proc. Natl. Acad. Sci. USA* **101,** 446–451.

Tan, X., Surovtsev, I. V., and Lindahl, P. A. (2006). Kinetics of CO insertion and acetyl group transfer steps, and a model of the acetyl-CoA synthase catalytic mechanism. *J. Am. Chem. Soc.* **128,** 12331–12338.

Tan, X., Martinho, M., Stubna, A., Lindahl, P. A., and Münck, E. (2008). Mössbauer evidence for an exchange-coupled [Fe4S4]1+ Nip1+ A-cluster in isolated alpha subunits of acetyl-coenzyme A synthase/carbon monoxide dehydrogenase. *J. Am. Chem. Soc.* **130,** 6712–6713.

Terlesky, K. C., Nelson, M. J., and Ferry, J. G. (1986). Isolation of an enzyme complex with carbon monoxide dehydrogenase activity containing corrinoid and nickel from acetate-grown *Methanosarcina thermophila*. *J. Bacteriol.* **168,** 1053–1058.

Thauer, R. K. (1988). Citric-acid cycle, 50 years on. Modifications and an alternative pathway in anaerobic bacteria. *Eur. J. Biochem.* **176,** 497–508.

Volbeda, A., Darnault, C., Tan, X., Lindahl, P. A., and Fontecilla-Camps, J. C. (2009). Novel domain arrangement in the crystal structure of a truncated acetyl-CoA synthase from *Moorella thermoacetica*. *Biochemistry* **48,** 7916–7926.

Xia, J., Sinclair, J. F., Baldwin, T. O., and Lindahl, P. A. (1996). Carbon monoxide dehydrogenase from *Clostridium thermoaceticum*: Quaternary structure, stoichiometry of its SDS-induced dissociation, and characterization of the faster-migrating form. *Biochemistry* **35,** 1965–1971.

Zou, X., Evans, D. R., and Brown, K. L. (1995). Efficient and convenient method for axial nucleotide removal from vitamin B_{12} and its derivatives. *Inorg. Chem.* **34,** 1634–1635.

CHAPTER ELEVEN

Acetate Kinase and Phosphotransacetylase

James G. Ferry

Contents

Abstract

Most of the methane produced in nature derives from the methyl group of acetate, the major end product of anaerobes decomposing complex plant material. The acetate is derived from the metabolic intermediate acetyl-CoA via the combined activities of phosphotransacetylase and acetate kinase. In *Methanosarcina* species, the enzymes function in the reverse direction to activate acetate to acetyl-CoA prior to cleavage into a methyl and carbonyl group of which the latter is oxidized providing electrons for reduction of the former to methane. Thus, phosphotransacetylase and acetate kinase have a central role in the conversion of complex organic matter to methane by anaerobic microbial food chains. Both enzymes have been purified from *Methanosarcina thermophila* and

Department of Biochemistry and Molecular Biology, Pennsylvania State University, University Park, Pennsylvania, USA

Methods in Enzymology, Volume 494
ISSN 0076-6879, DOI: 10.1016/B978-0-12-385112-3.00011-1

characterized. Both enzymes from *M. thermophila* have also been produced in *Escherichia coli* permitting crystal structures and amino acid variants, the kinetic and biochemical studies of which have lead to proposals for catalytic mechanisms. The high identity of both enzymes to paralogs in the domain *Bacteria* suggests ancient origins and common mechanisms.

1. Introduction

Phosphotransacetylase and acetate kinase, catalyzing reactions 1 and 2, are

$$CH_3COSCoA + HPO_4^{2-} \rightleftarrows CoASH + CH_3CO_2PO_3^{2-} \quad (1)$$

$$CH_3CO_2PO_3^{2-} + ADP \rightleftharpoons CH_3COO^- + ATP \quad (2)$$

widespread in anaerobic species of the domain *Bacteria* and a few species from the domain *Archaea*. Acetate kinase has also been identified in the domain *Eukarya* (Smith *et al.*, 2006). The enzymes play a universal role in the anaerobic decomposition of complex organic matter to methane by microbial food chains, an essential component of the global carbon cycle and a viable process for conversion of renewable plant biomass to methane as a biofuel. The conversion of cellulose to methane in freshwater environments is illustrated in Fig. 11.1. Fermentative anaerobes are the first link in the food chain degrading cellulose to glucose extracellularly that is fermented primarily to acetate plus lesser amounts of propionate, butyrate, H_2, and formate. The propionate and butyrate are further converted to acetate and H_2 by obligate proton-reducing acetogens. In some environments, H_2 and formate are electron donors for the reduction of carbon dioxide to acetate by homoacetogens. Acetyl-CoA, which is the central intermediate in the degradative pathways of both fermentatives and acetogens, is further metabolized to acetate and ATP by acetate kinase and phosphotransacetylase providing the major source of energy for biosynthesis. The methyl group of acetate is converted to methane and the carboxyl group to carbon dioxide by methanogenic species from the domain *Archaea*. The first step in the pathway of *Methanosarcina* species is activation of acetate to acetyl-CoA by reversal of reactions 1 and 2 catalyzed by acetate kinase and phosphotransacetylase. Thus, acetate kinase and phosphotransacetylase play a universal role in anaerobic microbial food chains illustrated in Fig. 11.1. In *Methanosaeta*, the only other acetotrophic genera, thiokinase, replaces acetate kinase and phosphotransacetylase (Kohler and Zehnder, 1984). In both genera, the C–C and C–S bonds of acetyl-CoA are cleaved by CO dehydrogenase/acetyl-CoA synthase yielding a methyl group that is ultimately reduced to methane with electrons derived from oxidation of the

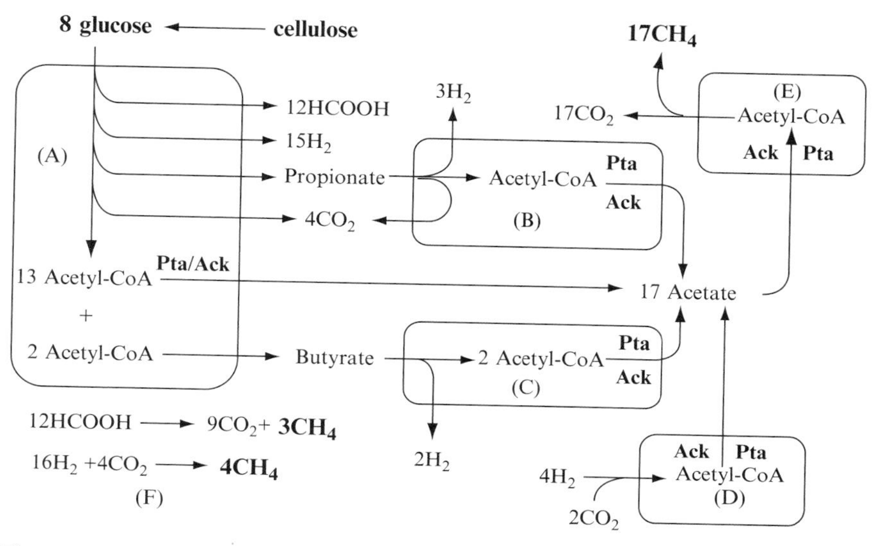

Figure 11.1 A model freshwater anaerobic microbial food chain converting cellulose to methane representing the major trophic groups and illustrating the central role of phosphotransacetylase and acetate kinase. (A) fermentatives, (B) propionate-utilizing obligate proton-reducing acetogens, (C) butyrate-utilizing obligate proton-reducing acetogens, (D) homoacetogens, (E) acetate-utilizing methanogens, and (F) obligate carbon dioxide-reducing methanogens.

carbonyl group to carbon dioxide. Most of the methane in freshwater environments is derived from acetate. In environments where homoacetogens are absent, H_2 and formate are electron donors for obligate carbon dioxide-reducing methane-producing species. Smaller amounts of methane are derived from the methyl groups of methanol, methylamines, and methylsulfide by methylotrophic species (not shown).

2. Acetate Kinase of *Methanosarcina thermophila*

2.1. Assay of activity

Two methods are used to assay acetate kinase in the forward direction. The enzyme-linked assay couples ADP formation to oxidation of NADH ($\varepsilon_{340} = 6.22\ \mathrm{m}M^{-1}\ \mathrm{cm}^{-1}$) through pyruvate kinase and lactic dehydrogenase (Allen *et al.*, 1964). The 5-ml assay mixture contains (final concentrations) 100 m*M* Tricine–KOH buffer (pH 8.2) (or 200 m*M* Tris–HCl buffer (pH 7.3) where noted), 200 m*M* potassium acetate, 1.5 m*M* ATP, 2 m*M*

$MgCl_2$, 2 m*M* phosphoenolpyruvate, 0.4 m*M* NADH, 2 m*M* dithiothreitol, 9 units of pyruvate kinase, and 26 units of lactic dehydrogenase. Assays are initiated by addition of acetate kinase. The hydroxamate assay is an adaptation of the method of (Lipmann and Tuttle (1945) and Rose *et al.* (1954) utilizing the reaction of acetyl phosphate with hydroxylamine to form acetyl hydroxamate which forms a colored complex with trivalent iron. The following components are combined in a volume of 333 μl (final concentrations): 145 m*M* Tris–HCl (pH 7.4), 200 m*M* potassium acetate, 10 m*M* $MgCl_2$, 10 m*M* ATP, and 705 m*M* hydroxylamine hydrochloride (neutralized with KOH before addition). The reaction is initiated with acetate kinase and stopped after 12 min by the addition of 333 μl of 10% trichloroacetic acid followed by the addition of 333 μl of 2.5% FeCl in 2.0 N HCl. The mixture is incubated for 5–30 min to allow color development, and the absorbance at 540 nm is measured. Acetate kinase is assayed in the reverse direction by linking ATP formation to the reduction of NADP through hexokinase and glucose-6-phosphate dehydrogenase similar to a previously described method (Bowman *et al.*, 1976). The reaction mixture (0.5 ml) contains (final concentrations) 100 m*M* Tris–HCl (pH 7.4), 5 m*M* ADP, 10 m*M* $MgCl_2$, 5.5 m*M* glucose, 1 m*M* NADP, 2 m*M* dithiothreitol, 6 units of hexokinase, and 3 units of glucose-6-phosphate dehydrogenase. Acetate kinase is added, and the reaction initiated by the addition of acetyl phosphate (final concentration 20 m*M*). Reduction of NADP is followed at 340 nm ($\varepsilon_{340} = 6.22\ mM^{-1}\ cm^{-1}$).

2.2. Purification from *M. thermophila*

Cells are grown as previously described with acetate as the growth substrate (Sowers *et al.*, 1984). Cell extract is prepared from frozen cell paste suspended 1:1 (w/v) in lysis buffer (50 m*M* potassium phosphate) containing 0.1 m*M* phenylmethylsulfonyl fluoride (PMSF) and 1 μg of DNase I ml^{-1}). The cells are lysed by passage of the suspension through a French pressure cell at 110 MPa. The lysate is centrifuged at 47,800×*g* for 20 min and the pellet washed with one-half of the original suspension volume of lysis buffer. The supernatant solutions are used as the starting material for purification. All steps for purification are performed aerobically at 4 °C. Extract (1.5 ml containing 60–70 mg of protein) is injected onto a Mono-Q HR 10/10 anion-exchange column (Pharmacia LKB Biotechnology Inc.) equilibrated with 50 m*M* TES buffer (pH 6.8) containing 10% (v/v) ethylene glycol, 10 m*M* $MgCl_2$, and 2 m*M* dithiothreitol (buffer A). A linear gradient from 0.0 to 1.0 *M* KCl is applied at 0.5 ml min^{-1} using an FPLC system (Pharmacia LKB Biotechnology Inc.) equipped with a model GP-250 gradient programmer. The active fractions from several Mono-Q separations of cell extract are pooled and loaded onto a 5.0 × 0.9-cm ATP affinity column previously equilibrated with 10 m*M*

Tris–acetate buffer (pH 7.4) containing 1.2 *M* KCI, 1 m*M* $MgCl_2$, and 2 m*M* dithiothreitol (buffer B) at a flow rate of 0.5 ml min^{-1}. Contaminating proteins and a small amount of acetate kinase are eluted with 15 ml of buffer B, followed by re-equilibration with 10 ml of buffer C (buffer B without KCl). The acetate kinase is then eluted at 0.1 ml min^{-1} by the application of 1 ml of buffer C containing 15 m*M* ATP followed by 10 ml of buffer C without ATP. The enzyme is stored in liquid nitrogen.

2.3. Purification of *M. thermophila* Acetate Kinase Produced in *Escherichia coli*

E. coli strain BL21(DE3) is transformed with the overexpression plasmid pML703 containing the *ack* gene (Latimer and Ferry, 1993). Transformants are grown at 37 °C in Luria-Bertani (LB) broth containing 100 μg ml^{-1} ampicillin and induced with 1% (final concentration) Bacto-lactose or 0.4 m*M* (final concentration) IPTG. All purification steps are performed aerobically at 4 °C. Approximately 40 g (wet weight) of cells are suspended in 130 ml (total volume) of 25 m*M* Bis–Tris buffer (pH 6.5) containing 2 m*M* DTT and lysed by French pressure cell disruption at 110 MPa. The lysate is centrifuged at 78,400×*g* for 25 min at 4 °C. Streptomycin sulfate is added to the supernatant (1% final concentration, w/v) and the mixture is centrifuged as above. The supernatant is applied to a 5 × 10-cm Q-Sepharose Fast Flow column previously equilibrated with two column volumes of 25 m*M* Bis–Tris buffer (pH 6.5) plus 2 m*M* DTT. The enzyme eluted from the column between 220 and 270 m*M* KCl using a 1800-ml linear gradient of 0–1 *M* KCl at 6 ml min^{-1}. Fractions containing the highest total activities are pooled and an equal volume of 1.8 *M* ammonium sulfate in 50 m*M* Tris (pH 7.2) plus 2 m*M* DTT is added. A protein sample (100 ml) is loaded onto a Phenyl-Sepharose HiLoad 26/10 column equilibrated with two column volumes of 900 m*M* ammonium sulfate in 50 m*M* Tris buffer (pH 7.2) containing 2 m*M* DTT. The acetate kinase elutes from the column at ~720 m*M* ammonium sulfate using a 600-ml decreasing linear gradient of 900–0 m*M* ammonium sulfate at 3 ml min^{-1}. The fractions containing the highest total activity are pooled, diluted 20-fold with 25 m*M* Tris (pH 7.6) plus 2 m*M* DTT, and loaded on the Mono-Q column equilibrated with five column volumes of 25 m*M* Tris buffer (pH 7.6) containing 2 m*M* DTT. The purified acetate kinase elutes at 230 m*M* KCl developed with a 200-ml 0–1 *M* KCl gradient at 2 ml min^{-1}. The enzyme is stored at −80 °C with minimal loss of activity over a period of 1 year.

Heterologous expression of the his-tagged *M. thermophila* acetate kinase is accomplished with the T7-based expression vector pET*ack* with a 60-nucleotide leader sequence containing six tandem histidine codons fused in-frame to the 5′ end of the wild-type *ack* gene (Singh-Wissmann *et al.*, 1998). LB medium containing 100 μg ml^{-1} of ampicillin is inoculated with *E. coli* BL21 (DE3) transformed with pET*ack* and cultured at 37 °C. IPTG is added at an O.D. of 0.7 to a final concentration of 1 m*M* and after 1.5–2.0 h cells are harvested by centrifugation at 13,000×*g* for 10 min at 4 °C. The cell pellet can be stored at −80 °C for several months without loss of enzyme activity. Approximately 20 g of thawed cells from 4 l of culture are resuspended in 50 m*M* HEPES buffer (pH 7.4) containing 500 m*M* NaCl, 20 m*M* imidazole, 2 m*M* $MgCl_2$, and 2 m*M* β-mercaptoethanol. DNase and 0.25 m*M* PMSF are added and the cells lysed by twice passage through a French pressure cell at 110 MPa. Cell debris and membranes are removed by centrifugation at 100,000×*g* for 45 min at 4 °C. The supernatant is loaded onto a Talon superflow metal affinity column (2.6 × 12 cm) (Clontech) equilibrated with resuspension buffer containing 40 m*M* imidazole. The enzyme is eluted with three column volumes of resuspension buffer containing 500 m*M* imidazole. The fractions containing activity are concentrated using a Centricon YM 10 (10 kDa cutsoff) membrane (Millipore) to a final volume of 15–30 ml. The concentrated solution is dialyzed overnight against 2 l of 50 m*M* HEPES (pH 7.4) containing 150 m*M* KCl, 2 m*M* $MgCl_2$, and 2 m*M* DTT. This protocol typically yields ~40,000 units of acetate kinase with a specific activity of ~600 units mg^{-1}. For long-term storage, Ficoll is added to the solution and the enzyme is lyophilized. The lyophilized enzyme can be stored for 1 year at 4 °C with loss of less than 10% activity.

2.4. Catalytic mechanism of acetate kinase from *M. thermophila*

Acetate kinase, discovered in 1944 by Lipmann (1944), is one of the earliest phosphoryl transfer enzymes recognized. Following the first purification in 1954 from *E. coli* (Rose *et al.*, 1954), the enzyme was the subject of investigations leading to two proposals for the catalytic mechanism: a direct in-line transfer of the γ-phosphoryl group of ATP to acetate (Blattler and Knowles, 1979; Skarstedt and Silverstein, 1976) or a triple-displacement mechanism involving two covalent phosphoenzyme intermediates (Spector, 1980). It was not until 1993 when acetate kinase from *M. thermophila* was overproduced in *E. coli* (Latimer and Ferry, 1993) leading to the first crystal structure (Buss *et al.*, 2001; Gorrell *et al.*, 2005). The crystal structure also revealed characteristics suggesting that acetate kinase is the ancestral enzyme of the ASKHA (acetate and sugar kinases/Hsc70/actin) superfamily of phosphotransferases (Buss *et al.*, 2001). The application of site-directed mutagenesis and biochemical characterization of variants revealed a direct in-line mechanism (Fig. 11.2) and the function of essential active-site residues (Gorrell *et al.*, 2005; Ingram-Smith *et al.*, 2000, 2005; Miles *et al.*, 2001; Singh-Wissmann *et al.*, 1998, 2000). Most compelling, the enzyme co-crystallized with ADP, $Al3^+$, F^-, and acetate shows an

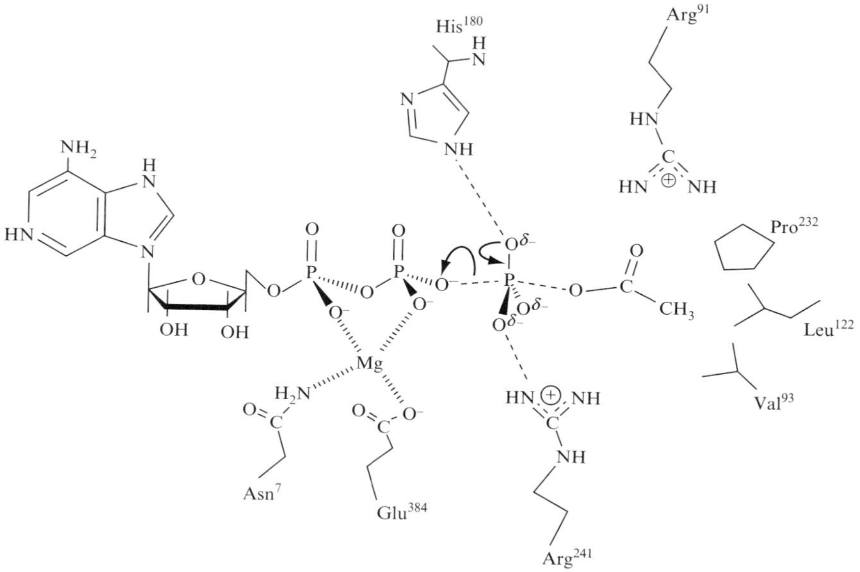

Figure 11.2 Postulated mechanism of *Methanosarcina thermophila* acetate kinase. Interactions are shown for the transition state in the direction of acetyl phosphate synthesis. Acetate is shown with the methyl group bound in the hydrophobic pocket formed by Val^{193}, Pro^{232}, and Leu^{122}. The mechanism is proposed to proceed via nucleophilic attack of the carboxyl group of acetate on the γ-phosphate of ATP. The transition state is stabilized through coordination of the equatorial oxygen atoms through interactions with His^{180} and Arg^{241}. Reprinted with permission Gorrell *et al.* (2005).

ADP-AlF_3-acetate linear array in the active-site cleft wherein AlF_3 is proposed to mimic the *m*-phosphate transition state (Gorrell *et al.*, 2005). Evidence is also reported for domain motion and that catalysis does not occur as two independent active sites of the homodimer but that the active-site activities are coordinated in a half-the-sites manner (Gorrell and Ferry, 2007). The *M. thermophila* acetate kinase has high identity to acetate kinases from the domain *Bacteria* suggesting an ancient origin and common mechanism (Ingram-Smith *et al.*, 2005).

2.5. Measurement of acetate in biological fluids

The measurement of acetate in biological fluids has both clinical and industrial applications. While traditional methods based on gas chromatography have a low detection limit, they require expensive equipment and specially trained personnel. On the other hand, acetate kinase-based assays utilize relatively inexpensive equipment and do not require advanced training. The E97D variant of acetate kinase from *M. thermophila* has been evaluated for detection of acetate in various biological fluids using the hydroxamate assay (Iyer and Ferry, 2005). Heterologous expression and facile purification of the his-tagged variant provide a low cost reagent. Furthermore, the variant acetate kinase has nearly an eightfold lower K_m for acetate (<3 m*M*) than the wild type (Singh-Wissmann *et al.*, 1998) which is lower than those reported for commercially available acetate kinase from *Bacillus stearothermophilus* (120 m*M*) (Nakajima *et al.*, 1978) or *E. coli* (7–300 m*M*) (Fox and Roseman, 1986; Nakajima *et al.*, 1978) increasing the sensitivity. The his-tagged E97D variant is overproduced from plasmid pET*ack* E97D (Singh-Wissmann *et al.*, 1998) and purified as described above for his-tagged wild type. The variant enzyme is equal to the wild type in stability. The variant enzyme has been evaluated by addition of acetate to various biological samples (Iyer and Ferry, 2005). Recovery is nearly quantitative between 0.1 and 1.0 m*M* acetate, and the method is sensitive to levels as low as 0.1 m*M* and linear up to 1 m*M* acetate without dilution.

3. Phosphotransacetylase of *M. thermophila*

3.1. Assay of activity

Activity is assayed by monitoring thioester bond formation at 233 nm ($\varepsilon_{233} = 5.55$ mM^{-1} cm^{-1}) similar to that described previously (Oberlies *et al.*, 1980). The assay mixture contains 50 m*M* Tris–HCl (pH 7.2), 20 m*M* KCl, 1 m*M* potassium lithium acetyl phosphate, 0.2 m*M* lithium CoASH, and 2 m*M* dithioerythritol. Specific activities are reported as micromoles of acetyl-CoA formed per minute per milligram of protein. The reaction is initiated by the addition of enzyme.

3.2. Purification from *M. thermophila*

Cell extract from acetate-grown cells is prepared as described above for purification of acetate kinase. All steps are performed aerobically at 4 °C. Buffer A is 50 m*M* potassium/TES, pH 7.2, unless otherwise indicated. The ammonium sulfate fractionation is performed at 4 °C and all other steps at 20–25 °C. Solid ammonium sulfate is added to cell extract to 45% saturation and the suspension centrifuged for 20 min at 47,800×*g*. The supernatant fluid is discarded and the pellet is dissolved in a minimal volume of buffer A (~5–10 ml), assayed, then diluted fivefold, and applied to a 2.5 × 9-cm DEAE-cellulose (Whatman DE52, Whatman LabSales, Hillsboro, OR) anion-exchange column equilibrated with buffer A. Protein is eluted with a 486-ml linear gradient of 0–1 *M* KCl in buffer A at 2 ml min^{-1} to resolve proteins for other uses. The bulk of phosphotransacetylase eluted in the pass-through fraction (50–60 ml) which also contains membranes. A small amount of activity (2%) elutes in ~250 m*M* KCI; however, the enzyme is of lower purity than that of the pass-through fractions and is discarded. Triton X-100 is added to the pass-through DE52 fraction to a final concentration of 2% (v/v). Ammonium sulfate is added to 70% saturation and the precipitate containing phosphotransacetylase activity is pelleted by centrifugation at 47,800 × g for 15 min. The pellet is dissolved in 3–6 ml of buffer A and assayed for activity. Activity is labile to dialysis; therefore, the enzyme solution is diluted fivefold in buffer A, directly applied to a Mono-Q 10/10 anion-exchange column (Pharmacia LKB Biotechnology Inc., Piscataway, NJ), and developed with a 360-ml linear gradient of 0–1 *M* KC1 in buffer A at 2 ml min^{-1}. Nearly all the protein is recovered in the pass-through fraction. The pass-through fraction containing enzyme activity is reapplied to the Mono-Q 10/10 and rechromatographed. Again, the majority of the activity of high purity eluted in the pass-through fractions with a significant increase in specific activity; a small amount of activity (~1%) of high purity elutes in 125 m*M* KCl. The phosphotransacetylase-containing fractions are pooled for affinity column chromatography. The Mono-Q fractions containing highest specific activity are combined. The solution is vacuum degassed with N_2 and 6 ml aliquots applied to an agarose-hexane-CoASH Type A (US Biochemical, Cleveland, OH) affinity column (1.2 × 2.5 cm) previously activated by equilibration with two bed volumes of degassed buffer B (50 m*M* potassium/TES, pH 8.0) containing 20 m*M* dithioerythritol followed by two bed volumes of degassed buffer B. After the protein is loaded, the column is washed with two bed volumes of degassed buffer B. Phosphotransacetylase is eluted with buffer B containing 5 m*M* CoASH and 2 m*M* dithioerythritol. The affinity chromatography fractions containing phosphotransacetylase are pooled and concentrated by application on a Mono-Q 10/10 column equilibrated with buffer A and batch eluted with 0.5 *M* KC1 in buffer A. Glycerol is added to a final concentration of 5% (v/v) and the purified protein is stored in liquid nitrogen.

3.3. Purification of *M. thermophila* phosphotransacetylase produced in *E. coli*

E. coli strain BL21(DE3) is transformed with plasmid pML702 (Latimer and Ferry, 1993). Transformants are grown at 37 °C in LB broth containing 100 μg ml^{-1} ampicillin and induced with 1% (final concentration) Bacto-lactose or 0.4 m*M* (final concentration) IPTG. All steps are performed aerobically at 4 °C. A 25-m*M* Tris (pH 7.2) buffer containing 2 m*M* DTT is used in all steps of the purification unless otherwise noted. Approximately 40 g (wet weight) of cells are suspended in 130 ml (total volume) of 25 m*M* Bis–Tris (pH 6.5) with 2 m*M* DTT and lysed by French pressure cell disruption at 110 MPa. The lysate is centrifuged at 4000×*g* for 25 min at 4 °C. The 4000×*g* pellet is dispersed in 60 ml of buffer and centrifuged as above. The washed pellet is dispersed in 10 ml of buffer to a final volume of 15 ml. One volume (15 ml) of 12 *M* urea is added to the suspension and the mixture incubated at 13 °C for 15 min. The solution is diluted to 300 ml with buffer and incubated at 13 °C for 5 h. The solution is loaded on a Q-Sepharose Fast Flow column (5 × 10 cm) equilibrated with two column volumes of buffer. The column is developed with a 1800-ml linear gradient of 0–1 *M* KCl at 6 ml min^{-1}. Fractions with the highest total activity are pooled, diluted with one volume 50 m*M* Tris (pH 7.6) plus 2 m*M* DTT, and loaded on a Mono-Q column equilibrated with five column volumes of 50 m*M* Tris (pH 7.6) with 2 m*M* DTT. The purified phosphotransacetylase elutes at 180 m*M* KCl using a 200 ml linear gradient from 0 to 1 *M* KCl at 2 ml min^{-1}. The enzyme can be stored at −80 °C with minimal loss of activity on one cycle of freeze-thawing for up to 1 year.

3.4. Catalytic mechanism of phosphotransacetylase from *M. thermophila*

Although phosphotransacetylase was first purified from the fermentative anaerobe *Clostridium kluyveri* in the early 1950s (Stadtman, 1951), it has been investigated in less detail than acetate kinase. Kinetic analyses of phosphotransacetylases from *C. kluyveri* and *Veillonella alcalescens* suggest a mechanism proceeding through formation of a ternary complex (Kyrtopoulos and Satchell, 1972; Pelroy and Whiteley, 1972). In an attempt to detect an acetyl-enzyme intermediate in the enzyme from *C. kluyveri*, no isotope exchange from labeled acetyl phosphate into either acetyl-CoA or inorganic phosphate was observed in the absence of free CoA. Further, attempts to isolate an acetyl-phosphotransacetylase intermediate were also unsuccessful, a result consistent with a ternary complex mechanism (Henkin and Abeles, 1976). Mechanistic analyses of the enzyme were abandoned until 1993 when heterologous expression of phosphotransacetylase from *M. thermophila* in *E. coli* (Latimer and Ferry, 1993) leads to the crystal structure joined by the structure of phosphotransacetylase from *Streptococcus pyogenes* (Iyer *et al.*, 2004; Xu *et al.*, 2004). Both structures form a homodimer with each monomer consisting of two α/β domains and a cleft along the domain boundary that is the proposed active site. The *M. thermophila* structure with HS-CoA bound (Lawrence *et al.*, 2006) combined with kinetic studies (Lawrence and Ferry, 2006) and biochemical analyses of site-directed variants (Iyer and Ferry, 2001; Iyer *et al.*, 2004; Lawrence *et al.*, 2006; Rasche *et al.*, 1997) suggest a mechanism proceeding through base-catalyzed abstraction of the thiol proton of HS-CoA followed by nucleophilic attack of the thiolate anion of HS-CoA on the carbonyl carbon of acetyl phosphate (Fig. 11.3). The *M. thermophila* phosphotransacetylase

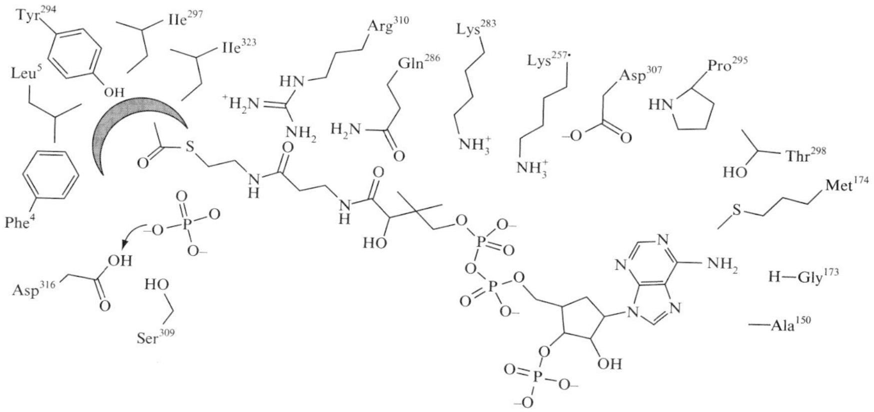

Figure 11.3 Postulated mechanism of *Methanosarcina thermophila* phosphotransacetylase. The crescent indicates the hydrophobic pocket interacting with the methyl group of acetyl phosphate. Acetyl phosphate is proposed to bind to the active site with the phosphate group coordinated by the catalytically essential Arg^{310}. The Asp^{316} is proposed to abstract a proton leading to the thiolate anion of HS-CoA attacking the carbonyl carbon of acetyl phosphate in the direction of acetyl-CoA synthesis. Hydrogen bonds between CoA and the protein are indicated. Reprinted with permission Lawrence *et al.* (2006).

has high identity to phosphotransacetylases from the domain *Bacteria* suggesting an ancient origin and common mechanism (Rasche *et al.*, 1997).

ACKNOWLEDGMENTS

Research in the author's laboratory was supported by the National Institutes of Health, National Science Foundation, and the Department of Energy.

REFERENCES

Allen, S. H. G., Kellermeyer, R. W., Stjernholm, R. L., and Wood, H. G. (1964). Purification and properties of enzymes involved in the propionic acid fermentation. *J. Bacteriol.* **87,** 171–187.

Blattler, W. A., and Knowles, J. R. (1979). Stereochemical course of phosphokinases. The use of adenosine [gamma-(S)-^{16}O ^{17}O ^{18}O] triphosphate and the mechanistic consequences for the reactions catalyzed by glycerol kinase, hexokinase, pyruvate kinase, and acetate kinase. *Biochemistry* **18,** 3927–3933.

Bowman, C. M., Valdez, R. O., and Nishimura, J. S. (1976). Acetate kinase from *Veillonella alcalescens.* Regulation of enzyme activity by succinate and substrates. *J. Biol. Chem.* **251,** 3117–3121.

Buss, K. A., Cooper, D. R., Ingram-Smith, C., Ferry, J. G., Sanders, D. A., and Hasson, M. S. (2001). Urkinase: structure of acetate kinase, a member of the ASKHA superfamily of phosphotransferases. *J. Bacteriol.* **183,** 680–686.

Fox, D. K., and Roseman, S. (1986). Isolation and characterization of homogeneous acetate kinase from *Salmonella typhimurium* and *Escherichia coli*. *J. Biol. Chem.* **261,** 13487–13497.

Gorrell, A., and Ferry, J. G. (2007). Investigation of the *Methanosarcina thermophila* acetate kinase mechanism by fluorescence quenching. *Biochemistry* **46,** 14170–14176.

Gorrell, A., Lawrence, S. H., and Ferry, J. G. (2005). Structural and kinetic analyses of arginine residues in the active-site of the acetate kinase from *Methanosarcina thermophila*. *J. Biol. Chem.* **280,** 10731–10742.

Henkin, J., and Abeles, R. H. (1976). Evidence against an acyl-enzyme intermediate in the reaction catalyzed by clostridial phosphotransacetylase. *Biochemistry* **15,** 3472–3479.

Ingram-Smith, C., Barber, R. D., and Ferry, J. G. (2000). The role of histidines in the acetate kinase from *Methanosarcina thermophila*. *J. Biol. Chem.* **275,** 33765–33770.

Ingram-Smith, C., Gorrell, A., Lawrence, S. H., Iyer, P., Smith, K., and Ferry, J. G. (2005). Identification of the acetate binding site in the *Methanosarcina thermophila* acetate kinase. *J. Bacteriol.* **187,** 2386–2394.

Iyer, P. P., and Ferry, J. G. (2001). Role of arginines in coenzyme A binding and catalysis by the phosphotransacetylase from *Methanosarcina thermophila*. *J. Bacteriol.* **183,** 4244–4250.

Iyer, P., and Ferry, J. G. (2005). Acetate kinase from *Methanosarcina thermophila*, a key enzyme for methanogenesis. *In* "Microbial Enzymes and Biotransformations," (J. L. Barredo, ed.), Vol. 17, pp. 239–246. Humana Press Inc., Totowa, NJ.

Iyer, P. P., Lawrence, S. H., Luther, K. B., Rajashankar, K. R., Yennawar, H. P., Ferry, J. G., and Schindelin, H. (2004). Crystal structure of phosphotransacetylase from the methanogenic archaeon *Methanosarcina thermophila*. *Structure* **12,** 559–567.

Kohler, H.-P. E., and Zehnder, A. J. B. (1984). Carbon monoxide dehydrogenase and acetate thiokinase in *Methanothrix soehngenii*. *FEMS Microbiol. Lett.* **21,** 287–292.

Kyrtopoulos, S. A., and Satchell, D. P. (1972). Kinetic studies with phosphotransacetylase. IV. Inhibition by products. *Biochim. Biophys. Acta* **276,** 383–391.

Latimer, M. T., and Ferry, J. G. (1993). Cloning, sequence analysis, and hyperexpression of the genes encoding phosphotransacetylase and acetate kinase from *Methanosarcina thermophila*. *J. Bacteriol.* **175,** 6822–6829.

Lawrence, S. H., and Ferry, J. G. (2006). Steady-state kinetic analysis of phosphotransacetylase from *Methanosarcina thermophila*. *J. Bacteriol.* **188,** 1155–1158.

Lawrence, S. H., Luther, K. B., Schindelin, H., and Ferry, J. G. (2006). Structural and functional studies suggest a catalytic mechanism for the phosphotransacetylase from *Methanosarcina thermophila*. *J. Bacteriol.* **188,** 1143–1154.

Lipmann, F. (1944). Enzymatic synthesis of acetyl phosphate. *J. Biol. Chem.* **155,** 55–70.

Lipmann, F., and Tuttle, L. C. (1945). *J. Biol. Chem.* **159,** 21–28.

Miles, R. D., Iyer, P. P., and Ferry, J. G. (2001). Site-directed mutational analysis of active site residues in the acetate kinase from *Methanosarcina thermophila*. *J. Biol. Chem.* **276,** 45059–45064.

Nakajima, H., Suzuki, K., and Imahori, K. (1978). Purification and properties of acetate kinase from *Bacillus stearothermophilus*. *J. Biochem.* **84,** 193–203.

Oberlies, G., Fuchs, G., and Thauer, R. K. (1980). Acetate thiokinase and the assimilation of acetate in methanobacterium thermoautotrophicum. *Arch. Microbiol.* **128,** 248–252.

Pelroy, R. A., and Whiteley, H. R. (1972). Kinetic properties of phosphotransacetylase from *Veillonella alcalescens*. *J. Bacteriol.* **111,** 47–55.

Rasche, M. E., Smith, K. S., and Ferry, J. G. (1997). Identification of cysteine and arginine residues essential for the phosphotransacetylase from *Methanosarcina thermophila*. *J. Bacteriol.* **179,** 7712–7717.

Rose, I. A., Grunberg-Manago, M., Korey, S. R., and Ochoa, S. (1954). Enzymatic phosphorylation of acetate. *J. Biol. Chem.* **211,** 737–756.

Singh-Wissmann, K., Ingram-Smith, C., Miles, R. D., and Ferry, J. G. (1998). Identification of essential glutamates in the acetate kinase from *Methanosarcina thermophila. J. Bacteriol.* **180,** 1129–1134.

Singh-Wissmann, K., Miles, R. D., Ingram-Smith, C., and Ferry, J. G. (2000). Identification of essential arginines in the acetate kinase from *Methanosarcina thermophila. Biochemistry* **39,** 3671–3677.

Skarstedt, M. T., and Silverstein, E. (1976). *Escherichia coli* acetate kinase mechanism studied by net initial rate, equilibrium, and independent isotopic exchange kinetics. *J. Biol. Chem.* **251,** 6775–6783.

Smith, C.-I., Martin, S. R., and Smith, K. S. (2006). Acetate kinase: Not just a bacterial enzyme. *Trends Microbiol.* **14,** 249–253.

Sowers, K. R., Nelson, M. J. K., and Ferry, J. G. (1984). Growth of acetotrophic, methane-producing bacteria in a pH auxostat. *Curr. Microbiol.* **11,** 227–230.

Spector, L. B. (1980). Acetate kinase: A triple-displacement enzyme. *Proc. Natl Acad. Sci. USA* **77,** 2626–2630.

Stadtman, E. R. (1951). The purification and properties of phosphotransacetylase. *J. Biol. Chem.* **196,** 527–534.

Xu, Q. S., Shin, D. H., Pufan, R., Yokota, H., Kim, R., and Kim, S. H. (2004). Crystal structure of a phosphotransacetylase from *Streptococcus pyogenes. Proteins* **55,** 479–481.

CHAPTER TWELVE

Sodium Ion Translocation and ATP Synthesis in Methanogens

Katharina Schlegel *and* Volker Müller

Contents

Molecular Microbiology & Bioenergetics, Institute of Molecular Biosciences, Goethe-University Frankfurt, Frankfurt, Germany

Methods in Enzymology, Volume 494
ISSN 0076-6879, DOI: 10.1016/B978-0-12-385112-3.00012-3

Abstract

Methanogens are the only significant biological producers of methane. A limited number of C_1 substrates, such as methanol, methylamines, methyl sulfate, formate, $H_2 + CO_2$ or CO, and acetate, serve as carbon and energy source. During degradation of these compounds, a primary proton as well as a primary sodium ion gradient is established, which is a unique feature of methanogens. This raises the question about the coupling ion for ATP synthesis by the unique A_1A_o ATP synthase. Here, we describe how to analyze and determine the Na^+ dependence of two model methanogens, the hydrogenotrophic *Methanothermobacter thermautotrophicus* and the methylotrophic *Methanosarcina barkeri*. Furthermore, the determination of important bioenergetic parameters like the ΔpH, $\Delta\Psi$, or the intracellular volume in *M. barkeri* is described. For the analyses of the A_1A_O ATP synthase, methods for measurement of ATP synthesis as well as ATP hydrolysis in *Methanosarcina mazei* Gö1 are described.

1. Introduction

Methanogenic archaea grow on a limited number of C_1 substrates, such as methanol, methylamines, formate, and some can also use acetate (Deppenmeier, 2002; Ferry, 1997; Thauer *et al.*, 2008). Central to all those pathways is the Wood–Ljundahl pathway (Ljungdahl, 1994; Ragsdale, 2008). During growth on $H_2 + CO_2$, CO_2 is first bound to methanofuran and thereby reduced to a formyl group (Fig. 12.1). This endergonic reaction is catalyzed by the formylmethanofuran dehydrogenase and driven by the electrochemical ion gradient across the membrane (Kaesler and Schönheit, 1989; Winner and Gottschalk, 1989). The formyl group is transferred to tetrahydromethanopterin (H_4MPT) and reduced to a methyl group. This methyl group is transferred to cofactor M (CoM-SH) by the methyl-H_4MPT:HS-CoM methyltransferase. This exergonic reaction ($\Delta G_0' = -29$ kJ/mol) is used to transfer 1.7 Na^+/CH_4 over the membrane and thereby generates a primary and electrogenic Na^+ gradient across the membrane (Becher *et al.*, 1992b; Gottschalk and Thauer, 2001; Lienard *et al.*, 1996; Müller *et al.*, 1987). In the next step, methyl-CoM undergoes a nucleophilic attack by the thiolate anion of HS-CoB (Coenzyme B, 7-thioheptanoyl-*o*-phosho-L-threonine), thus liberating CH_4 and generating a disulfide of CoM and CoB, the so-called *heterodisulfide*. The heterodisulfide is the terminal electron acceptor in methanogenesis and is reduced and thereby cleaved by a membrane-bound heterodisulfide reductase.

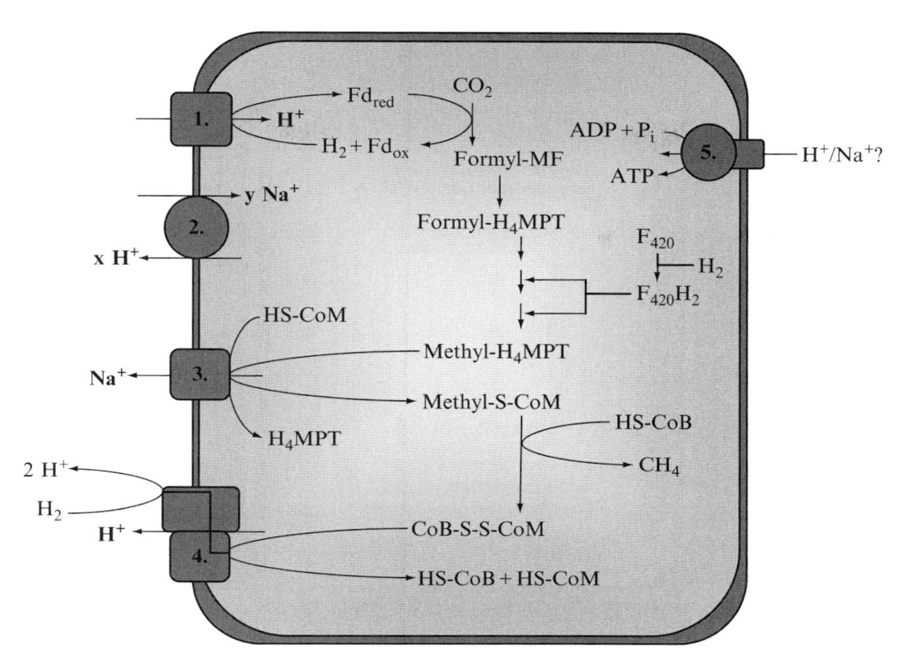

Figure 12.1 Ion currents during methanogenesis from $H_2 + CO_2$. Please note that this model only describes the pathway in *M. mazei* and *M. barkeri*. *M. acetivorans* does not contain hydrogenases. (1) Ech/Eha hydrogenase; (2) Na^+/H^+ antiporter; (3) methyl-H_4MPT coenzyme M methyltransferase; (4) H_2: heterodisulfide oxidoreductase system; (5) A_1A_O ATP synthase; MF, methanofuran; H_4MPT, tetrahydromethanopterin; CoM-SH, coenzyme M; CoB-SH, coenzyme B; Fd, ferredoxin.

The electrons necessary for this reaction are provided by F_{420} (during growth on methylated substrates), a membrane-bound hydrogenase (during growth on $H_2 + CO_2$) or ferredoxin (during growth on acetate) and transferred via an electron transport chain. In this reaction, three to four H^+ are translocated over the membrane (Blaut *et al.*, 1987; Deppenmeier *et al.*, 1990).

During growth on methyl-group containing substrates, the methyl groups are channeled to CoM-SH by specific methyltransferases. One quarter of the methyl group is oxidized to CO_2 to gain the reducing equivalents to reduce the other 75% to methane (Deppenmeier, 2002; van der Meijden *et al.*, 1983; Fig. 12.2). Methanogenesis from acetate starts with the activation of acetate to acetyl-CoA, followed by an oxidation of acetyl-CoA to CO_2 and a coenzyme-bound methyl group that is reduced to CH_4 with electrons gained during the oxidation reaction (Ferry, 1997; Terlesky and Ferry, 1988; Thauer, 1990).

A common feature of methanogenesis found in all methanogens investigated so far is its Na^+ requirement. Na^+ is required not only for growing

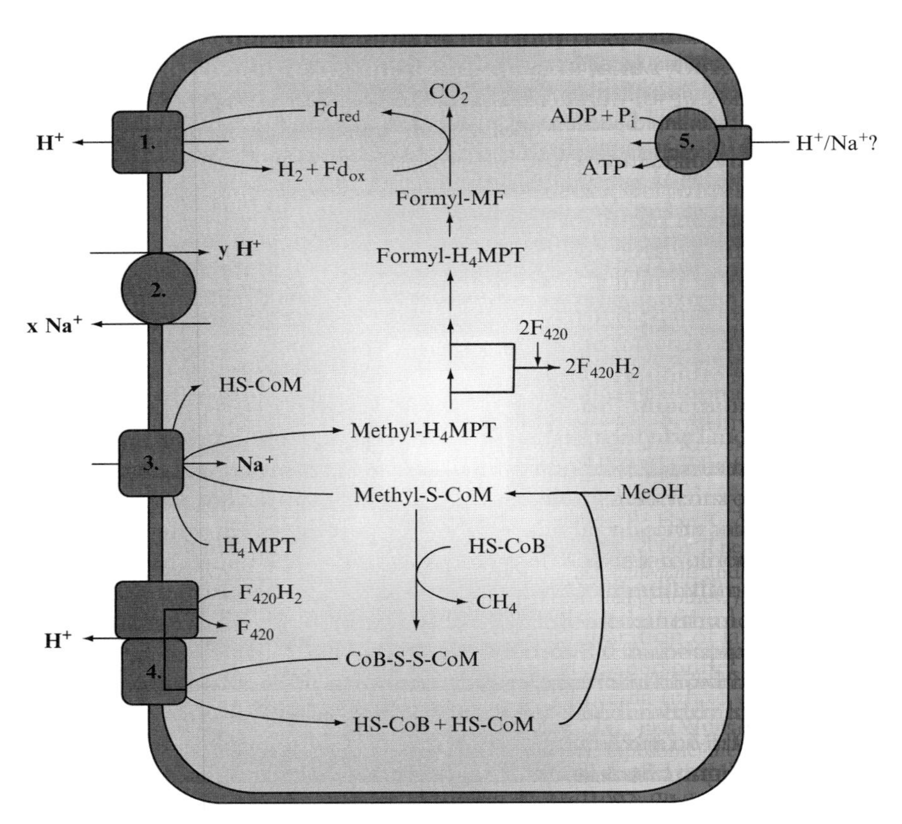

Figure 12.2 Ion currents during methanogenesis from methanol. Please note that this model only describes the pathway in *M. mazei* and *M. barkeri*. *M. acetivorans* does not contain an Ech hydrogenase but a Rnf complex. The electrons of the hydrogen produced by the Ech/Eha hydrogenase are finally transferred to the heterodisulfide reductase. It is still a matter of debate whether this occurs via a soluble F_{420} reducing hydrogenase or directly by the H_2: heterodisulfide oxidoreductase system. (1) Ech/Eha hydrogenase; (2) Na^+/H^+ antiporter; (3) methyl-H_4MPT-coenzyme M methyltransferase; (4) F_{420}: heterodisulfide oxidoreductase system; (5) A_1A_O ATP synthase; MF, methanofuran; H_4MPT, tetrahydromethanopterin; CoM-SH, coenzyme M; CoB-SH, coenzyme B; Fd, ferredoxin.

cells but also by resting cells for methanogenesis, indicating an involvement of Na^+ in one of the steps of the pathway (Müller *et al.*, 1986, 1988; Perski *et al.*, 1981, 1982). Indeed, it turned out that the methyl-H_4SPT: HS-CoM-methyltransferase is a primary Na^+ pump (Becher *et al.*, 1992a; Müller *et al.*, 1987). Apart from the Na^+ motive methyl-H_4MPT:HS-CoM-methyltransferase, the heterodisulfide reductase system is proton motive. Thus, methanogens are the only organisms that generate a primary proton and sodium ion gradient at the same time. How the two ion motive forces are connected to ATP synthesis is still a matter of debate, but the A_1A_O ATP synthase is essential for ATP synthesis (Deppenmeier and Müller, 2008; Müller and Grüber, 2003; Pisa *et al.*, 2007; Saum *et al.*, 2009).

In this chapter, we describe how Na^+ dependence of growth and methanogenesis is analyzed, how Na^+ transport is measured in cell suspensions, and how ATP synthesis as well as ATP hydrolysis is measured.

2. Analysis of the Na^+ Dependence of Growth and Methanogenesis

The first organism for which a Na^+ dependence was shown was *Methanothermobacter thermautotrophicus* (formerly, *Methanobacterium thermautotrophicus*) (Perski *et al.*, 1981), and later confirmed for many other methanogens such as *Methanobacterium bryantii, M. barkeri*, or *Methanobrevibacter hungatii* (Müller *et al.*, 1986; Perski *et al.*, 1982).

2.1. Na^+ dependence of growth

2.1.1. *M. thermautotrophicus* ΔH

M. thermautotrophicus ΔH (DSM 1053 = ATCC29096) is obtained from the German Collection of Microorganisms (DMSZ, Braunschweig, Germany). The organism is grown as described (Schönheit *et al.*, 1980). To analyze the sodium ion dependence of growth, the medium contains KH_2PO_4, 50 m*M*; NH_4Cl, 40 m*M*; nitrilotriacetate, 0.5 m*M*; $MgCl_2$, 0.2 m*M*; $FeSO_4$, 50 μ*M*; $CoCl_2$, 1 μ*M*; Na_2MoO_4, 1 μ*M*; $NiCl_2$, 5 μ*M*; and resazurin, 20 μ*M*. The Na^+ concentration is 0, 0.1, 0.2, 0.5, 1, or 5 m*M* (supplied as NaCl). The medium is flushed with 80% H_2/20% CO_2/0.2% H_2S for 15 min. To remove the remaining oxygen, the gas is passed over copper files heated at 370 °C. The pH is adjusted to 7 with NH_3 after gassing the medium. Cultures are incubated in 500-ml glass fermenters containing 250 ml of media at 65 °C. The cultures are continuously stirred at 1100 rpm and flushed with 80% H_2/20% CO_2/0.2% H_2S at a rate of 250 ml/min via a micro filter candle (porosity 3, Schott, Mainz, Germany). To monitor growth, the optical density at 578 nm is measured. According to Perski *et al.* (1981), an OD of 1 corresponds to a cell concentration of 460 mg/l.

2.1.2. *M. barkeri*

M. barkeri (DSM 804) is obtained from DSMZ (Braunschweig, FRG). The cells are cultivated in medium under a N_2 atmosphere, as described by Müller *et al.* (1986). The medium contains imidazole-HCl, 40 m*M*; KH_2PO_4, 0.5 m*M*; $K_2HPO_4{\cdot}3H_2O$, 0.5 m*M*; KCl, 50 m*M*; $MgCl_2{\cdot}6H_2O$, 1.7 m*M*; $CaCl_2{\cdot}2H_2O$, 2.0 m*M*; $(NH_4)_2{\cdot}6H_2O$, 0.1 m*M*; ammonium acetate, 10 m*M*; vitamin solution (Wolin *et al.*, 1964), 10 ml/l; trace element solution SL6 (Scherer and Sahm, 1981), 3 ml/l; K_2S, 2.6 m*M*; cysteine hydrochloride, 1.7 m*M*; and titanium citrate solution (Zehnder and Wuhrmann, 1976), 2 ml/l. The Na^+ concentrations (supplied as NaCl) used are 0, 5, 10, or 20 m*M*. The medium is titrated to a pH of 6.8 with 6 N HCl and prepared under an atmosphere of N_2. Reducing agents and methanol (25 *M*) are flushed with N_2 in aqueous stock solutions and added prior to inoculation. Methanol serves as a substrate in the final concentration of 200 m*M*. Growth is started by the addition of 10% inoculum. The cells are incubated at 37 °C and growth is analyzed by determining gas production over time.

2.1.3. Expected results

In both cases, the growth rate increases with increasing salinities in a Michaelis–Menten fashion. For *M. thermautotrophicus*, saturation is observed at 5 m*M* and the K_s is determined to 0.8 m*M* Na^+ (Perski *et al.*, 1981). For *M. barkeri*, the optimal growth is at 18 m*M* and the K_s is 4.3 m*M* NaCl (Müller *et al.*, 1986). Residual, contaminating Na^+ in the growth media has to be monitored by flame photometry or a Na^+-selective electrode and taken into account. Please note that the Na^+ dependence of growth of *M. barkeri* is better seen at high K^+ concentrations. Therefore, the medium contains extra 50 m*M* KCl.

2.2. Analysis of Na^+ dependence of methanogenesis

2.2.1. *M. thermautotrophicus* ΔH

For the preparation of a cell suspension, exponentially growing cultures are cooled down to 0 °C. During this rapid cooling, the cultures are continuously flushed with 80% H_2/20% CO_2/0.2% H_2S. Fifty milliliters of culture is transferred anaerobically in 120-ml centrifuge tubes and harvested at 2000×*g* for 10 min at 4 °C under an atmosphere of N_2. To remove the media, the cells are washed twice in a 50 m*M* KH_2PO_4/K_2HPO_4 buffer, pH 7.0, containing 2 m*M* $MgSO_4 \cdot 7H_2O$. The washing buffer contains different amounts of NaCl such as 0.15, 0.5, 1, or 5 m*M*. The cells are resuspended in the washing buffer to a final concentration of 2.5 mg dry weight per ml. To start the reaction, 1 ml of the cell suspension is diluted in 5 ml of washing buffer (50 m*M* KH_2PO_4/K_2HPO_4 buffer, pH 7.0, containing 2 m*M* $MgSO_4 \cdot 7H_2O$) in a 120-ml serum bottle (Miller and Wolin, 1974). The NaCl concentration is the same as in the washing buffers. The gas phase is 80% H_2/20% CO_2/0.2% H_2S (pressure 1.6 atm.). The serum bottle is rapidly shaken at 65 °C. Samples are taken as described below.

2.2.2. *M. barkeri*

The harvest is carried out in the late exponential growth phase after 70–80% of the total expected amount of gas produced from methanol is obtained. The cells are transferred to 300-ml centrifuge tubes in an anaerobic chamber (Coy Laboratory Products, Michigan, USA) filled with N_2 (97% N_2, 3% H_2, $O_2 > 10$ ppm) and centrifuged for 20 min at 10,000 rpm in a cooled Sorvall Centrifuge Type RC-5B (Du Pont Instruments, Newton, USA) at 4 °C. After washing the cells twice with 1 m*M* imidazole/HCl buffer, pH 6.7, containing 25 m*M* KCl, 4.4 μ*M* resazurin, and 2 ml/l titanium (III) citrate solution (Zehnder and Wuhrmann, 1976), the pellet is resuspended in the same buffer to a concentration of 10–20 mg protein/ml, filled in 16-ml Hungate tubes, and stored on ice. The gas phase is exchanged against N_2 by gassing the suspension for 10 min. An 18-ml anaerobic bottle is flushed with N_2 for 20 min prior to usage. Hundred microliters of concentrated cell suspension is added via a 250-μl Hamilton syringe anaerobically to the bottle filled with 1 ml of 200 m*M* imidazole/HCl buffer, pH 6.7, containing 20 m*M* $MgSO_4$. The Na^+ concentration is adjusted by adding NaCl from a stock solution anaerobically via a Hamilton syringe. The assay is preincubated at 37 °C on a rotary shaker for 15 min. After preincubation, 20 m*M* methanol is added to start methanogenesis. Again, a Hamilton syringe is used for addition. The production of methane must be measurable now.

2.3. Measurement of methane production

Samples (5 μl) of the head space are taken every 10 min over a period of 60 min. The measurements are made with a gas chromatograph (Bodenseewerk Perkin-Elmer, Üblingen, Germany) with a flame ionization detector. The chromatograph is connected with an integrator model 3370 (Hewlett Packard, USA). The separation is achieved with a 2 m×2 mm glass column filled with Porapak QS, 80–100 mesh (Sigma-Aldrich, Germany). The oven temperature is 75 °C, the temperature of the injector and detector is 200 °C. N_2 is used as a carrier gas with a flow rate of 30 ml/min.

3. Na^+ Transport in Cell Suspensions

Methanogenesis is not only dependent on Na^+, but also coupled to the generation of a transmembrane electrochemical Na^+ gradient (Müller *et al.*, 1987, 1988). This protocol describes the preparation of cell suspension and the measurement of $^{22}Na^+$ transport in metabolizing cells of *M. barkeri*.

3.1. Cultivation and preparation of cell suspension

M. barkeri (DSM 804) is cultivated in medium under a N_2 atmosphere as described (Müller *et al.*, 1986). The medium contains imidazole-HCl, 40 m*M*; $Na_2HPO_4 \cdot H_2O$, 0.4 m*M*; $NaH_2PO_4 \cdot H_2O$, 0.6 m*M*; $MgCl_2 \cdot 6H_2O$, 1.7 m*M*; $CaCl_2 \cdot 2H_2O$, 2 m*M*; KCl, 5.2 m*M*; NaCl, 34 m*M*; NH_4Cl, 9.4 m*M*; $(NH_4)_2Fe(SO_4)_2$, 0.1 m*M*; yeast extract, 2 g/l (Difco Laboratories, Detroit, USA); casitone, 2 g/l (Difco Laboratories); vitamin solution (Wolin *et al.*, 1964), 10 ml/l; trace element solution SL6, 3 ml/l; resazurin, 13 μ*M*; and H_2O bidest. The medium is titrated to a pH of 6.8 with 6 N HCl and prepared under an atmosphere of N_2. Reducing agents (cysteine–HCl·H_2O, 2.5 m*M*; $Na_2S \cdot 9H_2O$, 1.2 m*M*) and methanol (25 *M*) are flushed with N_2 in aqueous stock solutions and added prior to inoculation. Methanol serves as a substrate in a final concentration of 200 m*M*. Growth is started by the addition of 10% inoculum. The cells are incubated at 37 °C.

The cells are harvested in the late exponential growth phase after 70–80% of the total expected gas amount was produced. After washing the cells twice with a 20 m*M* imidazole/HCl buffer, pH 6.7, containing 25 m*M* KCl, 4.4 μ*M* resazurin, and 2 ml/l titanium (III) citrate solution (Zehnder and Wuhrmann, 1976), the pellet is resuspended in the same buffer to a concentration of 10–20 mg protein/ml, filled in 16-ml Hungate tubes, and stored on ice. The gas phase is exchanged against N_2 by gassing the suspension for 10 min.

3.2. Assay for determination of internal Na^+ concentration in the steady state of methanogenesis

1. An 18-ml anaerobic bottle is flushed with N_2 for 20 min prior to usage. Hundred microliters of concentrated cell suspension is added via a 250-μl Hamilton syringe anaerobically to the serum tube filled with 1 ml of 200 m*M* imidazole/HCl buffer, pH 6.7, containing 20 m*M* $MgSO_4$.
2. The assay is preincubated at 37 °C on a rotary shaker for 15 min.
3. After preincubation, 10 m*M* NaCl and 20 m*M* methanol are added to start methanogenesis. Again, a Hamilton syringe is used for addition. The production of methane must be measurable now.
4. The assay is incubated for an additional 30 min for equilibration.
5. The measurement is started by the addition of 10 μCi of $^{22}NaCl$ (carrier-free) into the assay. Samples (100 μl) are taken every 5 min. A 250-μl Hamilton syringe is used for sampling. The syringe has to be flushed with H_2O bidest. Thrice after every sample.

3.3. Assay for measurement of Na^+ efflux due to methanogenesis

1. An 18-ml anaerobic bottle is flushed with N_2 for 20 min prior to usage to remove the oxygen. 100 µl of a concentrated cell suspension is added via a 250-µl Hamilton syringe anaerobically to the serum tube filled with 1 ml of 200 m*M* imidazole/HCl buffer, pH 6.7, containing 20 m*M* $MgSO_4$.
2. The assay is preincubated at 37 °C on a rotary shaker for 15 min.
3. After preincubation, 10 m*M* NaCl and 10 µCi ^{22}NaCl (carrier-free) are added. It is important to incubate this assay for an additional 25 min to equilibrate internal and external radioactivity. It is possible to study the effect of inhibitors on the metabolism with this assay. If supplements are added, they have to incubate for another 30 min with the cells for complete activity.
4. To start Na^+ export, 20 m*M* of methanol is added with a 250-µl Hamilton syringe. Samples (100 µl) are taken every 5 min and cells are separated from the medium by filtration. The syringe has to be flushed thrice with H_2O bidest. After every sample.

3.4. Filtration

The aim of the filtration and washing is the removal of $^{22}Na^+$ not taken up by the cell and to remove Na^+ bound to the cell surface. Thus, only the internal Na^+ is quantified. As the $^{22}Na^+$ binds unspecific to the filter, the filter has to be pretreated. Therefore, cellulose nitrate membrane filters (diameter 25 mm, pore size 0.45 µm, Schleicher und Schüll, Dassel, Germany) are incubated with a 20-m*M* NaCl solution overnight and dried thereafter before usage. This procedure decreases the unspecific binding of $^{22}Na^+$ up to 70%. Still, it is useful to measure the unspecific binding by repeating the experiment using boiled cells and measure the radioactivity due to unspecific effects.

The filter is then put into a filter holder, the 100-µl sample is spotted into the middle of the filter, and the medium is sucked through the filter via underpressure (Fig. 12.3). Wash the cells with 10 times the volume of the samples for three times. The whole sampling and washing procedure should take not more than 15 s to avoid leakage of Na^+ out of the cells. The buffer used for washing is the same as in the experiment (20 m*M* imidazole/HCl buffer, pH 6.7, containing 20 m*M* $MgSO_4$, and 10 m*M* NaCl).

3.5. Counting radioactivity

The filters with the washed cells are placed for 1 min on absorptive paper to remove excess water and then put into a scintillation vial. The measurement is carried out in a gamma counter model PW 4800 (Philips, Hamburg,

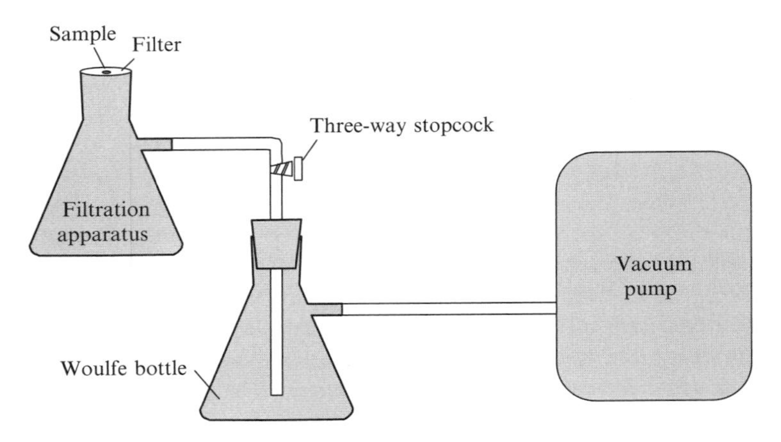

Figure 12.3 Filtration apparatus to separate cells from the medium. The filtration apparatus is connected to a Woulfe bottle that is connected to a vacuum pump. The vacuum pump is on all the time during the experiment, and the vacuum on the filtration apparatus is obtained or released by a three-way stopcock. Turn off the vacuum, put the filter on the filter holder, moisturize, and turn the vacuum on again. Apply the cells and take care that the cells stay just in the middle of the filter, otherwise the filter gets clumped and washing takes too long. After washing, release the vacuum and then remove the filter using forceps. Take care not to break the filter.

Germany). The positron radiation is counted for 10 min with an energy window of 460–640 keV. It is important to measure the radioactive background and deduct it from the counts of the sample. Counting can also be done in a liquid scintillation counter. Therefore, the filter is placed into 0.5 ml 3 N NaOH for 1 h. Then, 4 ml of Rotiszint eco plus (Roth, Karlsruhe, Germany) is added. Radioactivity can be counted after 60 min when chemiluminescence has stopped and the filter is completely dissolved (Schmidt *et al.*, 2007).

4. Determining Bioenergetic Parameters via Different Localization of Radioactively Labeled Substances: Intracellular Volume, ΔpH, and $\Delta\Psi$

4.1. Silicon oil centrifugation

To calculate the internal Na^+ concentration, the intracellular volume of the cells has to be known. In addition, for bioenergetic analyses of Na^+ transport, the electrical field ($\Delta\Psi$) produced by the Na^+ transport as well as changes in the ΔpH should be determined. These procedures have in common that the distribution of a labeled compound between the cell and the supernatant is analyzed. Therefore, cells have to be separated very fast and efficiently from the medium. Silicone oil centrifugation turned out to be very suitable for anaerobic conditions (Fig. 12.4).

The density of the silicone oil used depends on the ionic strength of the buffer used for the experiment. For the assay described below, oil with a density of 1.023 is applied. Oil of different densities can be produced by mixing oil with hexadecane.

1. Silicone oil (0.2 ml) is transferred into a 1.5-ml Eppendorf cup via a 1.5-ml plastic syringe. To remove the oxygen, the cups are transferred into the anaerobic chamber 12 h prior to the experiment.
2. Directly before the experiment, the cups are taken out of the anaerobic chamber.
3. To separate cells and medium, 0.5 ml of a sample like a cell suspension is added to the oil via a plastic syringe under anaerobic conditions.
4. Centrifuge the sample for 1 min in a Beckman microcentrifuge B (Beckmann, Fullerton, USA). There must be a clear separation of the supernatant (upper phase) and the silicone oil (lower phase) after centrifugation. The anaerobic conditions have to remain while centrifuging.
5. Fifty microliters of the supernatant are transferred into a 7-ml scintillation vial (Packard, Frankfurt, Germany) containing 0.5 ml of 3 *M* NaOH. The rest of the supernatant as well as the silicone oil is carefully removed.
6. The bottom of the Eppendorf cup containing the cell pellet is cut off with a razor blade ("Rotbart- extra thin," Wilkinson Sword, Solingen, Germany) and also transferred into a scintillation cup containing 0.5 ml of 3 *M* NaOH. To lyze the cells, the cup is incubated at 37 °C for 12 h and vigorously shaken several times.

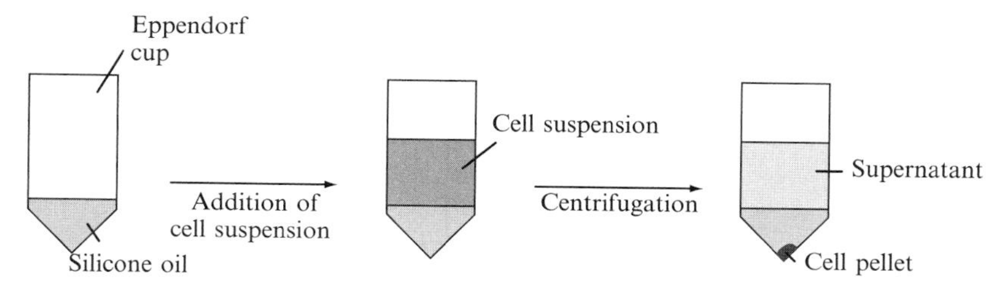

Figure 12.4 Separation of cells from the medium by silicone oil centrifugation. After centrifugation, check successful separation by visual inspection and microscopy of the supernatant–oil interface. Take care to sustain anaerobic conditions during centrifugation.

4.1.1. Counting radioactivity

To determine the radioactivity, 5 ml of Quickzint 2000 (Zinsser, Frankfurt, Germany) is added to the scintillation vials and placed into a LS 7500 Scintillation counter (Beckmann Instruments, Fullerton, USA). The "energy window" is set on 0–655. The samples are measured for 10 min.

4.2. Determining the intracellular volume

The general idea behind this method is that big molecules like sugar cannot penetrate the cellular membrane, whereas water can (Fig. 12.5). The distribution of [^{14}C] sucrose and ^3H_2O between the cell pellet and the supernatant is measured and gives a measure for the internal volume.

1. The preparation of a cell suspension is described in Section 3.1.
2. The experiments are carried out in 58-ml anaerobic bottles filled with 9–9.5 ml of imidazole/HCl buffer via a 10-ml plastic syringe. The gas atmosphere is oxygen-free N_2. Cell suspension (0.5–1 ml) is added via a plastic syringe to a final volume of 10 ml.
3. The measurements are started by the addition of 10 μCi ^3H_2O and 1 μCi [^{14}C] sucrose (final concentration 10 μ*M*). Samples (0.5 ml) are taken every 10 min. Cells and the supernatant are separated via silicon oil centrifugation and the radioactivity in each fraction is determined.

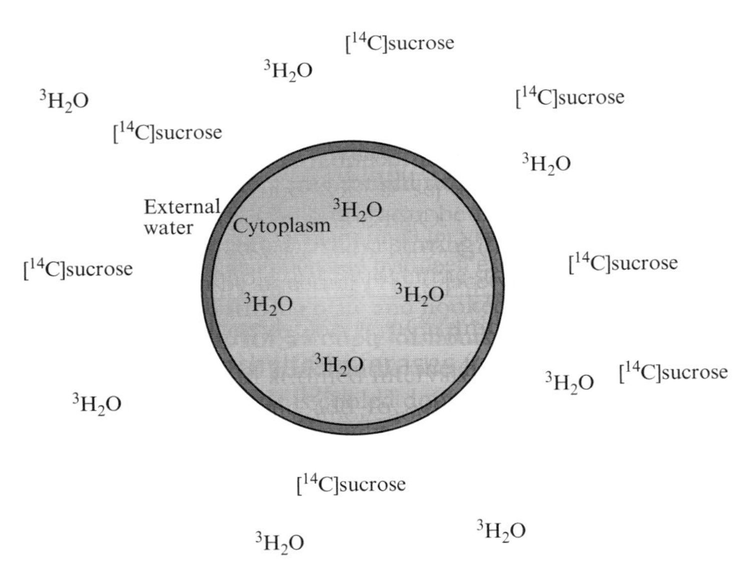

Figure 12.5 Principle to determine the intracellular volume. Cells are incubated with the radiotracers and then separated from the medium by silicone oil centrifugation.

Make sure that the settings of the liquid scintillation counter allow for the simultaneous and nonoverlapping counting of ^{3}H and ^{14}C.

The determination of the intracellular volume [μl/mg protein] is calculated as in Eqs. (12.1) and (12.2) (Rottenberg, 1979):

$$\nu_{\mathrm{e}} = \frac{[^{14}\mathrm{C}]\text{sugar in pellet}}{[^{14}\mathrm{C}]\text{sugar in supernatant}} \nu_{\mathrm{s}} \tag{12.1}$$

$$\nu_{\mathrm{i}} = \left(\frac{^{3}\mathrm{H_2O} \text{ in pellet}}{^{3}\mathrm{H_2O} \text{ in supernatant}} - \frac{[^{14}\mathrm{C}]\text{sugar in pellet}}{[^{14}\mathrm{C}]\text{sugar in supernatant}} \right) \nu_{\mathrm{s}} \tag{12.2}$$

where ν_e, external water; ν_i, intercellular water; ν_s, volume of the supernatant.

4.3. Determining the membrane potential of the cell

The principle of determining the membrane potential is also based on the diffusion of radioactively labeled substances over the cellular membrane (Fig. 12.6). The substance has to be freely diffusible, thus hydrophobic and carry a charge by which it is distributed across the energized membrane. A positively charged compound is used for cells in which the electric field is negative inside. The probe accumulates inside the cells according to the

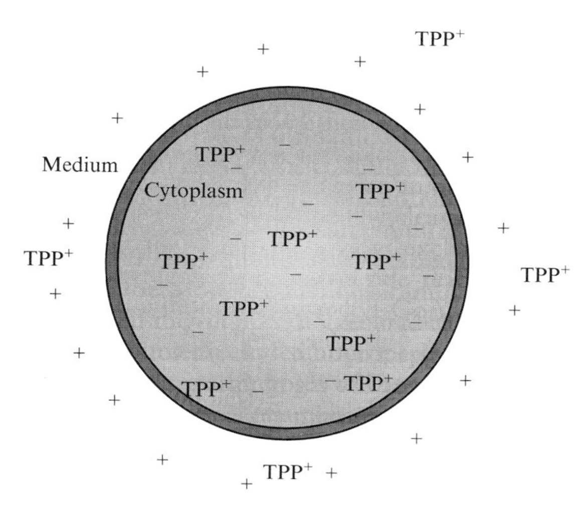

Figure 12.6 Principle to determine the membrane potential across the cytoplasmic membrane. Cells are incubated with the radiotracers and then separated from the medium by silicone oil centrifugation. The lipophilic cation TPP$^+$ (tetraphenylphosphonium) is membrane-permeable and distributes across the membrane according to the Nernst equation.

Nernst equation and the magnitude of accumulation reflects the size of the electrical gradient. It is important to use the probe in micromolar concentration to avoid an uncoupling effect by dissipation of the electrical field. The substance used in this test is $[^{14}C]$ TPP^+ (tetraphenylphosphonium bromide).

1. The preparation of a cell suspension is as described in Section 3.1.
2. The experiments are carried out in 58-ml anaerobic bottles filled with 9–9.5 ml of imidazole/HCl buffer via a 10-ml plastic syringe. The gas atmosphere is oxygen-free N_2. Cell suspension (0.5–1 ml) is added via a plastic syringe to a final volume of 10 ml.
3. To start the measurement, 1 μCi $[^{14}C]$ TPP^+ is added (final concentration of 10 μ*M*) and preincubated for 10 min. Post equilibration, 20 m*M* of methanol is added to induce methanogenesis. After that, 500 μl of sample is taken every 5 min. The cells and the supernatant are separated via silicon oil centrifugation (Section 4.1) and the radioactivity in the supernatant and the pellet is analyzed.

TTP^+ binds unspecifically to membrane components. As this simulates an uptake of TPP^+ in the experiment, the amount of TPP^+ bound to the cell surface has to be determined and deducted from the other values. To do so, the cells are deenergized with the protonophore TCS (4,5,4′,5′-tetrachlorosalicylanilide), incubated with 1 μCi TPP^+, and samples of 500 μl analyzed by silicone oil centrifugation (Section 4.1) and radioactive measurements (Section 4.1.1). The measured radioactivity represents the unspecifically bound TPP^+.

The membrane potential is calculated as in Eq. (12.3):

$$\Delta\psi = \frac{RT}{F}\ln\frac{[^{14}C]TPP_e^+ - [^{14}C]TPP_d^+}{[^{14}C]TPP_s^+ v_i}\frac{v_s}{p} \qquad (12.3)$$

where $[^{14}C]TPP_e^+$, radioactivity of energized cells in the pellet; $[^{14}C]TPP_d^+$, radioactivity of nonenergized cells in the pellet (unspecifically bound TPP^+); $[^{14}C]TPP_s^+$, radioactivity of the supernatant of the energized cells of the chosen aliquot; v_i, intracellular waterspace; v_s, volume of the supernatant of the chosen aliquot; and p, protein amount in the pellet (mg).

4.4. Determining the pH gradient of the cell

The third method using a different ratio of radioactively labeled substances in and outside of the cells is the determination of the intracellular pH. The substance used for this analysis is a radioactively labeled weak acid, which has to meet the following requirements:

1. The free acid in its undissociated, uncharged form must be able to move freely through the cellular membrane.

2. The dissociation constant of the acid must be the same inside and outside of the cell.

The second requirement results in the following assumption (Eq. (12.4)):

$$K_s = \frac{(H^+)_i(A^-)_i}{(AH)_i} = \frac{(H^+)_e(A^-)_e}{(AH)_e} \tag{12.4}$$

Following the first requirement, the concentration of free acid inside and outside of the cell equals: $AH_i = AH_e$. The conversion of this equation provides the basis for ΔpH determination (Eq. (12.5)):

$$\Delta pH = pH_i - pH_e = \log\frac{(A^-)_i}{(A^-)_e} \tag{12.5}$$

If the external pH is more than one unit higher than the pK_s-value of the according weak acid, the amount of undissociated acid is negligible. If the difference is less, the undissociated acid has to be taken into account. In this case, pH_i is calculated as in Eq. (12.6):

$$pH_i = \log\left(\frac{(A_t)_i}{(A_t)_e}(10^{pH_a - pK_s} + 1) - 1\right) + pK_s \tag{12.6}$$

As there is also extracellular water space inside the cell pellet, it has also to be taken into account (Section 4.2). Using the concentrations of the indicator acid inside the pellet and the supernatant as well as the intracellular and extracellular water space, the following relation can be set up (Eq. (12.7)):

$$\frac{(A_t)_i}{(A_t)_e} = \frac{[^{14}C]A_{tp}/{}^3H_2O_p}{[^{14}C]A_{ts}/{}^3H_2O_s}\left(1 + \frac{\nu_e}{\nu_i}\right) - \frac{\nu_e}{\nu_i} \tag{12.7}$$

After insertion of this equation into the former one, the internal pH can be calculated as in Eq. (12.8) (Waddell and Butler, 1959):

$$pH_i = \log\left[\left(\frac{[^{14}C]A_{tp}/{}^3H_2O_p}{[^{14}C]A_{ts}/{}^3H_2O_s}\left(1 + \frac{\nu_e}{\nu_i}\right) - \frac{\nu_e}{\nu_i}\right)(10^{pH_a - pK_s} + 1) - 1\right] + pK_s \tag{12.8}$$

For the following assay, benzoic acid is used as a weak acid (Fig. 12.7).

1. The preparation of a cell suspension is as described in Section 3.1.
2. The experiments are carried out in 58-ml anaerobic bottles filled with 9–9.5 ml of imidazole/HCl buffer via a 10-ml plastic syringe. The gas atmosphere is oxygen-free N_2. Cell suspension (0.5–1 ml) of *M. barkeri* is added via a plastic syringe to a final volume of 10 ml.

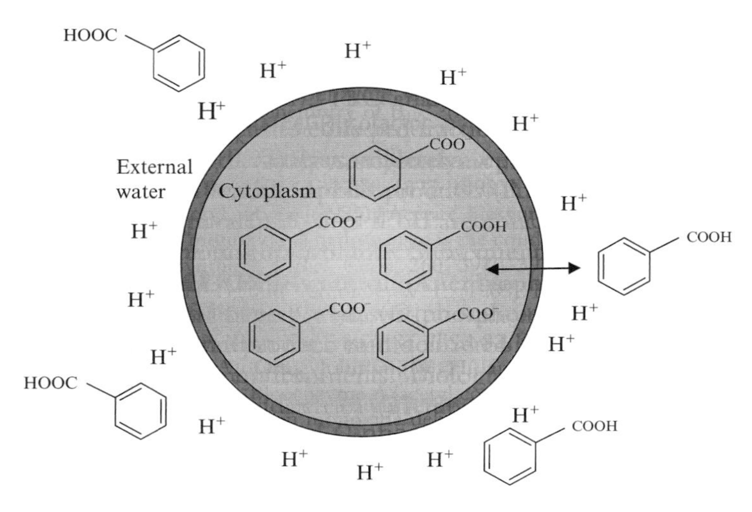

Figure 12.7 Principle to determine the pH gradient across the cytoplasmic membrane. Cells are incubated with the radiotracers and then separated from the medium by silicone oil centrifugation. [^{14}C] benzoic acid diffuses across the membrane, dissociates into protons and benzoate and thus, the intracellular benzoate concentration is a function of the internal pH. For investigations at different pH, other weak acids or bases can be used, such as acetate at low pH or trimethylamine at high pH.

3. For ΔpH determination, 10 μCi ^3H_2O and 1 μCi [7-^{14}C] benzoic acid (final concentration: 27 μ*M*) are used. Post equilibration (10 min), 20 m*M* of methanol is added to induce methanogenesis. Samples of 500 μl are taken every 5 min and analyzed by silicon oil centrifugation, as described previously (Section 4.1).

5. ATP Synthesis

The model organism well established for the analyses of the A_1A_O ATP synthase is *Methanosarcina mazei* Gö1. This mesophilic and halotolerant organism is easy to cultivate and offers the opportunity to do bioenergetic studies using an inverted vesicle system. ATP is determined by the luciferin–luciferase system (Imkamp and Müller, 2002; Kimmich *et al.*, 1975).

5.1. Cell cultivation

M. mazei Gö1 (DSM 7222) is obtained from the DMSZ (Braunschweig, Göttingen). The medium DSMZ 120 without yeast extract or casitone is used for cultivation. The medium contains K_2HPO_4, 2 m*M*; KH_2PO_4,

1.7 m*M*; NH_4Cl, 9.3 m*M*; $MgSO_4 \cdot 7H_2O$, 2 m*M*; $CaCl_2 \cdot 2H_2O$, 2 m*M*; NaCl, 38.5 m*M*; $FeSO_4 \cdot 7H_2O$, 7 μ*M*; vitamin solution (DSM 141), 10 ml/l; trace element solution (SL-10), 1 ml/l; resazurin, 4.4 μ*M*; and $NaHCO_3$, 34 m*M*. The medium is flushed with 80% N_2/20% CO_2 for 20 min until the pH has reached a value between 6.8 and 7. To remove the remaining oxygen, the gas is passed over copper files heated at 370 °C. The reducing agents, cysteine–HCl (2.5 m*M*), $Na_2S \cdot 9H_2O$ (1.2 m*M*), and methanol (150 m*M*), are added from anaerobic stock solutions prior to inoculation. Growth is started by the addition of 10% inoculum. The cells are incubated at 37 °C without shaking. Growth is monitored by measuring the OD at 578 nm.

5.2. Cell suspension

For the preparation of cell suspension, the cells are harvested under strictly anaerobic conditions at the late exponential growth phase ($OD_{578} = 0.8$). Typically, 500-ml cultures are grown.

The cells are filled into 300-ml PPCO centrifuge tubes (Laborgeräte Beranek, Weinheim, Germany) in an anaerobic chamber filled with 95% N_2/5% H_2 (Coy Laboratory Products). For harvest, the cells are centrifuged for 20 min at 8000 rpm at 4 °C in an Avanti® J-25 centrifuge (Beckmann, Munich, Germany). The cells are washed twice with a washing buffer containing 2 m*M* K_2HPO_4, 2 m*M* KH_2PO_4, 10 m*M* NH_4Cl, 2 m*M* $MgSO_4 \cdot 7H_2O$, 2 m*M* $CaCl_2 \cdot 2H_2O$, 38.5 m*M* NaCl, 7 μ*M* $FeSO_4 \cdot 7H_2O$, and 4.4 μ*M* resazurin. The pH of the buffer is adjusted to 6.8 with KOH. To remove the oxygen, the buffer is flushed with N_2 for 20 min. 10 m*M* DTE (Dithioerythritol) is added after flushing to remove the remaining oxygen. After washing the cells, they are resuspended in 10 ml of the same buffer and filled in 16-ml Hungate tubes and stored on ice. The gas phase is exchanged against N_2 by gassing the suspension for 10 min. The protein concentration is determined as described (Schmidt *et al.*, 1963) and typically around 15 mg/ml.

5.3. Assay for determination of ATP synthesis coupled to methanogenesis

1. A 50-ml anaerobic bottle is flushed with N_2 for 20 min to remove the oxygen. Five hundred microliters of concentrated cell suspension is transferred via a 250-μl Hamilton syringe anaerobically to the anaerobic bottle filled with 9.5 ml of buffer containing 2 m*M* K_2HPO_4, 2 m*M* KH_2PO_4, 10 m*M* NH_4Cl, 2 m*M* $MgSO_4 \cdot 7H_2O$, 2 m*M* $CaCl_2 \cdot 2H_2O$, 38.5 m*M* NaCl, 7 μ*M* $FeSO_4 \cdot 7H_2O$, and 4.4 μ*M* resazurin. The pH of the buffer is adjusted to 6.8 with KOH. The assay is incubated at 37 °C for 10 min prior to the experiment on a rotary shaker.

2. After preincubation, 500-μl samples are taken after 0, 5, 8, 10, 12, 14, 16, 18, 20, 25, and 30 min. After 9.5 min, 20 m*M* of methanol is added. The samples are transferred into the tubes containing 200 μl 4 *M* TCA and incubated for 60 min. During this time, they have to be shaken occasionally. It is important that each sample is incubated with TCA for the same time (60 min).
3. After cell lyses, 100 μl of TES-Buffer containing 0.4 *M* Na-TES (*N*-tris (hydroxymethyl)methyl-2-aminoethanesulfonic acid) buffer, pH 7.4, are added and the pH is adjusted to 7 with a saturated solution of K_2CO_3. The pH adjustment is an important step, because of acid-catalyzed hydrolysis of ATP.
4. To remove the originated $KClO_4$, the samples are centrifuged for 10 min at 14,000 rpm (EBA 20, Hettich, Tuttlingen, Germany).
5. The supernatant is removed and stored on ice.

5.4. ATP synthesis by an artificial ΔpH in whole cells

ATP synthesis can be driven by artificial ΔpH's across the membrane (Fig. 12.8). The procedure is given below.

5.4.1. Cell suspension

The cell suspension is produced as described (Section 5.2), with the exception of the washing buffer used in this assay: 20 m*M* imidazole/HCl buffer, pH 6.8, containing 20 m*M* $MgSO_4$ and 6 m*M* DTT. The cells are resuspended in this buffer to a final protein concentration of 30–40 mg/ml. The protein concentration is determined as described (Schmidt *et al.*, 1963).

5.5. ΔpH-driven ATP-synthesis

In this assay, the cells are equilibrated to a neutral pH. To impose a ΔpH across the membrane, HCl is added to different final pH values, creating different ΔpH over the membrane. The H^+ gradient can be used by the ATP synthase for ATP synthesis.

1. A 50-ml anaerobic bottle is flushed with N_2 for 20 min to remove the oxygen. The bottle is filled with 10 ml of 20 m*M* imidazole/HCl buffer, pH 7.0, containing 20 m*M* $MgSO_4$ and 10 m*M* potassium citrate. The buffer is flushed with N_2 for 20 min and reduced with 6 m*M* DTT.
2. The cell suspension is diluted in the buffer to a final protein concentration of 1 mg/ml.
3. The assay is preincubated at 37 °C on a rotary shaker (70 rpm) for 10 min.
4. After preincubation, 500-μl samples are taken after 0, 5, 8, 10, 12, 14, 16, 18, 20, 25, and 30 min. After 9.5 min, HCl is added. It is important

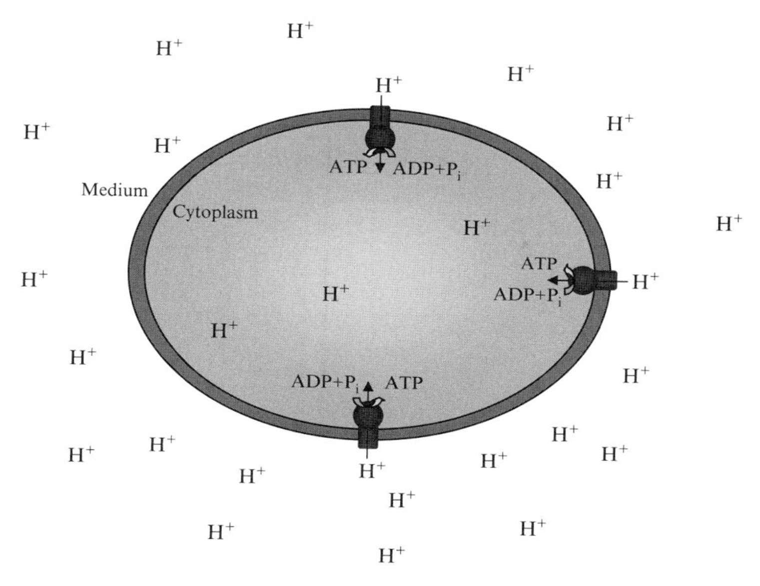

Figure 12.8 Principle of ΔpH-driven ATP synthesis in whole cells. The medium pH is lowered by the addition of HCl. Please note that Na^+ ATP synthases can also use H^+ in the absence of Na^+.

to impose different ΔpH values in this experiment. This can be achieved by adding different amounts of HCl to the assay and using different buffers to adjust different pH values. The final pH has to be determined for every test.

5. Handling of the samples is as described before.

5.6. Measurement of ATP

1. For measurement of ATP, 5–15 μl of sample is placed into a Lumacuvette (Celsis-Lumac, Landgraaf, The Netherlands), containing 250 μl of an ATP determination buffer (5 m*M* $NaHAsO_4$, 4 m*M* $MgSO_4$, 20 m*M* glycylglycine, pH 8).
2. Twenty microliters of firefly crude extract (Sigma, Taufkirchen, Germany) is added and the mixture is vortexed vigorously.
3. The measurement is performed with a Luminometer Junior LB 9509—standard model (Berthold Technologies GmbH & Co. KG, Bad Wildbad, Germany). The luminescence is measured for 10 s. To calculate the total amount of ATP, a standard curve from 2.5 to 40 pmol of ATP per assay is made.

6. ATP Hydrolysis

ATP hydrolysis is measured at membrane preparations of *M. mazei* by determining the amount of P_i produced from ATP over time (Heinonen and Lahti, 1981).

6.1. Cell cultivation and preparation of membranes

The cells are cultivated as described (Section 5.1), and harvested and washed as described previously (Section 5.2). The NaCl of the washing buffer is replaced by 0.1 *M* sucrose to reduce the amount of Na^+. The preparation can be performed aerobically without loss of the enzyme function.

1. Add 100 μ*M* of PMSF (phenylmethanesulfonyl fluoride) and a few crystals of DNase (Roche Diagnostic GmbH, Penzberg, Germany) to the cell suspension prior to disruption.
2. The cells are disrupted by passing them through the French press at 400 PsiG for three times. The solution must be darker afterward.
3. Remove the cell debris by centrifugation for 10 min at 4000×*g*.
4. To separate the membrane fraction, ultracentrifuge the sample for 1 h at 135,000×*g* at 4 °C and wash the preparation twice in 50 m*M* Tris buffer, pH 7.5, containing 5 m*M* $MgCl_2$, 1 m*M* DTE, 0.1 m*M* PMSF, and 10% (v/v) glycerol to remove the phosphate ions of the washing buffer. The membranes are resuspended carefully with a brush in the same buffer to prevent the loss of the A_1 part of the ATP synthase. The protein concentration is determined as described (Bradford, 1976) and the concentration should be around 20 mg/ml.

6.2. Assay for ATP hydrolysis

1. Sixty microliters of the membrane preparation is diluted in 1 ml of 100 m*M* morpholinoethanesulfonic acid (MES) buffer, pH 5.2, containing 40 m*M* Na-acetate, 40 m*M* $Na_2S_2O_5$, 10 m*M* $MgSO_4$, and 10% (v/v) glycerol.
2. This assay is preincubated for 10 min at 37 °C.
3. To start the reaction, 2.5 m*M* of Na_2-ATP is added to the test.
4. Two hundred microliters of sample is taken every 5 min and mixed with 40 μl of 30% (v/v) TCA to stop the reaction.
5. To remove the cell debris, the samples are centrifuged for 3 min at 13,000×*g* at room temperature.
6. Two hundred microliters of the supernatant is transferred into 1 ml of AAM reagent that contains 10 m*M* $(NH_4)_6Mo_7O_{24}$, 5 N H_2SO_4, and

acetone in a ratio of 1:1:2. When preparing the AAM reagent, it is important to mix the acetone with the H_2SO_4 prior to addition of $(NH_4)_6Mo_7O_{24}$.

7. After 10 min of incubation at room temperature, the samples are measured at a wavelength of 355 nm.
8. For determination of phosphate, a standard curve with 0–200 nmol of phosphate per assay is produced.

ACKNOWLEDGMENTS

Financial support of Katharina Schlegel by Biodiversity and Climate Research Centre (BiK-F), Frankfurt a.M., supported by the research funding program "LOEWE—Landes-Offensive zur Entwicklung Wissenschaftlich-ökonomischer Exzellenz" of Hesse's Ministry of Higher Education, Research, and the Arts is gratefully acknowledged. The authors are also grateful to the DFG and Bik-F for financial support of former and present projects.

REFERENCES

Becher, B., Müller, V., and Gottschalk, G. (1992a). The methyltetrahydromethanopterin: Coenzyme M methyltransferase of *Methanosarcina* strain Gö1 is a primary sodium pump. *FEMS Microbiol. Lett.* **91,** 239–244.

Becher, B., Müller, V., and Gottschalk, G. (1992b). N^5-methyl-tetrahydromethanopterin: Coenzyme M methyltransferase of *Methanosarcina* strain Gö1 is a Na^+ translocating membrane protein. *J. Bacteriol.* **174,** 7656–7660.

Blaut, M., Müller, V., and Gottschalk, G. (1987). Proton translocation coupled to methanogenesis from methanol + hydrogen in *Methanosarcina barkeri*. *FEBS Lett.* **215,** 53–57.

Bradford, M. M. (1976). A rapid and sensitive method for the quantification of microgram quantities of protein utilizing the principle of proteine-dye-binding. *Anal. Biochem.* **72,** 248–254.

Deppenmeier, U. (2002). The unique biochemistry of methanogenesis. *Prog. Nucleic Acid Res. Mol. Biol.* **71,** 223–283.

Deppenmeier, U., and Müller, V. (2008). Life close to the thermodynamic limit: How methanogenic archaea conserve energy. *Results Probl. Cell Differ.* **45,** 123–152.

Deppenmeier, U., Blaut, M., Mahlmann, A., and Gottschalk, G. (1990). Reduced coenzyme F_{420}:Heterodisulfide oxidoreductase, a proton-translocating redox system in methanogenic bacteria. *Proc. Natl. Acad. Sci. USA* **87,** 9449–9453.

Ferry, J. G. (1997). Enzymology of the fermentation of acetate to methane by *Methanosarcina thermophila*. *Biofactors* **6,** 25–35.

Gottschalk, G., and Thauer, R. K. (2001). The Na^+-translocating methyltransferase complex from methanogenic archaea. *Biochim. Biophys. Acta* **1505,** 28–36.

Heinonen, J. E., and Lahti, R. J. (1981). A new and convenient colorimetric determination of inorganic orthophosphate and its application to the assay of inorganic pyrophosphatase. *Anal. Biochem.* **113,** 313–317.

Imkamp, F., and Müller, V. (2002). Chemiosmotic energy conservation with Na^+ as the coupling ion during hydrogen-dependent caffeate reduction by *Acetobacterium woodii*. *J. Bacteriol.* **184,** 1947–1951.

Kaesler, B., and Schönheit, P. (1989). The role of sodium ions in methanogenesis. Formaldehyde oxidation to CO_2 and $2H_2$ in methanogenic bacteria is coupled with primary electrogenic Na^+ translocation at a stoichiometry of 2–3 Na^+/CO_2. *Eur. J. Biochem.* **184,** 223–232.

Kimmich, G. A., Randles, J., and Brand, J. S. (1975). Assay of picomole amounts of ATP, ADP and AMP using the luciferase enzyme system. *J. Bacteriol.* **158,** 844–848.

Lienard, T., Becher, B., Marschall, M., Bowien, S., and Gottschalk, G. (1996). Sodium ion translocation by N^5-methyltetrahydromethanopterin:Coenzyme M methyltransferase from *Methanosarcina mazei* Gö1 reconstituted in ether lipid liposomes. *Eur. J. Biochem.* **239,** 857–864.

Ljungdahl, L. G. (1994). The acetyl-CoA pathway and the chemiosmotic generation of ATP during acetogenesis. *In* "Acetogenesis," (H. L. Drake, ed.), pp. 63–87. Chapman & Hall, New York.

Miller, T. L., and Wolin, M. J. (1974). A serum bottle modification of the Hungate technique for cultivating obligate anaerobes. *Appl. Microbiol.* **27,** 985–987.

Müller, V., and Grüber, G. (2003). ATP synthases: Structure, function and evolution of unique energy converters. *Cell. Mol. Life Sci.* **60,** 474–494.

Müller, V., Blaut, M., and Gottschalk, G. (1986). Utilization of methanol plus hydrogen by *Methanosarcina barkeri* for methanogenesis and growth. *Appl. Environ. Microbiol.* **52,** 269–274.

Müller, V., Blaut, M., and Gottschalk, G. (1987). Generation of a transmembrane gradient of Na^+ in *Methanosarcina barkeri*. *Eur. J. Biochem.* **162,** 461–466.

Müller, V., Blaut, M., and Gottschalk, G. (1988). The transmembrane electrochemical gradient of Na^+ as driving force for methanol oxidation in *Methanosarcina barkeri*. *Eur. J. Biochem.* **172,** 601–606.

Perski, H. J., Moll, J., and Thauer, R. K. (1981). Sodium dependence of growth and methane formation in *Methanobacterium thermoautotrophicum*. *Arch. Microbiol.* **130,** 319–321.

Perski, H. J., Schönheit, P., and Thauer, R. K. (1982). Sodium dependence of methane formation in methanogenic bacteria. *FEBS Lett.* **143,** 323–326.

Pisa, K. Y., Weidner, C., Maischak, H., Kavermann, H., and Müller, V. (2007). The coupling ion in methanoarchaeal ATP synthases: H^+ versus Na^+ in the A_1A_O ATP synthase from the archaeon *Methanosarcina mazei* Gö1. *FEMS Microbiol. Lett.* **277,** 56–63.

Ragsdale, S. W. (2008). Enzymology of the Wood-Ljungdahl pathway of acetogenesis. *Ann. N. Y. Acad. Sci.* **1125,** 129–136.

Rottenberg, H. (1979). The measurement of membrane potential and ΔpH in cells, organelles and vesicles. *Methods Enzymol.* **55,** 547–569.

Saum, R., Schlegel, K., Meyer, B., and Müller, V. (2009). The F_1F_O ATP synthase genes in *Methanosarcina acetivorans* are dispensable for growth and ATP synthesis. *FEMS Microbiol. Lett.* **300,** 230–236.

Scherer, P., and Sahm, H. (1981). Influence of sulphur-containing compounds on the growth of *Methanosarcina barkeri* in defined medium. *Eur. J. Appl. Microbiol. Biotechnol.* **12,** 28–35.

Schmidt, K., Liaanen-Jensen, S., and Schlegel, H. G. (1963). Die Carotinoide der *Thiorodaceae*. *Arch. Mikrobiol.* **46,** 117–126.

Schmidt, S., Pflüger, K., Kögl, S., Spanheimer, R., and Müller, V. (2007). The salt-induced ABC transporter Ota of the methanogenic archaeon *Methanosarcina mazei* Gö1 is a glycine betaine transporter. *FEMS Microbiol. Lett.* **277,** 44–49.

Schönheit, P., Moll, J., and Thauer, R. K. (1980). Growth parameters (K_s, μ_{max}, Y_s) of *Methanobacterium thermoautotrophicum*. *Arch. Microbiol.* **127,** 59–65.

Terlesky, K. C., and Ferry, J. G. (1988). Ferredoxin requirement for electron transport from the carbon monoxide dehydrogenase complex to a membrane-bound hydrogenase in acetate-grown *Methanosarcina thermophila*. *J. Biol. Chem.* **263,** 4075–4079.

Thauer, R. K. (1990). Energy metabolism of methanogenic bacteria. *Biochim. Biophys. Acta* **1018,** 256–259.

Thauer, R. K., Kaster, A. K., Seedorf, H., Buckel, W., and Hedderich, R. (2008). Methanogenic archaea: Ecologically relevant differences in energy conservation. *Nat. Rev. Microbiol.* **6,** 579–591.

van der Meijden, P., Heythuysen, H. J., Pouwels, F. P., Houwen, F. P., van der Drift, C., and Vogels, G. D. (1983). Methyltransferase involved in methanol conversion by *Methanosarcina barkeri. Arch. Microbiol.* **134,** 238–242.

Waddell, W. J., and Butler, T. C. (1959). Calculation of intracellular pH from the distribution of 5,5-dimethyl-2,4-oxazolidinedione (DMO). Application to skeletal muscle of the dog. *J. Clin. Invest.* **38,** 720–729.

Winner, C., and Gottschalk, G. (1989). H_2 and CO_2 production from methanol or formaldehyde by the methanogenic bacterium strain Gö1 treated with 2-bromoethansulfonic acid. *FEMS Microbiol. Lett.* **65,** 259–264.

Wolin, E. A., Wolfe, R. S., and Wolin, M. J. (1964). Viologen dye inhibition of methane formation by *Methanobacillus omelianskii. J. Bacteriol.* **87,** 993–998.

Zehnder, A. J., and Wuhrmann, K. (1976). Titanium (III) citrate as a nontoxic oxidation-reduction buffering system for the culture of obligate anaerobes. *Science* **194,** 1165–1166.

CHAPTER THIRTEEN

Proton Translocation in Methanogens

Cornelia Welte *and* Uwe Deppenmeier

Contents

Abstract

Methanogenic archaea of the genus *Methanosarcina* possess a unique type of metabolism because they use $H_2 + CO_2$, methylated C_1-compounds, or acetate as energy and carbon source for growth. The process of methanogenesis is fundamental for the global carbon cycle and represents the terminal step in the anaerobic breakdown of organic matter in freshwater sediments. Moreover, methane is an important greenhouse gas that directly contributes to climate change and global warming. *Methanosarcina* species convert the aforementioned substrates to CH_4 via the CO_2-reducing, the methylotrophic, or the aceticlastic pathway. All methanogenic processes finally result in the oxidation of two thiol-containing cofactors (HS-CoM and HS-CoB), leading to the formation of the so-called heterodisulfide (CoM-S-S-CoB) that contains an intermolecular disulfide bridge. This molecule functions as the terminal electron acceptor of a branched respiratory chain. Molecular hydrogen, reduced coenzyme F_{420}, or reduced ferredoxin are used as electron donors. The key enzymes of the

Institute of Microbiology and Biotechnology, University of Bonn, Bonn, Germany

Methods in Enzymology, Volume 494
ISSN 0076-6879, DOI: 10.1016/B978-0-12-385112-3.00013-5

respiratory chain (Ech hydrogenase, F_{420}-nonreducing hydrogenase, $F_{420}H_2$ dehydrogenase, and heterodisulfide reductase) couple the redox reactions to proton translocation across the cytoplasmic membrane. The resulting electrochemical proton gradient is the driving force for ATP synthesis. Here, we describe the methods and techniques of how to analyze electron transfer reactions, the process of proton translocation, and the formation of ATP.

1. Introduction

Methanogenic archaea produce methane as the major metabolic end product and combine the pathway of methanogenesis with a unique energy-conserving system. Biological methane formation is one of the most important processes for the maintenance of the global carbon flux, because these organisms perform the terminal step in the mineralization of organic material in many anaerobic environments. In addition, methane is an important greenhouse gas that significantly contributes to climate change and global warming. In this context, it is important to note that the atmospheric methane concentration increased more than 2.5-fold during the past 150 years due to anthropogenic effects including expanded cultivation of rice and ruminant life stock (Jain *et al.*, 2004; Khalil and Rasmussen, 1994). Thus, it is essential to consider the future atmospheric CH_4 budgets for the stabilization of greenhouse gas concentrations. However, the formation of biogenic methane as part of the so-called biogas is of great economical importance. In recent decades, the combustion of biogas, produced by fermenting biomass, to generate electricity in combined heat and power stations, has largely expanded (Kashyap *et al.*, 2003).

Most methanogenic archaea can use $H_2 + CO_2$ as substrates. However, only methylotrophic methanogens (e.g., the genus *Methanosarcina*) are able to produce methane from methylated C_1-compounds such as methanol, methylamines, or methylthiols. Furthermore, only the genera *Methanosarcina* (*Ms.*) and *Methanosaeta* are able to use acetic acid for methanogenesis and growth (Boone *et al.*, 1993). Acetate is converted to methane and CO_2 and creates approximately 70% of the biologically produced methane (Ferry and Lessner, 2008). All methanogenic substrates are formed by a complex degradation process of organic matter as catalyzed by fermentative and synthrophic bacteria (McInerney *et al.*, 2009). The substrates are converted to CH_4 via the CO_2-reducing, the methylotrophic, or the aceticlastic pathway of methanogenesis (Deppenmeier, 2002b). The biochemistry of methanogenesis clearly indicates that there is no site for ATP regeneration by substrate-level phosphorylation. Therefore, it has been assumed for quite some time that membrane-bound electron transport might be coupled to ion translocation and that the ion motive force generated might be used for ATP synthesis (Thauer *et al.*, 1977).

The process of methanogenesis is very complex and involves many unusual enzymes and cofactors (Deppenmeier and Müller, 2008; Ferry, 1999; Thauer, 1998; Thauer *et al.*, 2008; Wolfe, 1985). However, all catabolic processes ultimately lead to the formation of methyl-2-mercaptoethanesulfonate (methyl-S-CoM; Fig. 13.1), which is then reduced to CH_4 by the catalytic activity of the methyl-S-CoM reductase. The enzyme uses *N*-7-mercaptoheptanoyl-L-threonine phosphate (HS-CoB; Fig. 13.1) as the electron donor; hence, the final product of this reaction is the heterodisulfide of HS-CoM and HS-CoB (CoM-S-S-CoB). This compound functions as electron acceptor of the methanogenic respiratory chain (Fig. 13.2). Molecular hydrogen, reduced coenzyme F_{420} (Fig. 13.1), or reduced ferredoxin are used as electron donors. The redox reactions, catalyzed by the membrane-bound electron transport chains, are coupled to proton translocation across the cytoplasmic membrane (Fig. 13.2). The resulting electrochemical proton gradient is the driving force for ATP synthesis as catalyzed by an A_1A_O-type ATP synthase (Deppenmeier and Müller, 2008; Pisa *et al.*, 2007).

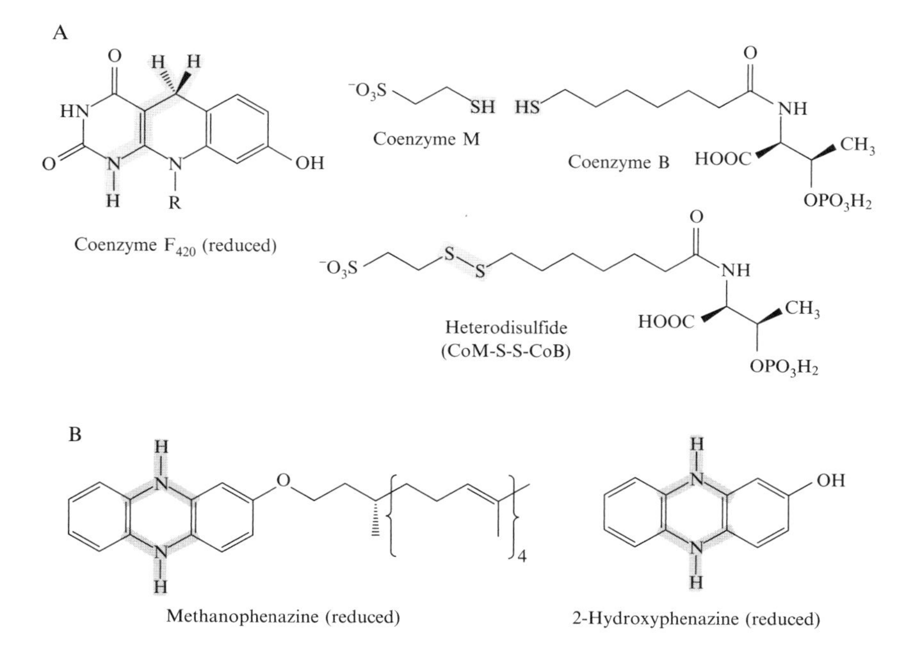

Figure 13.1 Electron carriers in *Ms. mazei.*

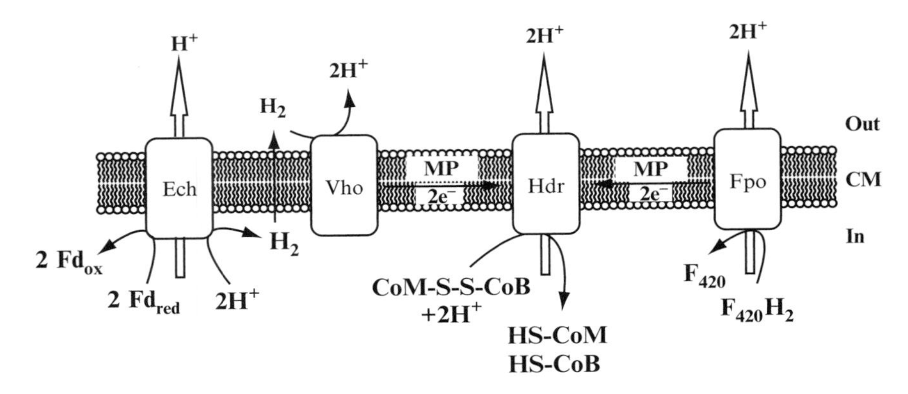

Figure 13.2 Membrane-bound electron transfer system of *Ms. mazei.* Ech, Ech hydrogenase; Vho, F_{420}-nonreducing hydrogenase; Fpo, $F_{420}H_2$ dehydrogenase; Hdr, heterodisulfide reductase; MP, methanophenazine.

2. Analysis of Membrane-Bound Electron Transport in *Methanosarcina mazei*

The question as to the mechanism of ATP synthesis in methanogenic archaea has been investigated in whole cells (Blaut and Gottschalk, 1984; Blaut *et al.*, 1987; Peinemann *et al.*, 1988) or in protoplasts (Peinemann *et al.*, 1989). The results obtained were in accordance with a chemiosmotic mechanism of ATP synthesis. Further details of the energy metabolism were obtained when reduction of CoM-S-S-CoB with H_2 was studied using subcellular membrane preparations of *Ms. mazei* (Deppenmeier *et al.*, 1992).

2.1. Preparation of cytoplasmic membranes

In *Ms. mazei*, all enzymes involved in energy transduction are tightly membrane-bound (Fig. 13.2). Hence, for the analysis of electron transfer events, the preparation of cytoplasmic membranes is necessary. Washing steps of the membrane fraction are performed to remove all soluble cellular components.

2.1.1. Procedure

Ms. mazei is grown anaerobically at 37 °C on the media described previously, supplemented with 1 g L^{-1} of sodium acetate (Hippe *et al.*, 1979). The cultures are typically grown with trimethylamine as substrate (50 m*M* final concentration; doubling time 7.5 h). Other carbon and energy sources can be used at the following final concentrations: methanol (150 m*M*, doubling time 13 h), dimethylamine (100 m*M*), monomethylamine (150 m*M*), or sodium acetate (100 m*M*, doubling time 28 h).

Subsequent preparation steps are performed anaerobically in an anaerobic chamber (Coy Laboratory Products, Michigan, USA) under a 97% N_2/3% H_2 atmosphere. Exponentially grown cells are filled in gas-tight centrifugation bottles (500 mL) and harvested by centrifugation (3000×*g*, 30 min). The cell pellet is resuspended in a small volume 40 m*M* potassium phosphate (K-phosphate) buffer, pH 7.0, 5 m*M* dithioerythritol, and 1 μg mL^{-1} resazurin. In the course of resuspension, the cells lyze because the protein cell wall is not stabile in this buffer. After incubated with DNase I for 30 min at 4 °C, the cell lysate is ultracentrifuged (90 min, 150,000×g) and the supernatant is carefully removed and stored at −20 °C (e.g., for a subsequent isolation of coenzyme F_{420}). The membrane pellet is homogenized in 20 m*M* K-phosphate buffer (pH 7.0) containing 20 m*M* $MgSO_4$, 500 m*M* sucrose, 5 m*M* dithioerythritol, and 1 μg mL^{-1} resazurin. The ultracentrifugation step is repeated, the supernatant discarded, and the membrane pellet again homogenized in the previously mentioned buffer (300 μL L^{-1} culture). The membrane fraction is finally checked for remaining contamination of soluble compounds of the cells by measuring the activity of the CO dehydrogenase, which is only present in the cytoplasm of *Ms. mazei*.

To assay for CO dehydrogenase activity, an anaerobic rubber-stoppered glass cuvette containing 700 μL K-phosphate buffer (40 m*M*, pH 7.0, 5 m*M* dithioerythritol, 1 μg mL^{-1} resazurin) is flushed with 5% CO/95% N_2. Methyl viologen is added to a final concentration of 5 m*M* ($\varepsilon_{604} = 13.6\ cm^{-1}\ mM^{-1}$). Upon addition of the prepared washed membrane fraction, no activity should occur if the membrane fraction is indeed cytoplasm-free.

2.2. Analysis of the H_2-dependent electron transport chain

The source of reducing equivalents necessary for the reduction of CoM-S-S-CoB depends on the growth substrate. If molecular hydrogen is present, a membrane-bound F_{420}-nonreducing hydrogenase channels electrons via b-type cytochromes to the heterodisulfide reductase, which reduces the terminal electron donor (Fig. 13.2) (Deppenmeier *et al.*, 1992). This electron transport system was referred to as H_2:heterodisulfide oxidoreductase. Later on it was found that the electron transport between the enzymes is mediated by the novel redox active cofactor methanophenazine (Figs. 13.1 and 13.2; Abken *et al.*, 1998; Beifuss and Tietze, 2000, 2005). With the identification of methanophenazine as an electron carrier in the membrane of *Ms. mazei*, it became evident that the key enzymes of the membrane-bound electron transfer systems were able to interact with 2-hydroxyphenazine which is a water-soluble homolog of methanophenazine

(Bäumer *et al.*, 2000). To analyze the key enzymes of this system, the following procedure can be followed.

2.2.1. F_{420}-nonreducing hydrogenase

Several types of nickel–iron hydrogenases were described in methanogenic archaea (Vignais *et al.*, 2001). Among them is the F_{420}-nonreducing hydrogenase, which is composed of a small electron transfer subunit and a large catalytic subunit harboring a nickel–iron center. Both subunits are located at the outer leaflet of the cytoplasmic membrane. It has been shown that the metals in the bimetallic reaction center are directly involved in H_2 cleavage to $2e^-$ and $2H^+$. The small subunit contains three Fe/S clusters and is responsible for electron transport from the catalytic center to the third subunit, which is a b-type cytochrome. This subunit forms the membrane anchor and transfers electrons to methanophenazine (Deppenmeier, 2002a).

2.2.1.1. Procedure The membrane-bound hydrogenase is assayed in 1.5-mL glass cuvettes gassed with H_2 and closed with rubber stoppers. After gassing for 5 min, the cuvette is filled with 1 mL anaerobic 40 m*M* K-phosphate buffer, pH 7.0, containing 5 m*M* dithiothreitol and 1 μg mL^{-1} resazurin, and 10–15 μg of the membrane fraction is added. The reaction is started by addition of 10 m*M* methyl viologen ($\varepsilon_{604} = 13.6\ cm^{-1}\ mM^{-1}$). The H_2-dependent 2-hydroxyphenazine reduction is followed photometrically at 425 nm ($\varepsilon = 4.5\ mM^{-1}\ cm^{-1}$) by adding 4 μL of 2-hydroxyphenazine from a 20 m*M* stock solution in ethanol. The activity should be between 1.5 and 2 U mg^{-1} membrane protein.

2.2.2. Heterodisulfide reductase

The heterodisulfide reductase catalyzes the reduction of CoM-S-S-CoB, which is the final step in the anaerobic respiratory chain of *Ms. mazei* and other *Methanosarcina* strains (Hedderich *et al.*, 2005; Heiden *et al.*, 1994; Künkel *et al.*, 1997). *In vivo* electrons for CoM-S-S-CoB reduction derive from the reduced form of methanophenazine (Figs. 13.1 and 13.2). *In vitro*, the enzyme accepts electrons from the artificial electron donors, benzyl viologen and 2-hydroxyphenazine (Fig. 13.2).

2.2.2.1. Procedure The benzyl viologen-dependent heterodisulfide reductase activity is determined at room temperature in 1 mL 40 m*M* K-phosphate buffer, pH 7.0, preflushed with N_2. After addition of 2 μL 1 *M* benzyl viologen, 3 μL 50 m*M* Na-dithionite, and 1–5 μL of the membrane preparation, the reaction is started by addition of CoB-S-S-CoM to a final concentration of 180 μ*M* and followed at 575 nm ($\varepsilon_{benzyl\ viologen} = 8.9\ mM^{-1}\ cm^{-1}$). Typical activities range between 1 and 2 U mg^{-1} membrane protein.

2-hydroxyphenazine (0.25 mM final concentration) is reduced in 25 mM MOPS buffer (pH 7.0) containing 2 mM dithioerythritol, 1 μg mL^{-1} resazurin containing platinum(VI)-oxide (2 mg/40 mL) under a hydrogen atmosphere overnight. After reduction is completed, the catalyst is removed by centrifugation in an anaerobic chamber and the resulting solution is kept under nitrogen. For the determination of membrane-bound heterodisulfide reductase activity, 25 mM MOPS buffer (pH 7.0) containing 2 mM dithioerythritol, 1 μg mL^{-1} resazurin, and the reduced 2-hydroxyphenazine solution are mixed in a 1:1 ratio (final volume 1 mL). After washed membranes are added, the reaction is started by addition of CoB-S-S-CoM to a final concentration of 180 μM and followed at 425 nm ($\varepsilon_{2\text{-hydroxyphenazine}} = 4.5$ mM^{-1} cm^{-1}). Typically, the activity is in the range of 2 U mg^{-1} membrane protein.

2.2.3. H_2:Heterodisulfide oxidoreductase (coupled reaction of F_{420}-nonreducing hydrogenase and heterodisulfide reductase)

The entire electron transport within the H_2:heterodisulfide oxidoreductase system (Fig. 13.2) cannot be followed by photometric methods. However, the end product of the reaction, the thiols HS-CoM and HS-CoB (Fig. 13.2), can be determined using the so-called Ellman's test in which 5,5′-dithiobis(2-nitrobenzoic acid) (DTNB, also known as Ellman's reagent) reacts with mercaptans, like HS-CoM and HS-CoB, under alkaline conditions (Ellman, 1958; Zahler and Cleland, 1968). In the course of the reaction, the yellow *p*-nitrobenzenethiol anion is released, which has a molar extinction coefficient of 13.6 mM^{-1} cm^{-1} at 412 nm.

2.2.3.1. *Procedure*

1. Rubber-stoppered glass vials are flushed with N_2, and 250 μL K-phosphate buffer (40 mM, pH 7.0, 1 μg mL^{-1} resazurin, flushed with N_2) is added. The solution is reduced until it just becomes colorless (using titanium(III) citrate). Approximately 150 μg washed *Ms. mazei* membranes are added, and the whole vial is flushed with 100% H_2.
2. The reaction is started by addition of 400 μM heterodisulfide and the first sample is directly taken: a 20 μL sample of the reaction mixture is mixed with 950 μL Tris–HCl (150 mM, pH 8.1) and 100 μL DTNB (5 mM in 50 mM Na-acetate, pH 5.0) and the absorption at 412 nm is immediately measured.
3. Samples are continuously retrieved until the heterodisulfide is completely reduced ($\Delta A \sim 0.2$). Typically about 300 nmol thiols are produced per minute per milligram membrane protein.

Note: A buffer without mercaptan-containing reducing agent is needed for the assay, thus the buffer must not be reduced with dithioerythritol or dithiothreitol. Instead, titanium(III)citrate (Zehnder and Wuhrmann, 1976) or other mercaptan-free reducing agent can be used. The titanium (III)citrate solution is prepared as follows: 15 mmol potassium citrate is solvated in 48.75 mL H_2O and gassed with N_2 for 10 min. Carefully, 9.38 mL of a 12% Ti(III)chloride solution (in 12% HCl) are added under nitrogen gassing. The pH is adjusted to 7 with a saturated K_2CO_3 solution.

2.2.4. Preparation of the heterodisulfide (CoM-S-S-CoB)

The heterodisulfide CoM-S-S-CoB is not commercially available and needs to be synthesized by condensation of HS-CoB and HS-CoM. HS-CoB has to be synthesized *de novo* (Ellermann *et al.*, 1988), whereas HS-CoM (2-mercaptoethanesulfonate) is available from various suppliers.

A. *Synthesis of* N*-hydroxysuccinimidester according to Noll* et al. *(1987)*

1. In a 500-mL Erlenmeyer flask, 200 mL pure ethanol, 17.8 g 7-bromoheptanoic acid (85.3 mmol), and 32.4 g thiourea (426.5 mmol) are added. The mixture is refluxed at 78 °C for 16 h.
2. The solution is cooled to room temperature and alkalized with 50 mL NaOH (60%, w/v). The mixture is again refluxed at 78 °C for 16 h. After cooling to room temperature, the solution is acidified with HCl to a pH < 1.
3. The solution is extracted three times with 50 mL chloroform (total 150 mL). The chloroform phases are combined and extracted with 1 *M* aqueous Na_2CO_3. The aqueous phase is acidified with 50 mL HCl (37%) and extracted with chloroform (2×100 mL). An aqueous solution of iodine (10%, w/v)/potassium iodide (20%, w/v) is added until a light brown color appears.
4. The aqueous phase is removed, and the chloroform phase is washed with water three times. Then, the solution is dried with anhydrous $MgSO_4$ and concentrated under reduced pressure.
5. The resulting reddish-brown substance is solvated in benzene/pentane (1:1, v/v) under gentle warming, and the substance is allowed to crystallize at 4 °C. The resulting crystals are filtered and recrystallized using the same solvents. Light yellow dithiodiheptanoic acid crystals form. The yield should be in the range of 4–5 g.
6. The disulfide product is activated by synthesis of its *N*-hydroxysuccinimide ester using dicyclohexylcarbodiimide (DCCD). In all, 2 g dithiodiheptanoic acid, 1.5 g *N*-hydroxysuccinimide, and 2.6 g DCCD are solvated in 60 mL anhydrous 1,4-dioxane and incubated for 20 h at room temperature under stirring. The suspension is filtered, the solid matter (dicyclohexylurea) is discarded, and the filtrate is dried under vacuum to give a white solid. The white solid is recrystallized twice using 2-propanol to yield approximately 2 g *N*-hydroxysuccinimidester.

B. *Synthesis of HS-CoB according to Kobelt* et al. *(1987)*

1. L-Threonine phosphate (546 mg) (2.74 mmol) is solvated in 5 mL H_2O including 760 μL triethylamine. The mixture is stirred and 672 mg (1.30 mmol) *N*-hydroxysuccinimide ester solvated in 40 mL tetrahydrofuran and 14 mL acetonitrile are added. The mixture is stirred for 36 h under argon and then concentrated using a rotary evaporator. The volume is brought to approximately 40 mL with H_2O. The pH is adjusted to 6.7 with 1 *M* NaOH or HCl and the solution gassed with N_2. The disulfide compounds are reduced by the addition of $NaBH_4$. The pH is titrated to pH 0 with 1 *M* HCl.
2. For a hydrophobic interaction chromatography, an XAD-2 column (2.5×8 cm) is prepared. The column is washed with water and equilibrated with 1 *M* HCl before the reaction mixture is applied. The column is then washed with 100 mL HCl (1 *M*) and 140 mL H_2O.
3. HS-CoB is eluted using a linear gradient of 0–50% methanol/H_2O (100 mL). The thiol-containing elution fractions are pooled and concentrated under vacuum. To analyze the thiol content, the following test is performed: Mix 950 μL Tris–HCl (150 m*M* pH 8.1), 50–100 μL sample, and a few grains of $NaBH_4$. Let the mixture stand open for 10 min, add 100 μL acetone and 100 μL DTNB solution (5 m*M* in 50 m*M* Na-acetate pH 5.0), and measure at 412 nm against a control without sample (Ellmann, 1958; Zahler and Cleland, 1968).

C. *Synthesis of the heterodisulfide of HS-CoB and 2-mercaptoethanesulfonate (HS-CoM) according to Ellermann* et al. *(1988)*

1. The pH of the HS-CoB solution (Coenzyme B solution) is adjusted to 6.7 with 1 *M* NaOH/HCl and reduced with a few grains of $NaBH_4$ under a N_2 atmosphere. The approximate content of HS-CoB is calculated using the thiol test as described above (MW (HS-CoB) = 395 g mol^{-1}) and a fivefold access of HS-CoM is added. The pH is adjusted to 9 (1 *M* NaOH) and the reaction mixture is gassed with air for 16 h at room temperature to allow the formation of the mixed disulfide.
2. After adjustment to pH 0 (1 *M* HCl), the preparation is applied to a XAD-2 column, equilibrated as described above. The column is washed with 150 mL 1 *M* HCl, 250 mL H_2O (elution of HS-CoM and CoM-S-S-CoM). The heterodisulfide CoM-S-S-CoB is eluted with 250 mL 20% methanol, and subsequently concentrated under vacuum, frozen, and lyophilized.
3. The content of biologically active heterodisulfide is determined with a biological test: 700 μL K-phosphate buffer (pH 7.0, 5 m*M* dithioerythritol, 1 μg mL^{-1} resazurin), 5 m*M* methyl viologen (ε_{604} = 13.6 mM^{-1} cm^{-1}), and 50–100 μg *Ms. mazei* cytoplasmic membrane preparation (see above)

are mixed in an anaerobic rubber-stoppered cuvette. The reaction mixture is titrated with Na-dithionite till an absorption of about 2 at 604 nm is reached. The test is started with a small amount of the synthesized heterodisulfide (solvated in H_2O). From the change of absorbance at 604 nm, the approximate content can be calculated. The typical yield is 15–50 μmol heterodisulfide.

2.2.5. Preparation of 2-hydroxyphenazine according to Ott (1959)

It was found that the condensation of *p*-benzoquinone with o-phenylenediamine in ethanolic solution leads directly to 2-hydroxyphenazine.

1. Five hundred milligrams (4.6 mmol, 1.0 equiv.) of 1,4-benzoquinone is solvated in 5 mL of dry ethanol by slight warming.
2. Cool to room temperature, add 500 mg (4.6 mmol, 1.0 equiv.) of o-phenylenediamine, and incubate for 1 h at ambient temperature.
3. Heat to 75 °C and incubate for another 30 min.
4. Cool to room temperature and add 1 mL of H_2O; incubate for 8 h at 4 °C.
5. Filter the cooled solution and solvate the brownish-black crystals in 0.5 *M* NaOH.
6. Filter again and acidify the filtrate with concentrated acetic acid. A brownish compound will precipitate.
7. Repeat the filtration and dry the precipitation under vacuum.
8. Purify the compound by silica gel chromatography (silica gel 60, mesh 0.05–0.20 mm, *Merck*) with a linear gradient of the solvent system diethylether/petroleum ether (3:1, v/v)/diethylether.

2-Hydroxyphenazine is a slight yellow solid. The yield is about 10%.

The natural electron carrier methanophenazine can also be chemically synthesized (Beifuss and Tietze, 2000; Beifuss *et al.*, 2000; Beifuss and Tietze, 2005) It is a multistep chemical synthesis and is not described here.

2.3. Analysis of the $F_{420}H_2$-dependent electron transport chain

During methanogenesis from methylated C_1-compounds in the absence of H_2, one quarter of the methyl groups is oxidized to CO_2 and reducing equivalents are in part transferred to the central electron carrier F_{420}. In the reductive branch of the pathway, three of four methyl groups are transferred to coenzyme M. Methyl-S-CoM is reductively cleaved by methyl-S-CoM reductase, forming methane and CoB-S-S-CoM (Deppenmeier, 2002a). Thus, $F_{420}H_2$ and CoB-S-S-CoM are generated as electron donor and acceptor for the electron transport chain of the membrane-bound $F_{420}H_2$:heterodisulfide oxidoreductase system. Reduced F_{420} is oxidized by $F_{420}H_2$

dehydrogenase and CoB-S-S-CoM is reduced by heterodisulfide reductase, respectively (Fig. 13.2). These two enzymes are interconnected by the membrane-soluble electron carrier methanophenazine (Deppenmeier, 2002a).

2.3.1. Photometric analysis

$F_{420}H_2$ dehydrogenase activity is determined at room temperature in glass cuvettes under N_2. The 700 μL anaerobic assay mixture routinely contains 40 m*M* K-phosphate buffer, pH 7.0, 20 m*M* $MgSO_4$, 0.5 *M* sucrose, 1 μg mL^{-1} resazurin, 10 m*M* dithioerythritol, and 16 μ*M* $F_{420}H_2$. The $F_{420}H_2$ dehydrogenase reaction is started with 0.5 m*M* metronidazole and 0.3 m*M* methyl viologen. Metronidazole reoxidizes reduced methyl viologen, which is necessary because the concentration of reduced methyl viologen has to be kept low to prevent the spectroscopic interference with F_{420}. The oxidation of $F_{420}H_2$ is followed at 420 nm ($\varepsilon = 40\ mM^{-1}\ cm^{-1}$). The activity is about 50 mU mg^{-1} membrane protein.

$F_{420}H_2$:heterodisulfide oxidoreductase activity is determined similarly to the $F_{420}H_2$ dehydrogenase activity, except that metronidazole and methyl viologen are replaced by the intrinsic electron acceptor heterodisulfide to a final concentration of 20 μ*M*. Typical activities range between 50 and 200 mU mg^{-1} membrane protein. As the $F_{420}H_2$ dehydrogenase interacts with the membrane-soluble methanophenazine pool, the terminal electron acceptor heterodisulfide can be replaced by the intermediary electron acceptor 2-hydroxyphenazine, which is used at a final concentration of 50 μ*M*. An activity of about 200 mU mg^{-1} membrane protein is to be expected.

2.3.2. Purification and reduction of coenzyme F_{420}

The ultracentrifugation supernatants from membrane or vesicle preparations contain all cytoplasmic compounds and can be used for the preparation of F_{420}. Alternatively, cells of *Ms. barkeri* or *Ms. mazei* can be used. The preparations are suspended under N_2 in 40 m*M* NH_4–acetate buffer, pH 4.8, containing 10 m*M* 2-mercaptoethanol. The preparations are boiled in a water bath for 30 min under an atmosphere of argon, sedimented by centrifugation at 25,000×*g* for 30 min, and resuspended in the same buffer. Boiling and centrifugation is repeated, and the supernatants of both extractions are combined and applied to a DEAE-Sephadex A25 anion exchange column, which has been equilibrated with a 40 m*M* NH_4–acetate buffer, pH 4.8. A linear 1 L gradient (0–1 *M* NH_4Cl in NH_4–acetate buffer) is applied and fractions containing F_{420} are detected by their absorption at 420 nm. Pooled fractions are diluted 10-fold with 50 m*M* Tris–HCl buffer, pH 7.5, and loaded onto a second DEAE-Sephadex column (8×4 cm). The cofactor is eluted with a 0–1 *M* NaCl gradient in the same buffer. The salt content of the green fluorescent solution of F_{420} is reduced by diluting with 50 m*M* Tris–HCl buffer, pH 7.5, and the preparation is placed on a QAE-Sephadex column (2.6×7 cm) that has been equilibrated with the same buffer. A 0.5 L gradient of 0–2 *M* NaCl is passed through the column. Finally, the F_{420}-containing fractions are adjusted to pH 1 with 6 *M* HCl and applied to a XAD-4 column (2.6×5 cm). The preparation is desalted by applying 100 mL 1 *M* HCl and F_{420} is eluted with a linear methanol gradient (0–100%). The F_{420}-containing fractions are combined, neutralized with NaOH, and concentrated by rotary evaporation.

Coenzyme F_{420} is reduced under a flow of N_2 by adding $NaBH_4$ and incubating at room temperature until the solution turns colorless. Excess $NaBH_4$ is removed by adding anaerobic HCl to a final pH of 0. Anaerobic NaOH is used to readjust to pH 7.0. As the volume of this reaction is rather low, care must be taken not to lose significant amounts of material due to monitoring pH. Ideally, only 1 μL of the solution is applied to an appropriate pH test strip. After readjustment of the pH, the vial is sealed with a rubber stopper and is used the same day. The $F_{420}H_2$ solution is not suitable for storage and will slowly oxidize.

2.4. Analysis of the ferredoxin-dependent electron transport chain

The major part of methane produced by the anaerobic breakdown of organic matter derives from the methyl group of acetate (Ferry and Lessner, 2008). The process is referred to as the aceticlastic pathway of methanogenesis. Only species of the genera *Methanosarcina* and *Methanosaeta* are known to produce CH_4 and CO_2 from acetate. *Methanosarcina* species activate acetate by conversion to acetyl-CoA. The key enzyme in the aceticlastic pathway is the CO-dehydrogenase/acetyl-CoA synthase which is responsible for the cleavage of the C–C and C–S bonds in acetyl-CoA (Gencic *et al.*, 2010; Gong *et al.*, 2008; Ragsdale, 2008). This enzyme also catalyzes the oxidation of CO to CO_2 and the electrons are used for ferredoxin reduction (Clements *et al.*, 1994). The methyl group of acetyl-CoA is transferred to H_4SPT, forming methyl-H_4SPT (Fischer and Thauer, 1989), which is converted to methane by the Na^+-translocating methyl-H_4MPT:HS-CoM methyltransferase (Gottschalk and Thauer, 2001) (see Chapter 12 of this volume) and the methyl-S-CoM reductase as described above (Ebner *et al.*, 2010). Again, the heterodisulfide CoM-S-S-CoB is formed and is reduced by a membrane-bound electron transport system (Fig. 13.2). In *Ms. mazei* and *Ms. barkeri*, the aceticlastic respiratory chain comprises the three transmembrane proteins: Ech hydrogenase (Hedderich and Forzi, 2005; Meuer *et al.*, 1999, 2002), F_{420}-nonreducing hydrogenase (Ide *et al.*, 1999), and heterodisulfide reductase (Bäumer *et al.*, 2000; Fig. 13.2). The oxidation of reduced ferredoxin is catalyzed by Ech hydrogenase, resulting in the release of molecular hydrogen (Meuer *et al.*, 1999) that is then reoxidized by the F_{420}-nonreducing hydrogenase, and the electrons are channeled via methanophenazine to the heterodisulfide reductase (Welte *et al.*, 2010a,b). The entire electron transport system is referred to as ferredoxin:heterodisulfide oxidoreductase. In addition to aceticlastic methanogenesis, reduced ferredoxin is produced in the oxidative branch of methylotrophic methanogenesis. As previously described in Section 2.3, three-fourths of the methyl groups are used to generate the heterodisulfide with methane as a side product. A quarter of the methyl groups are oxidized to yield reducing equivalents in the form of two molecules of $F_{420}H_2$, and two reduced ferredoxins per oxidized methyl group. The enzyme responsible for the reduction of ferredoxin is the formyl-methanofuran dehydrogenase that oxidizes a formyl group to CO_2 (Bartoschek *et al.*, 2000).

Ferredoxin-mediated electron transport can be analyzed using washed cytoplasmic membranes that comprise all essential proteins involved in aceticlastic energy conservation (Welte *et al.*, 2010a). However, these membranes do not contain a system to generate reduced ferredoxin. Therefore, a system of *Moorella thermoacetica* CO-dehydrogenase/acetyl-CoA synthase (CODH/ACS) and *Clostridium pasteurianum* ferredoxin (Fd_{Cl}) is used. In this small electron transport chain, CO is oxidized by CODH/ACS, which transfers electrons to ferredoxin. Clostridial ferredoxin (Fd_{Cl}) can specifically donate electrons to Ech hydrogenase and initiate electron transport in the ferredoxin:heterodisulfide oxidoreductase system (Welte *et al.*, 2010b).

2.4.1. Analysis of the Fd:heterodisulfide oxidoreductase (coupled reaction of Ech hydrogenase, F_{420}-nonreducing hydrogenase and heterodisulfide reductase)

The Fd:heterodisulfide oxidoreductase system can be analyzed using the Ellman's test, as described in Section 2.2.3.

1. Rubber-stoppered glass vials are flushed with N_2. 250 μL K-phosphate buffer (40 m*M*, pH 7.0, 1 μg mL^{-1} resazurin, flushed with N_2) is added and reduced until the solution just becomes colorless (using titanium(III) citrate). Approximately 150 μg washed *Ms. mazei* membranes are added and the whole vial is flushed with 5% CO/95% N_2. 30 μg Fd_{Cl} and 70 μg CODH/ACS are also added.
2. The reaction is started by addition of 400 μ*M* heterodisulfide and the first sample is directly taken: a 20 μL sample of the reaction mixture is mixed with 950 μL Tris–HCl (150 m*M*, pH 8.1) and 100 μL DTNB (5 m*M* in 50 m*M* Na-acetate, pH 5.0) and the absorption at 412 nm is immediately measured.
3. Samples are taken until the heterodisulfide is completely reduced ($\Delta A \sim 0.2$). Normally, 70 nmol thiols are produced per minute per milligram membrane protein.

2.4.2. Ech hydrogenase: Quantification of molecular hydrogen

The activity of Ech hydrogenase can be determined independently of the complete Fd:heterodisulfide oxidoreductase. Ech hydrogenase catalyzes H_2 formation in the process of ferredoxin oxidation (Meuer *et al.*, 1999). So, when washed cytoplasmic membranes of *Ms. mazei* are provided with reduced ferredoxin, they will produce molecular hydrogen. To measure H_2 production rates, rubber-stoppered glass vials are filled with 500 μL K-phosphate buffer (40 m*M*, pH 7.0, 1 μg mL^{-1} resazurin, 5 m*M* dithioerythritol), 5% CO/95% N_2 in the 1.5 mL headspace, 500–700 μg washed membrane preparation, 33.5 μg Fd_{Cl}, and 20 μg CODH/ACS. The vials are shaken at 37 °C, and 10 μL of the headspace is injected into a gas chromatograph connected to a thermal conductivity detector with argon as carrier gas at various time points. The resulting peak areas can be compared with standard curves with defined hydrogen content. Expected activities range between 30 and 60 nmol H_2/mg membrane protein.

2.4.3. Preparation of ferredoxin

Fd_{Cl} can be easily prepared following a protocol developed by Mortenson (1964). The steps can be performed aerobically. However, if Fd_{Cl} is stored for a prolonged time under the influence of oxygen, the iron–sulfur centers will decompose and the protein will lose its color and function.

For the preparation of Fd_{Cl}, 2–4 L *C. pasteurianum* DSM 525 culture is pelleted and resuspended in water. The cells are lyzed by lyzozyme treatment and sonication (to improve lysis, the cell suspension can be frozen or lyophilized prior to sonication). The volume of the crude extract is brought to 150–200 mL with ice-cold water. An equal volume of ice-cold acetone is added and the solution is stirred for 15 min in an ice bath. The solution is centrifuged for 15 min at 7000×*g* and 4 °C, and the supernatant is applied onto an anion exchange column (DEAE cellulose, 2.5×8 cm, equilibrated with 50% (v/v) acetone). Ferredoxin directly becomes visible on top of the chromatography column as a dark brown ring. The column is washed with 200 m*M* Tris–HCl pH 8.0 until the bright yellow flavins no longer elute from the column (~100 mL). Fd_{Cl} is eluted with 500 m*M* Tris–HCl, pH 8.0. The brown fractions are concentrated via ultrafiltration and stored aerobically at −70 °C in aliquots.

2.4.4. Preparation of CO-dehydrogenase/acetyl-CoA synthase

CODH/ACS can be purified from *M. thermoacetica* ATCC 39073, as described by Ragsdale *et al.* (1983), with modifications as specified (Welte *et al.*, 2010a). All steps should be carried out anaerobically.

1. *M. thermoacetica* cells are lyzed (suspended in 50 m*M* Tris–HCl, pH 7.5, 2 m*M* Na-dithionite) by French press treatment.

2. The lysate is applied to a DEAE cellulose column and eluted with a step gradient of NaCl from 0.1 to 0.5 *M*, CODH/ACS elutes at 0.3 *M* NaCl. To screen for the CODH/ACS-containing fractions, the following enzymatic test is used: 700 μL K-phosphate buffer (40 m*M*, pH 7.0, 1 μg mL^{-1} resazurin, and 5 m*M* dithioerythritol), 5 m*M* methyl viologen, and 5% CO in the headspace. The reaction can be monitored at 600 nm at room temperature.
3. The enzyme containing fractions are applied to a Q-sepharose anion exchange column (elution at ~0.4 *M* NaCl) and a phenyl–sepharose hydrophobic interaction column (elution at ~0.2 *M* $(NH_4)_2SO_4$).
4. The fractions from the phenyl–sepharose column that contain CODH/ACS are concentrated and the buffer is exchanged into 50 m*M* Tris–HCl (pH 7.6) by ultrafiltration.

3. Measurement of Proton Translocation

As described above, the membrane-bound electron transfer of *Ms. mazei* has been analyzed in detail. To elucidate the coupling between electron transport and ion translocation, a subcellular system of this organism was developed that consisted of washed inverted membrane vesicles free from cytoplasmic components (Deppenmeier *et al.*, 1988; Mayer *et al.*, 1987). The advantage is that enzymes located at the inner leaflet of the membrane are accessible to the highly charged substrates, F_{420} and CoM-S-S-CoB, because of the inverted orientation of these vesicles. Using these vesicle preparations, it has been shown that electron transport as catalyzed by the Ech hydrogenase, the $F_{420}H_2$: heterodisulfide oxidoreductase system, and the H_2:heterodisulfide oxidoreductase system, is accompanied by proton transfer across the cytoplasmic membrane into the lumen of the inverted vesicles (Deppenmeier, 2004; Welte *et al.*, 2010b).

3.1. Preparation of washed inverted vesicles

In contrast to other *Methanosarcina* species, *Ms. mazei* lacks a heteropolysaccharide layer and only possesses a proteinaceous cell wall (Jussofie *et al.*, 1986). By treatment with proteases, the protein layer can be digested, resulting in the formation of protoplasts. Gentle French pressure treatment and centrifugation results in the formation of cell-free lysate containing membrane vesicles with 90% inside-out orientation (Fig. 13.3). Inside-out oriented membrane vesicles have the great advantage that the active sites of most energy-transducing enzymes face the outside and are accessible for extraneously added electron donors or acceptors (e.g., $F_{420}H_2$, Fd_{red}, heterodisulfide). Thus, inside-out oriented membrane vesicles offer a variety of

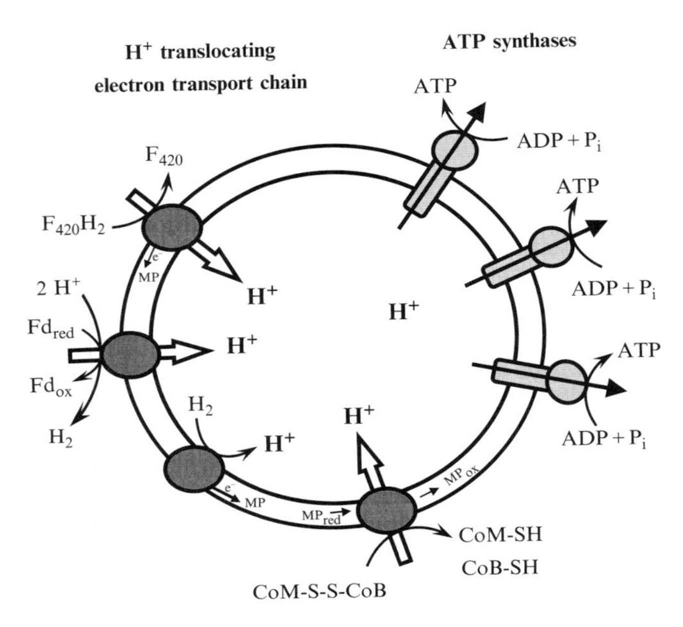

Figure 13.3 Scheme of inside-out vesicles.

possibilities to investigate membrane-bound energy transduction mechanisms. To produce inverted membrane vesicles, the following procedure can be used:

1. Grow at least 4 L of *Ms. mazei* culture. All steps should be performed anaerobically. Harvest the cells in the late exponential growth phase (OD_{600} = 0.8–1) and resuspend the cells in 20–50-mL vesicle buffer (20 m*M* K-phosphate buffer, pH 7.0, 20 m*M* $MgSO_4$, 500 m*M* sucrose, 10 m*M* dithioerythritol, and 1 μg mL^{-1} resazurin). Gently homogenize the cells.
2. For protoplast formation, add 2 mg pronase per liter culture volume. Incubate at 37 °C until protoplasts appear (this can be visualized by light microscopy; the cells become perfectly globe-shaped), and do not exceed 30 min. Stop the reaction by addition of approximately 50 μL PMSF (100 m*M* in isopropanol), and directly put the protoplast vial on ice. Always handle protoplasts (and vesicles) very gently!
3. Centrifuge the protoplast suspension at 12,000×*g* for 10 min at 4 °C. Resuspend the pellet in 30 mL vesicle buffer containing DNase. Apply the preparation to a French press cell within an anaerobic chamber and connect the outlet with a N_2-gassed and rubber-stoppered serum bottle by using a short tube ending with a needle which is injected into the serum flask. Take the cell out and French-press the washed protoplast

suspension at 600 psi. The preparation is released into the anaerobic serum bottle (caution: omit overpressure by inserting a second needle!). The result is a crude inverted vesicle preparation with 90% inside-out orientation, which is then transferred to appropriate centrifugation bottles within the anaerobic chamber. Spin the crude vesicle preparation at $1300 \times g$ for 10 min at 4 °C. The supernatant contains the crude vesicles and should be turbid. *Note*: sometimes, the vesicles form a slight brown pellet and should be included in the ultracentrifugation (see below)!

4. Ultracentrifuge the crude vesicle preparation at $150{,}000 \times g$ for 60–90 min at 4 °C. The brown pellet consists of the inverted membrane vesicles. Often, a black pellet underneath is observed, which contains sulfidic contaminants and should be discarded. Carefully remove the cytoplasmic supernatant (be careful not to remove the pellet: it is easily disturbed) and resuspend the pellet in the vesicle buffer. Transfer the vesicle suspension to a new ultracentrifugation vial and recentrifuge (30–60 min is sufficient) until the supernatant becomes colorless. *Note*: care must be taken in the initial centrifugation step, as membrane vesicles will not pellet if they are too concentrated. In this case, the vesicles must be diluted.
5. Finally, resuspend the washed inverted membrane vesicle preparation in a 500–2000-μL vesicle buffer (final protein concentration $\sim$25 mg mL^{-1}), and aliquot and freeze them anaerobically at −70 °C under an H_2 atmosphere.

3.2. Experimental determinations of the $H^+/2e^-$ ratio

In most organisms, electron flow through the respiratory chain is directly coupled to the formation of an electrochemical proton gradient (Δp) that is formed by protein-catalyzed proton transfer across biological phospholipid bilayers. This process is also performed by the methanogenic archaeon *Ms. mazei* and is coupled by the electron transfer from the electron input modules F_{420}-nonreducing hydrogenase and $F_{420}H_2$ dehydrogenase to the heterodisulfide reductase (Deppenmeier *et al.*, 1990, 1992). Furthermore, it was shown that the Ech hydrogenase also contributes to the electrochemical proton gradient (Welte *et al.*, 2010a,b). One important aspect of this phenomenon is the stoichiometric relationship between proton translocation and electron transport, which is expressed by the $H^+/2e^-$ ratio.

3.2.1. Procedure

Proton translocation can be followed in a glass vessel (5.5 mL total volume) filled with 2.5 mL 40 m*M* potassium thiocyanate, 0.5 *M* sucrose, 0.1 μg mL^{-1} resazurin, and 10 m*M* dithioerythritol (Fig. 13.4A). The thiocyanate ion is membrane-permeable and prevents the formation of a membrane potential to ensure electroneutral movement of protons. A sensitive

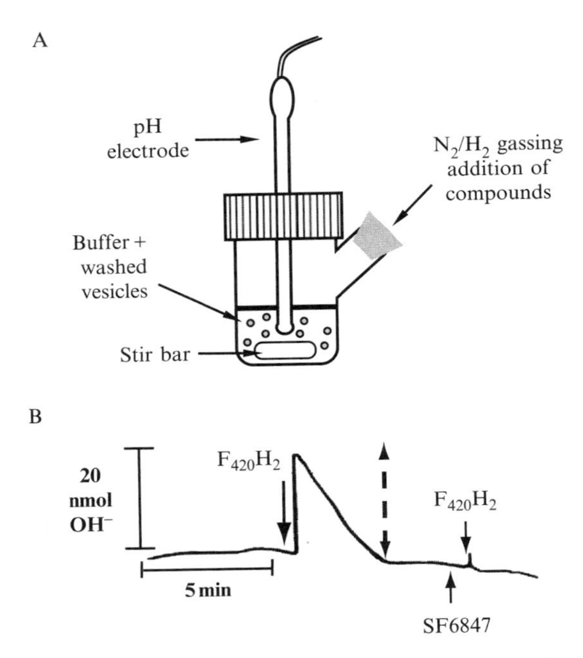

Figure 13.4 Analysis of proton translocation. (A) Device for continuous determination of the pH value, (B) flow chart of proton translocation, black arrows indicate the addition of $F_{420}H_2$ or the protonophor SF6847, and the broken arrow shows the extent of alkalinization.

pH electrode (e.g., model 8103 Ross, Orion research, Küsnacht, Switzerland) is inserted into the vessel from the top through a rubber stopper. Connect the electrode with a pH meter (e.g., Orion model EA 920) and a chart recorder. The vessel is subsequently gassed for 10 min with N_2 by means of two needles inserted from the side through a rubber stopper (Fig. 13.4A). In case the solution is not totally colorless, add up to 1 μL titanium(III) citrate for complete reduction. Then, 100 μL washed vesicles (0.35–0.5 mg protein/assay) are added, resulting in a final K-phosphate concentration of 1.5 m*M* in the reaction mixture. The preparation is continuously stirred and the pH is adjusted to 6.8–6.9. Additions are made with a microliter syringe from the side arm.

As mentioned above, proton translocation is due to electron transfer from an electron donor (H_2, $F_{420}H_2$, reduced 2-hydroxyphenazine) to an acceptor (2-hydroxyphenazine or CoM-S-S-CoB). In principle, one substrate is added in excess, whereas the second substrate is injected in low amounts (pulse injection; usually 2–10 nmol) that allows only a short period of electron transfer events. The pulse injection will lead to alkalinization of the medium as is apparent from the recorded time course of this process (Fig. 13.4B). A short period of alkalinization is then followed by a longer period of acidification until a stable baseline (pH) is reached again. The alkalinization is due to proton movement from the medium into the lumen of the inverted vesicles energized by the electron transfer reactions. The extent of alkalinization is dependent on the amount of the pulsed substrate. Changes in the pH value are converted into proton equivalents by double titration with standard solutions of HCl and KOH (Fig. 13.4B). The amount of H^+ translocated is calculated from the difference between the maximum of alkalinization and the final baseline after reacidification (Fig. 13.4B). The ratio of translocated protons and transferred electron can be calculated since the amount of pulsed substrate is known.

3.2.1.1. Reaction conditions

- H_2: heterodisulfide oxidoreductase: Use an H_2 atmosphere and pulse with CoM-S-S-CoB (2–10 nmol in oxygen-free H_2O).
- F_{420}-nonreducing hydrogenase: Use an H_2 atmosphere and pulse with hydroxyphenazine (2–10 nmol from a 1 m*M* stock solution solvated in anoxic ethanol).
- Heterodisulfide reductase: N_2 atmosphere in the reaction vessel; add 250 nmol CoM-S-S-CoB and pulse with reduced hydroxyphenazine (2–10 nmol from a 1 m*M* stock solution solvated in anoxic ethanol).
- $F_{420}H_2$ dehydrogenase: N_2 atmosphere in the reaction vessel; add 250 nmol hydroxyphenazine and pulse with $F_{420}H_2$ (2–10 nmol).
- $F_{420}H_2$: heterodisulfide oxidoreductase: N_2 atmosphere in the reaction vessel; add 250 nmol CoM-S-S-CoB and pulse with reduced F_{420} (2–10 nmol).

In our studies, the reversible alkalinization was used to calculate from more than 60 experiments an average of 1.9 protons translocated per $F_{420}H_2$ oxidized in the $F_{420}H_2$: heterodisulfide oxidoreductase assay. The observed alkalinization was specifically coupled with the $F_{420}H_2$-dependent CoM-S-S-CoB reduction. Alkalinization was not observed if $F_{420}H_2$ or CoM-S-S-CoB was omitted or if $F_{420}H_2$ was replaced by F_{420}. Similar values were obtained when the H_2:heterodisulfide oxidoreductase system was analyzed. Keeping in mind that about 50% of the membrane structures present in the vesicle preparations catalyze electron transport but are unable to establish a proton gradient (Deppenmeier *et al.*, 1996), the $H^+/2e^-$ values of the overall reaction increases to 3.8, which is in agreement with values found for whole cell preparations of *Ms. barkeri* (Blaut *et al.*, 1987).

The aforementioned washed inverted vesicles were found to couple electron transfer processes with the transfer of about four protons across the cytoplasmic membrane. Furthermore, it was shown that 2-hydroxyphenazine is reduced by molecular hydrogen, as catalyzed by the F_{420}-nonreducing hydrogenase, and the membrane-bound heterodisulfide reductase was able to use reduced 2-hydroxyphenazine as the electron donor for the reduction of CoM-S-S-CoB. In addition, the $F_{420}H_2$-dehydrogenase could use hydroxyphenazine as electron acceptor (Bäumer *et al.*, 2000; Ide *et al.*, 1999). All reactions allowed the translocation of protons with a stoichiometry of about $2H^+/2e^-$. Hence, there are two proton-translocating segments present in the $F_{420}H_2$:CoM-S-S-CoB and the H_2:CoM-S-S-CoB oxidoreductase system, respectively. The first one involves the electron-donating enzymes (F_{420}-nonreducing hydrogenase or $F_{420}H_2$ dehydrogenase) and the second one the heterodisulfide reductase (Fig. 13.2). Control experiments showed that the formation of a proton gradient was not possible in the presence of a protonophor, such as SF6847 (10 nmol mg^{-1} protein; Fig. 13.4B). In contrast, H^+ translocation was not affected by an inhibitor of the ATP synthase (e.g., DCCD; 400 μM) or in the presence of a Na^+ ionophor (e.g., ETH157; 30 μM).

4. Analysis of ATP Formation in Vesicle Preparation

When protons are translocated into the inverted membrane vesicles by enzymes comprising the methanogenic respiratory chain, and when ADP and P_i are present, the A_1A_O ATP synthase can phosphorylate ADP in the process of proton extrusion. ATP content can be measured using the luciferin/luciferase assay (Kimmich *et al.*, 1975). The reaction of luciferase with its substrate luciferin is strictly ATP-dependent and produces light, and the amount of ATP is directly proportional to the amount of emitted light.

4.1. Procedure

The rubber-stoppered, N_2-flushed glass reaction vial contains 500 μL vesicle buffer, 600 μM ADP, 300 μM AMP (to inhibit membrane-bound adenylate kinase), and 250–700-μg inverted membrane vesicles. Depending on the system to be observed, different electron donors or acceptors need to be used. For the H_2:heterodisulfide oxidoreductase, a H_2 atmosphere and 200–500 μM heterodisulfide as electron acceptor is used. For the $F_{420}H_2$: heterodisulfide oxidoreductase, a N_2 atmosphere, 10 μL $F_{420}H_2$ (final concentration 15 μM) as electron donor and 200–500 μM heterodisulfide as electron acceptor is used. For measuring the Fd:heterodisulfide oxidoreductase, a 5% CO/95% N_2 atmosphere is applied to the reaction vial. Furthermore, 70 μg CO-dehydrogenase and 30 μg Fd are added to generate Fd_{red}, and 200–500 μM heterodisulfide are added as terminal electron acceptor. If Ech hydrogenase as part of the Fd:heterodisulfide oxidoreductase is to be measured independently, the heterodisulfide can be omitted. Electrons are transferred to protons as terminal electron acceptors and form H_2.

The reaction vials are shaken in a 37 °C water bath and 10 μL samples are taken at various time points with a microliter syringe. The sample is mixed with 700 μL ATP determination buffer (4 m*M* $MgSO_4$, 20 m*M* glycylglycine, pH 8.0) and 100 μL firefly lantern extract (Sigma-Aldrich, Germany). The emitted light is quantified with a luminometer at 560 nm after a lag phase of 10 s. The values are compared to a standard curve generated with 1–100 pmol ATP. For the H_2:heterodisulfide oxidoreductase and Ech hydrogenase, an activity of about 1.5–2 nmol ATP mg^{-1} membrane protein, and for the complete Fd:heterodisulfide oxidoreductase, an activity of 3–4 nmol ATP mg^{-1} membrane protein, is expected.

ACKNOWLEDGMENTS

We thank Elisabeth Schwab for technical assistance and Paul Schweiger for critical reading of the chapter. Many thanks also go to Gunes Bender and Steve Ragsdale, Department of Biological Chemistry, University of Michigan Medical School for providing the CODH/ACS from *Moorella thermoacetica*. This work was supported by the Deutsche Forschungsgemeinschaft (Grant De488/9-1).

REFERENCES

Abken, H. J., Tietze, M., Brodersen, J., Bäumer, S., Beifuss, U., and Deppenmeier, U. (1998). Isolation and characterization of methanophenazine and function of phenazines in membrane-bound electron transport of *Methanosarcina mazei* Gö1. *J. Bacteriol.* **180**(8), 2027–2032.

Bartoschek, S., Vorholt, J. A., Thauer, R. K., Geierstanger, B. H., and Griesinger, C. (2000). N-carboxymethanofuran (carbamate) formation from methanofuran and CO_2 in methanogenic archaea. Thermodynamics and kinetics of the spontaneous reaction. *Eur. J. Biochem.* **267**(11), 3130–3138.

Bäumer, S., Ide, T., Jacobi, C., Johann, A., Gottschalk, G., and Deppenmeier, U. (2000). The $F_{420}H_2$ dehydrogenase from *Methanosarcina mazei* is a redox-driven proton pump closely related to NADH dehydrogenases. *J. Biol. Chem.* **275**(24), 17968–17973.

Beifuss, U., and Tietze, M. (2000). On the total synthesis of (S)-methanophenazine and the formal synthesis of (R)-methanophenazine from a common precursor. *Tetrahedron Lett.* **41**(50), 9759–9763.

Beifuss, U., and Tietze, M. (2005). Methanophenazine and other natural biologically active phenazines. *Top. Curr. Chem.* **244,** 77–113.

Beifuss, U., Tietze, M., Bäumer, S., and Deppenmeier, U. (2000). Methanophenazine: Structure, total synthesis, and function of a new cofactor from methanogenic archaea. *Angew. Chem. Int. Ed.* **39**(14), 2470–2472.

Blaut, M., and Gottschalk, G. (1984). Coupling of ATP synthesis and methane formation from methanol and molecular hydrogen in *Methanosarcina barkeri*. *Eur. J. Biochem.* **141**(1), 217–222.

Blaut, M., Müller, V., and Gottschalk, G. (1987). Proton translocation coupled to methanogenesis from methanol + hydrogen in *Methanosarcina barkeri*. *FEBS Lett.* **215**(1), 53–57.

Boone, D. R., Whitman, W. B., and Rouvière, P. E. (1993). Diversity and taxonomy of methanogens. *In* "Methanogenesis: Ecology, Physiology, Biochemistry and Genetics," (J. G. Ferry, ed.), pp. 35–80. Chapman & Hall, New York.

Clements, A. P., Kilpatrick, L., Lu, W. P., Ragsdale, S. W., and Ferry, J. G. (1994). Characterization of the iron-sulfur clusters in ferredoxin from acetate-grown *Methanosarcina thermophila*. *J. Bacteriol.* **176**(9), 2689–2693.

Deppenmeier, U. (2002a). Redox-driven proton translocation in methanogenic archaea. *Cell. Mol. Life Sci.* **59**(9), 1513–1533.

Deppenmeier, U. (2002b). The unique biochemistry of methanogenesis. *Prog. Nucleic Acid Res. Mol. Biol.* **71,** 223–283.

Deppenmeier, U. (2004). The membrane-bound electron transport system of *Methanosarcina* species. *J. Bioenerg. Biomembr.* **36**(1), 55–64.

Deppenmeier, U., and Müller, V. (2008). Life close to the thermodynamic limit: How methanogenic archaea conserve energy. *Results Probl. Cell Differ.* **45,** 123–152.

Deppenmeier, U., Blaut, M., Jussofie, A., and Gottschalk, G. (1988). A methyl-CoM methylreductase system from methanogenic bacterium strain Gö 1 not requiring ATP for activity. *FEBS Lett.* **241**(1–2), 60–64.

Deppenmeier, U., Blaut, M., Mahlmann, A., and Gottschalk, G. (1990). Reduced coenzyme F_{420}: Heterodisulfide oxidoreductase, a proton- translocating redox system in methanogenic bacteria. *Proc. Natl. Acad. Sci. USA* **87**(23), 9449–9453.

Deppenmeier, U., Blaut, M., Schmidt, B., and Gottschalk, G. (1992). Purification and properties of a F_{420}-nonreactive, membrane-bound hydrogenase from *Methanosarcina* strain Gö1. *Arch. Microbiol.* **157**(6), 505–511.

Deppenmeier, U., Müller, V., and Gottschalk, G. (1996). Pathways of energy conservation in methanogenic archaea. *Arch. Microbiol.* **165**(3), 149–163.

Ebner, S., Jaun, B., Goenrich, M., Thauer, R. K., and Harmer, J. (2010). Binding of coenzyme B induces a major conformational change in the active site of methyl-coenzyme M reductase. *J. Am. Chem. Soc.* **132**(2), 567–575.

Ellermann, J., Hedderich, R., Bocher, R., and Thauer, R. K. (1988). The final step in methane formation. Investigations with highly purified methyl-CoM reductase (component C) from *Methanobacterium thermoautotrophicum* (strain Marburg). *Eur. J. Biochem.* **172** (3), 669–677.

Ellman, G. L. (1958). A colorimetric method for determining low concentrations of mercaptans. *Arch. Biochem. Biophys.* **74**(2), 443–450.

Ferry, J. G. (1999). Enzymology of one-carbon metabolism in methanogenic pathways. *FEMS Microbiol. Rev.* **23**(1), 13–38.

Ferry, J. G., and Lessner, D. J. (2008). Methanogenesis in marine sediments. *Ann. N. Y. Acad. Sci.* **1125,** 147–157.

Fischer, R., and Thauer, R. K. (1989). Methyltetrahydromethanopterin as an intermediate in methanogenesis from acetate in *Methanosarcina barkeri*. *Arch. Microbiol.* **151**(5), 459–465.

Gencic, S., Duin, E. C., and Grahame, D. A. (2010). Tight coupling of partial reactions in the acetyl-CoA decarbonylase/synthase (ACDS) multienzyme complex from *Methanosarcina thermophila*: Acetyl C-C bond fragmentation at the A cluster promoted by protein conformational changes. *J. Biol. Chem.* **285**(20), 15450–15463.

Gong, W., Hao, B., Wei, Z., Ferguson, D. J., Tallant, T., Krzycki, J. A., and Chan, M. K. (2008). Structure of the alpha(2)epsilon(2) Ni-dependent CO dehydrogenase component of the *Methanosarcina barkeri* acetyl-CoA decarbonylase/synthase complex. *Proc. Natl. Acad. Sci. USA* **105**(28), 9558–9563.

Gottschalk, G., and Thauer, R. K. (2001). The Na^+-translocating methyltransferase complex from methanogenic archaea. *Biochim. Biophys. Acta* **1505**(1), 28–36.

Hedderich, R., and Forzi, L. (2005). Energy-converting [NiFe] hydrogenases: More than just H_2 activation. *J. Mol. Microbiol. Biotechnol.* **10**(2–4), 92–104.

Hedderich, R., Hamann, N., and Bennati, M. (2005). Heterodisulfide reductase from methanogenic archaea: A new catalytic role for an iron-sulfur cluster. *Biol. Chem.* **386** (10), 961–970.

Heiden, S., Hedderich, R., Setzke, E., and Thauer, R. K. (1994). Purification of a 2-subunit cytochrome-b containing heterodisulfide reductase from methanol-grown *Methanosarcina barkeri*. *Eur. J. Biochem.* **221**(2), 855–861.

Hippe, H., Caspari, D., Fiebig, K., and Gottschalk, G. (1979). Utilization of trimethylamine and other N-methyl compounds for growth and methane formation by *Methanosarcina barkeri*. *Proc. Natl. Acad. Sci. USA* **76**(1), 494–498.

Ide, T., Bäumer, S., and Deppenmeier, U. (1999). Energy conservation by the H_2:heterodisulfide oxidoreductase from *Methanosarcina mazei* Gö1: Identification of two proton-translocating segments. *J. Bacteriol.* **181**(13), 4076–4080.

Jain, N., Pathak, H., Mitra, S., and Bhatia, A. (2004). Emission of methane from rice fields—A review. *J. Sci. Ind. Res.* **63**(2), 101–115.

Jussofie, A., Mayer, F., and Gottschalk, G. (1986). Methane formation from methanol and molecular hydrogen by protoplasts of new methanogenic isolates and inhibition by dicyclohexylcarbodiimide. *Arch. Microbiol.* **146**(3), 245–249.

Kashyap, D. R., Dadhich, K. S., and Sharma, S. K. (2003). Biomethanation under psychrophilic conditions: A review. *Bioresour. Technol.* **87**(2), 147–153.

Khalil, M. A. K., and Rasmussen, R. A. (1994). Global emissions of methane during the last several centuries. *Chemosphere* **29**(5), 833–842.

Kimmich, G. A., Randles, J., and Brand, J. S. (1975). Assay of picomole amounts of ATP, ADP, and AMP using the luciferase enzyme system. *Anal. Biochem.* **69**(1), 187–206.

Kobelt, A., Pfaltz, A., Ankelfuchs, D., and Thauer, R. K. (1987). The L-form of N-7-mercaptoheptanoyl-O-phosphothreonine is the enantiomer active as component B in methyl-CoM reduction to methane. *FEBS Lett.* **214**(2), 265–268.

Künkel, A., Vaupel, M., Heim, S., Thauer, R. K., and Hedderich, R. (1997). Heterodisulfide reductase from methanol-grown cells of *Methanosarcina barkeri* is not a flavoenzyme. *Eur. J. Biochem.* **244**(1), 226–234.

Mayer, F., Jussofie, A., Salzmann, M., Lubben, M., Rohde, M., and Gottschalk, G. (1987). Immunoelectron microscopic demonstration of ATPase on the cytoplasmic membrane of the methanogenic bacterium strain Gö1. *J. Bacteriol.* **169**(5), 2307–2309.

McInerney, M. J., Sieber, J. R., and Gunsalus, R. P. (2009). Syntrophy in anaerobic global carbon cycles. *Curr. Opin. Biotechnol.* **20**(6), 623–632.

Meuer, J., Bartoschek, S., Koch, J., Kunkel, A., and Hedderich, R. (1999). Purification and catalytic properties of Ech hydrogenase from *Methanosarcina barkeri*. *Eur. J. Biochem.* **265** (1), 325–335.

Meuer, J., Kuettner, H. C., Zhang, J. K., Hedderich, R., and Metcalf, W. W. (2002). Genetic analysis of the archaeon *Methanosarcina barkeri* Fusaro reveals a central role for Ech hydrogenase and ferredoxin in methanogenesis and carbon fixation. *Proc. Natl. Acad. Sci. USA* **99**(8), 5632–5637.

Mortenson, L. E. (1964). Purification and analysis of ferredoxin from *Clostridium pasteurianum*. *Biochim. Biophys. Acta* **81**(1), 71–77.

Noll, K. M., Donnelly, M. I., and Wolfe, R. S. (1987). Synthesis of 7-Mercaptoheptanoylthreonine phosphate and its activity in the methylcoenzyme-M methylreductase system. *J. Biol. Chem.* **262**(2), 513–515.

Ott, R. (1959). Untersuchungen über Chinone. 2. Mitt.: Azinbildung durch Umsetzung von p-Chinonen mit o-Diaminen. *Monatsh. Chem.* **90**(6), 827–838.

Peinemann, S., Müller, V., Blaut, M., and Gottschalk, G. (1988). Bioenergetics of methanogenesis from acetate by *Methanosarcina barkeri*. *J. Bacteriol.* **170**(3), 1369–1372.

Peinemann, S., Blaut, M., and Gottschalk, G. (1989). ATP synthesis coupled to methane formation from methyl-CoM and H_2 catalyzed by vesicles of the methanogenic bacterial strain Gö1. *Eur. J. Biochem.* **186**(1–2), 175–180.

Pisa, K. Y., Weidner, C., Maischak, H., Kavermann, H., and Müller, V. (2007). The coupling ion in the methanoarchaeal ATP synthases: H(+) vs. Na(+) in the A(1)A(o) ATP synthase from the archaeon *Methanosarcina mazei* Gö1. *FEMS Microbiol. Lett.* **277**(1), 56–63.

Ragsdale, S. W. (2008). Catalysis of methyl group transfers involving tetrahydrofolate and B12. *Vitam. Horm.* **79,** 293–324.

Ragsdale, S. W., Ljungdahl, L. G., and Der Vartanian, D. V. (1983). Isolation of carbon monoxide dehydrogenase from *Acetobacterium woodii* and comparison of its properties with those of the *Clostridium thermoaceticum* enzyme. *J. Bacteriol.* **155**(3), 1224–1237.

Thauer, R. K. (1998). Biochemistry of methanogenesis: A tribute to Marjory Stephenson. *Microbiology* **144,** 2377–2406.

Thauer, R. K., Jungermann, K., and Decker, K. (1977). Energy conservation in chemotrophic anaerobic bacteria. *Bacteriol. Rev.* **41**(1), 100–180.

Thauer, R. K., Kaster, A. K., Seedorf, H., Buckel, W., and Hedderich, R. (2008). Methanogenic archaea: Ecologically relevant differences in energy conservation. *Nat. Rev. Microbiol.* **6**(8), 579–591.

Vignais, P. M., Billoud, B., and Meyer, J. (2001). Classification and phylogeny of hydrogenases. *FEMS Microbiol. Rev.* **25**(4), 455–501.

Welte, C., Kallnik, V., Grapp, M., Bender, G., Ragsdale, S., and Deppenmeier, U. (2010a). Function of Ech hydrogenase in ferredoxin-dependent, membrane-bound electron transport in *Methanosarcina mazei*. *J. Bacteriol.* **192**(3), 674–678.

Welte, C., Krätzer, C., and Deppenmeier, U. (2010b). Involvement of Ech hydrogenase in energy conservation of *Methanosarcina mazei*. *FEBS J.* **277,** 3396–3403.

Wolfe, R. S. (1985). Unusual coenzymes of methanogenesis. *Trends Biochem. Sci.* **10**(10), 396–399.

Zahler, W. L., and Cleland, W. W. (1968). A specific and sensitive assay for disulfides. *J. Biol. Chem.* **243**(4), 716–719.

Zehnder, A. J. B., and Wuhrmann, K. (1976). Titan(III) citrate a nontoxic oxidation-reduction buffering system for the culture of obligate anaerobes. *Science* **194,** 1165–1166.

CHAPTER FOURTEEN

Measuring Isotope Fractionation by Autotrophic Microorganisms and Enzymes

Kathleen M. Scott,* Gordon Fox,* *and* Peter R. Girguis†

Contents

Abstract

Physical, chemical, and biological processes commonly discriminate among stable isotopes. Therefore, the stable isotope compositions of biomass, growth substrates, and products often carry the isotopic fingerprints of the processes that shape them. Therefore, measuring isotope fractionation by enzymes and cultures of autotrophic microorganisms can provide insights at many levels, from metabolism to ecosystem function. Discussed here are considerations relevant to measuring isotope discrimination by enzymes as well as intact cells, with an emphasis on stable one-carbon isotopes and autotrophic microorganisms.

* Department of Integrative Biology, University of South Florida, Tampa, Florida, USA
† Department of Organismal and Evolutionary Biology, Harvard University, Cambridge, Massachusetts, USA

Methods in Enzymology, Volume 494
ISSN 0076-6879, DOI: 10.1016/B978-0-12-385112-3.00014-7

1. Introduction

Many biologically relevant elements occur as multiple stable (nonradioactive) isotopes in nature. Carbon, which is the focus of this chapter, exists primarily as ^{12}C, with approximately 1% as ^{13}C. Given that both isotopes have the same valence electrons, both can form the same compounds and are readily assimilated into biological materials. However, covalent bonds with ^{12}C atoms are typically more labile than those with ^{13}C, due to lower zero-point energies for bonds with the heavier isotope, which result in larger activation energies (Cook, 1998; Melander and Saunders, 1980). The relative rates of ^{12}C and ^{13}C reaction vary subtly from process to process, and can be used to identify as well as quantify biological and abiotic activities (Hayes, 1993).

Studies using ^{12}C and ^{13}C to describe processes use a reasonably standard terminology to describe isotopic compositions, as well as the relative rates of their reactions. The relative amounts of ^{12}C and ^{13}C are described as isotope ratios ($R = {}^{13}C/{}^{12}C$). Given that R values are cumbersome (e.g., currently, $R_{atmCO_2} \sim 0.011145$; CDIAC, ORNL) and typically differ by <0.001, stable carbon isotopic compositions are usually reported as $\delta^{13}C$ values:

$$\delta^{13}C = \left(\frac{R_{sample}}{R_{std}} - 1\right) \times 10^3‰ \quad (14.1)$$

(Hayes, 1993), in which R_{std} is the isotope ratio of the PeeDee Belemnite standard. Differences in relative rates of reaction of ^{12}C and ^{13}C are expressed as kinetic isotope effects ($\alpha = {}^{12}k/{}^{13}k$). For most processes, α values are close to 1; as with R values, kinetic isotope effects are usually reported as fractionation factors (ε values) to magnify their differences from one another:

$$\varepsilon = (\alpha - 1) \times 10^3‰ \quad (14.2)$$

Enzymatic reactions fractionate carbon to varying degrees. For example, carboxylation by Calvin–Benson–Bassham (CBB) cycle carboxylase ribulose 1,5-bisphosphate carboxylase/oxygenase (RubisCO) has a rather large fractionation factor compared to other carboxylases. RubisCO from spinach has an ε value of 29‰ (Roeske and O'Leary, 1984), while maize phosphoenol carboxylase fractionates substantially less ($\varepsilon = 2.9$‰; O'Leary *et al.*, 1981). It is important to note that enzymes isolated from different organisms can have markedly different fractionation factors. For example, RubisCO from other organisms fractionate less than the spinach enzyme (18–24‰; Guy *et al.*, 1993; Robinson *et al.*, 2003; Roeske and O'Leary, 1985; Scott, 2003; Scott *et al.*, 2007).

Enzyme fractionation factors affect the isotopic composition of an organism's biomass, and can be used to infer metabolism *in situ*. Due to the relatively large fractionation factors by RubisCO enzymes, microorganisms using the CBB cycle for carbon fixation often have ^{13}C-depleted biomass. Cultures of these organisms have biomass δ^{13}C values of -11 to -26‰ relative to dissolved inorganic carbon (DIC $= CO_2 + HCO_3^- + CO_3^{-2}$; Madigan *et al.*, 1989; Pardue *et al.*, 1976; Quandt *et al.*, 1977; Ruby *et al.*, 1987; Sakata *et al.*, 2008). Autotrophic methanogens using the Wood–Ljungdahl pathway for CO_2 reduction and uptake fractionate even more than CBB autotrophs, with biomass values that are -22 to -38‰ relative to DIC (Fuchs *et al.*, 1979; Londry *et al.*, 2008). Autotrophic microorganisms using either the reductive citric acid cycle or hydroxypropionate cycle fractionate CO_2 to a lesser degree, with δ^{13}C values that are -1 to -12‰ relative to DIC (Hügler *et al.*, 2005; Quandt *et al.*, 1977; vanderMeer *et al.*, 2001a,b; Williams *et al.*, 2006). Given the differences in isotope discrimination by cultured autotrophs that use different pathways, it is sometimes possible to infer the autotrophic pathway that microorganisms are using *in situ* based on biomass δ^{13}C values. For example, Yellowstone mats with a mixed chloroflexus/cyanobacterial community have δ^{13}C values that are more positive than would be expected for CBB autotrophs, consistent with organic carbon input from the hydroxypropionate cycle present in chloroflexi (vanderMeer, 2007).

2. Enzyme-Level Studies

2.1. General assay considerations

To measure an enzyme's kinetic isotope effect, the enzyme and substrate(s) are incubated together under physiologically relevant conditions of pH, temperature, and cofactor presence, and the reaction is monitored by recording the stable isotope composition, as well as the concentration, of substrate(s) and/or product(s). Either a single-timepoint or a multiple-timepoint approach can be taken. For the single-timepoint approach, the enzyme is incubated with a vast overabundance of substrate and the reaction is terminated before the concentration of substrate is significantly diminished, or its isotope composition significantly impacted. The stable isotope compositions of the substrate and product are compared and used to calculate the kinetic isotope effect directly:

$$\alpha = \frac{R_s}{R_p} \tag{14.3}$$

where R_s and R_p are the isotope ratios of the substrate and product, respectively.

For the multiple-timepoint approach, the enzyme incubation is assembled as a closed system with respect to substrates and products. The reaction is sampled over a timecourse and both the concentrations and stable isotope compositions of target compounds are measured; kinetic isotope effects are calculated using the Rayleigh distillation equation (RDE) to account for the cumulative effect of isotope discrimination on the composition of the reaction substrate and product (Fig. 14.1). If the isotope composition and concentration of the substrate are being measured, the appropriate version of the RDE is

$$\frac{R_{st}}{R_{si}} = \left(\frac{C_{st}}{C_{si}}\right)^{1/\alpha-1} \tag{14.4}$$

where R_{st} is the isotope composition and C_{st} is the concentration of substrate at a particular timepoint t, and R_{si} and C_{si} describe these quantities at the beginning of the reaction. The corresponding RDE when measuring the isotope ratio of accumulated product instead of substrate is

$$\frac{R_{pt}}{R_{si}} = \frac{\left[1 - (C_{pt}/C_{si})^{1/\alpha}\right]}{\left[1 - C_{pt}/C_{si}\right]} \tag{14.5}$$

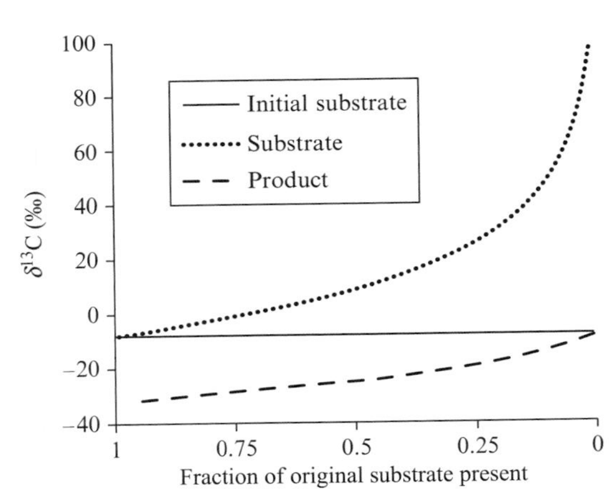

Figure 14.1 Isotopic composition of the substrate and the product as a reaction with a kinetic isotope effect progresses as a closed system. For this example, $\varepsilon = 25‰$ ($\alpha = 1.025$). Changes in the δ^{13}C value of the substrate and the product were modeled with the Rayleigh distillation equation (Eq. (14.4)). As the reaction progresses, due to selective consumption of the lighter isotope (^{12}C in this case), the remaining substrate is ^{13}C-enriched (and therefore has a more positive δ^{13}C value).

where R_{pt} and C_{pt} are the isotope ratio and concentration of accumulated product at timepoint t, and C_{si} is the initial concentration of substrate (see (Mariotti *et al.*, 1981) for review and algebra, but note that he calculates α as R_p/R_s).

The multiple-timepoint approach has the advantage of providing an estimate of α with associated error from a single incubation as it is sampled at multiple timepoints, which can provide "quality control" for reactions in which it is anticipated that isotope fractionation will be constant despite changes in C_{st}. Under these conditions, should isotope fractionation change over the timecourse of the reaction, it will suggest the necessity of further scrutiny either of reaction conditions or the assumption of a constant fractionation factor; such a "checkpoint" is absent from a single-timepoint incubation. Though this approach provides an estimate of error for α, it must be noted that this error estimate is with respect to this single experiment. Replicate incubations should be run, and the estimate of error should be based on α values from independent incubations (see below for calculations).

2.2. Setting up the enzyme assay

It is necessary to prepare the target enzyme in such a manner that its activity is as high as possible; high enzyme activity will result in large changes in substrate concentration and isotope ratios, which in turn yield data with more favorable signal-to-noise ratios. Ideally, purified enzyme will be used, as this will minimize the chance that the observed isotope fractionation is due to enzymes other than the one under study. However, in many cases, enzymes lose activity during the purification process and it is necessary to use partially purified enzymes or cell extracts. If this is the case, it is necessary to run control experiments to ensure that the observed reaction is indeed catalyzed by the target enzyme. For example, parallel incubations can be run, in which the target enzyme is inactivated (via the absence of a substrate or critical cofactor), and activity, if any is observed, is quantified.

With respect to the reaction vessel and conditions, a few commonsense considerations must apply. Given that incubation duration can sometimes extend to hours to allow the reaction to proceed sufficiently, it is necessary for the incubation to be sterile. The reaction vessel must be well mixed over the course of the reaction, as diffusion itself has a kinetic isotope effect (for CO_2, $\varepsilon = 0.7$‰ at 25 °C; O'Leary, 1984). Since the kinetic isotope effect for the enzyme is calculated from some combination of the isotope ratios of the substrate, product, and their concentrations, the reaction must be a closed system with respect to substrate (and product, if measured). Addition of substrate to supplement an ongoing reaction must be avoided, as it will likely change the isotope ratio of the substrate pool available to the enzyme, which in turn will impact the isotope ratio of the product, and therefore the α value calculated from that experiment.

Incubations with gaseous substrates and/or products must be undertaken with particular care. In this case, given the necessity of a closed system, as well as the kinetic isotope effect associated with diffusion, the reaction vessel must be fabricated from materials that are impermeable to the substrate or product (whichever is being measured to estimate the kinetic isotope effect). Furthermore, single-phase incubations (e.g., an aqueous assay without gaseous headspace) are optimal because two-phase incubations require an assumption of chemical and isotope equilibrium between the gas and aqueous phases, quantification of target compound in both phases, and application of the equilibrium isotope effect for dissolution (for CO_2, $\varepsilon = 1.1‰$ at 25 °C; Mook *et al.*, 1974) when calculating the kinetic isotope effect for the enzyme.

If the kinetic isotope effect of an enzyme will be calculated from the concentration and isotopic composition of one of the components of the DIC system, it is absolutely essential to keep the system in chemical and isotopic equilibrium. Abiotic interconversion of CO_2 and HCO_3^- is relatively slow, as full chemical and isotopic equilibration takes a few minutes at room temperature (Zeebe and Wolf-Gladrow, 2003). The interconversion of CO_2 and HCO_3^- also has an equilibrium effect of

$$\varepsilon_{b,c} = \left[\frac{R_{CO_2}}{R_{bic}} - 1\right] \times 10^3 = \left(\frac{-9.866 \times 10^3}{T}\right) + 24.12 \qquad (14.6)$$

(Mook *et al.*, 1974), in which T is the temperature in Kelvin. This equilibrium isotope effect results in dissolved HCO_3^- having a $\delta^{13}C$ value approximately 9‰ more enriched than dissolved CO_2 at 25 °C. When assaying for inorganic carbon assimilation or production, it is optimal to add carbonic anhydrase at sufficient activity to substantially outpace DIC consumption by the target enzyme. This will keep the DIC system in chemical and isotopic equilibrium.

2.3. Analytical concerns

Given the emphasis here on autotrophic processes (including methanogenesis), this section will focus on quantification issues with respect to DIC and methane, as well as the multicarbon products resulting from biological fixation of one-carbon compounds.

2.3.1. Dissolved inorganic carbon (CO_2, HCO_3^-, CO_3^{-2})

DIC can be quantified readily via infrared gas analyzer (Scott *et al.*, 2004b), gas chromatography (Dobrinski *et al.*, 2005), or membrane-inlet mass spectrometry (Girguis *et al.*, 2000, 2002; McNevin *et al.*, 2006) if samples are acidified to quantitatively convert all HCO_3^- and CO_3^{-2} to CO_2. Solid-state sensors also exist for quantifying CO_2, though many sensors are plagued by a lack of substrate specificity, and cross-react with other compounds (Fergus, 2007, 2008; Stetter and Li, 2008). DIC can also be quantified enzymatically via a coupled phosphoenolpyruvate carboxylase/malate dehydrogenase system (Arnelle and O'Leary, 1992), but this method is less reliable at DIC concentrations below approximately 1 m*M* (K. Scott, unpublished data).

Typically, enzymes will consume or produce only one species of DIC, which presents special issues in substrate or product quantification. It is possible to quantify CO_2 independently from HCO_3^- and CO_3^{-2} via membrane-inlet mass spectrometry and simultaneously quantify the abundances of $^{12}CO_2$ and $^{13}CO_2$ (McNevin *et al.*, 2006). Often, however, the concentrations of CO_2, HCO_3^-, and CO_3^{-2} are calculated from the DIC concentration and pH, using the appropriate dissociation constants and the Henderson–Hasselbach equation.

It is particularly important to select dissociation constants for the carbonate system carefully, as they are very sensitive to temperature, pressure, and ionic strength (Zeebe and Wolf-Gladrow, 2003). At the low-to-moderate ionic strengths present in most enzyme assays, the dissociation constant for the protonation of bicarbonate can be calculated from

$$\begin{aligned} pK_a = -\log\frac{[H^+][HCO_3^-]}{[CO_2]} &= \left[0.175\log\left(\frac{1}{I}\right)\right] \\ &+ \left[0.2957\log\left(\frac{1}{°C}\right)\right] + 6.3572 \end{aligned} \tag{14.7}$$

where I is the ionic strength of the incubation and °C is the incubation temperature in Celsius (Yokota and Kitaoka, 1985). If incubations have ionic strengths and compositions similar to seawater, the CO2SYS program can be used to calculate pK values (Lewis and Wallace, 1998). CO2SYS is available in a variety of formats from Oak Ridge National Laboratories at http://cdiac.ornl.gov/ftp/co2sys/.

Isotope ratios of DIC can be measured via conversion of DIC to CO_2, usually by adding phosphoric acid. Once the conversion is complete, the isotope ratio of the CO_2 can be determined via isotope ratio mass spectrometry, provided the instrument is sufficiently sensitive to measure natural abundance levels of ^{13}C. Newer technologies, for example, tunable laser diode systems and integrated off-axis absorption spectroscopy, provide resolution comparable to isotope ratio mass spectrometers (ca. 0.2‰; Baer *et al.*, 2002; Mihalcea *et al.*, 1997, 1998; Nagali *et al.*, 1996; Webber *et al.*, 2000). Because of the low cost and high performance of these instruments, they may soon displace isotope ratio mass spectrometers as the standard analytical methodology.

To calculate the isotopic composition of a particular component of DIC, it is necessary to take both the pK_a and the equilibrium fractionation factor for the interconversion of the different forms into account. For example, to calculate the isotope ratio of CO_2 from the isotope ratio of DIC at a pH where the dominant form of DIC is HCO_3^-, one would use Eq. (14.6), with the approximation that $R_{DIC} \sim R_{HCO_3^-}$.

Prolonged waiting periods between sample collection and DIC quantification and/or isotope ratio determination should be avoided if possible. This is particularly true if the samples have been acidified to quench the reaction and convert the DIC to CO_2. If the sample container is permeable to CO_2, such as might occur in plastic vessels, the sample will become ^{13}C-enriched due to the kinetic isotope for CO_2 diffusion (O'Leary, 1984), and/or will become contaminated with atmospheric CO_2. This can also be problematic when using glass anaerobic serum vials or autosampler analytical vials (e.g., exetainers), as the rubber stoppers used with these vials can retain inorganic carbon, causing carryover if the septa are reused. Thus, if analysis must be delayed, control experiments should be run with a standard to check whether the storage protocol significantly affects DIC concentration or isotopic composition. One fail-safe storage protocol for CO_2 is cryodistillation with a vacuum line and sealing in glass ampoules (Scott *et al.*, 2004b).

2.3.2. Methane

Unlike DIC, there is one chemical species of methane. As a consequence of its inherent stability (five atoms joined by covalent bonds), there is no abiotic oxidation or reduction to other compounds at typical laboratory conditions. For example, the abiotic oxidation of methane to carbon dioxide without ignition typically requires metal catalysts at elevated temperatures (Foger and Ahmed, 2005; Park *et al.*, 2000). Unlike DIC, methane is not highly soluble in water, nor can it be "trapped" by altering chemical conditions, for example, pH. It too can be absorbed into polymers and, for similar reasons as described above, methane stored for later analyses must be kept in appropriate vessels. Storage within glass ampoules that use low permeability rubber stoppers is acceptable in the short term. As with DIC, cryodistillation and storage in glass ampoules are preferred. Thus, storing methane requires one to consider its modest reactivity, limited solubility, and rapid diffusivity.

Methane concentrations can be readily quantified via infrared gas analyzer (Bartlome *et al.*, 2007; Borjesson *et al.*, 2007; Court and Sephton, 2009; Griffith *et al.*, 2008; Kassi *et al.*, 2008; Kim *et al.*, 2009; Uotila and Kauppinen, 2008), gas chromatography (Behrens *et al.*, 2008; Bock *et al.*, 2010; Court and Sephton, 2009; Fisher *et al.*, 2006; Harrison *et al.*, 2000; Jacq *et al.*, 2008; Kampbell and Vandegrift, 1998; Pedersen *et al.*, 2005; Tang *et al.*, 2006; Valentin *et al.*, 1985), solid-state amperometric gas sensors (Fergus, 2007, 2008; Stetter and Li, 2008), and membrane-inlet or isotope ratio mass spectrometry (Beckmann and Lloyd, 2001; Benstead and Lloyd, 1994; Fisher *et al.*, 2006; Girguis *et al.*, 2003, 2005; Hemond *et al.*, 2008; Lloyd *et al.*, 2002; Mastepanov and Christensen, 2008; Panikov *et al.*, 2007; Schluter and Gentz, 2008; Schluter *et al.*, 2009; Thomas *et al.*, 1995; Tortell and Long, 2009), without acidification or any pretreatment. Some analytical methods, for example, infrared spectroscopy, are best conducted on dry gas to avoid the interference of water with the absorption signal. No widespread biological assays for rapidly quantifying methane concentrations exist, though some studies have used biological methane oxidation as an index for methanogenesis (Fitzgerald, 1996; Owen *et al.*, 1979).

Methane isotope ratios are measured in the same manner as for DIC. Unlike DIC, there is no cause for concern regarding pH, as it has no measureable effect of methane at the relevant time scales. In some instances, such as when measuring trace amounts of methane, the gas is combusted in a high-temperature column with oxygen and a catalyst, and the resulting DIC is "base-trapped" and concentrated for isotope ratio determination. However, the advent of higher sensitivity continuous flow isotope ratio mass spectrometers now enables parts per billion measurements of atmospheric methane with automated chromatographic preconcentration (e.g., Brass and Röckmann, 2010).

It is worth nothing that biological methane oxidation (aerobic and anaerobic) is often quantified by measuring the production of labeled DIC from labeled methane. In such cases, accurate quantification is best enabled by "base-trapping" the DIC (as carbonate) and sparging the solution with high-purity nitrogen, argon, or helium to eliminate any dissolved methane. Vaccum may also be used, but air should never be used to sparge, as atmospheric carbon dioxide will accumulate within the fluids and potentially influence the measurements. In such cases, all the aforementioned concerns regarding DIC concentration and stable isotopic measurements must be considered, in particular, pH and temperature. As above, two-phase systems should be avoided to minimize complexity.

While radioisotopic DIC is commonly available, radioisotopic methane is more difficult to commercially acquire. Many laboratories produce their own radioisotopic methane from biological methanogenesis. If such is the case, care must be taken to purify the resulting methane prior to use as a tracer. Residual labeled biological and abiotic constituents (e.g., volatile fatty acids, DIC) can confound the data, yielding artifically inflated rates of methane usage. Here again, distillation of methane or impurities using chemical or physical traps is recommended prior to use in experimentation.

2.3.3. Multicarbon compounds

If measuring an enzyme kinetic isotope effect requires determining the isotope ratio of the single-carbon unit removed or added to a multiple-carbon compound, simple combustion of this multiple-carbon compound can result in less precise results, as was the case for early attempts to measure isotope fractionation by RubisCO. RubisCO adds a CO_2 molecule to the five-carbon sugar ribulose 1,5-bisphosphate, generating a transient six-carbon intermediate, which spontaneously converts into two molecules of phosphoglycerate. Isotope fractionation was initially calculated by comparing the isotope ratio of the CO_2 substrate to that of the phosphoglycerate produced. Since only one of the six carbons in the product resulted from RubisCO activity, it was necessary to subtract the isotope ratio of five-sixth of the sample to obtain the isotope ratio of the carbon of interest. This difficulty was remedied by purifying the phosphoglyceric acid product, enzymatically releasing the carbon atom fixed by the carboxylase, and measuring its isotope ratio (Roeske and O'Leary, 1984, 1985). If this approach is used, it is critical that the product be quantitatively decarboxylated, as many decarboxylases also have kinetic isotope effects.

3. Culture/Cell-Level Studies

To measure isotope discrimination by intact cells, many of the same considerations apply as for enzyme-level studies. However, one really critical consideration is that, if the reaction of interest occurs within the cell, or in a semi-enclosed extracellular space (e.g., a bacterial periplasm), it is likely that the isotopic composition of the substrate differs from that of the bulk medium. Accordingly, it is important when publishing values from whole-cell assays to acknowledge and explore this possibility when comparing values from enzyme-level assays.

3.1. General assay considerations

Isotope discrimination by cell cultures is measured by growing cells under conditions favorable to the process being studied, and monitoring the stable isotope composition and concentration of target compounds. As for enzyme-level studies, either a single-timepoint or multiple-timepoint approach is possible, and Eqs. (14.3)–(14.5) can be used to describe these systems. For a single-timepoint assay, a culture would be grown under conditions where the concentration and isotope composition of substrate were not substantially diminished, and α would be calculated directly from Eq. (14.3). For this approach, the best system to use would be a chemostat, as cells are growing under steady-state conditions (substrate concentration and composition will be constant). Such measurements are possible in batch culture (e.g., in a flask), but conditions (and therefore the physiological state of the cells) are continually changing in such cultures. Given that cellular responses to changing conditions include changes in nutrient demand, transporter arsenal, and metabolic pathways, isotope discrimination is likely to vary considerably over time in a batch culture. Indeed, carbon isotope fractionation by methanogens is inversely proportional to demand (Londry *et al.*, 2008; Penning *et al.*, 2005; Valentine *et al.*, 2004).

Though problematic for single-timepoint studies, batch culture could be used for multiple-timepoint assays. If, as for enzyme isotope discrimination experiments, the culture is a closed system with respect to the target compound(s) being analyzed, it is possible to apply the RDE to such a system. This approach would be particularly attractive if isotope discrimination is small; tracking the cumulative effect of a small isotope effect as substrate is consumed will magnify the difference between the substrate and the product. As described above, however, such an approach must be used with caution, as physiological changes in the cells resulting from changing culture conditions (e.g., nutrient depletion or end-product accumulation) are likely to result in changing isotope fractionation.

3.2. Setting up the culture assay

It is ideal for cultures used for cell-level studies to be axenic, to ensure that the measured activity is due to the target organism. However, mixed-culture studies are of course more appropriate for processes that require crossfeeding.

When designing the experiment, it is helpful to consider all possible sources and sinks of the target compounds to be measured. For example, the growth vessel should be selected with care to ensure that the target compounds are not contaminated with atmospheric inputs, and conversely cannot be depleted via diffusion from the growth vessel. Additionally, inocula should be small to minimize the effect of prior cultivation conditions on the isotopic composition of substrate and biomass. Furthermore, if isotope discrimination is to be estimated by comparing the isotope composition of substrate to products, it is important to capture, quantify, and analyze all substrates and products (e.g., both biomass and secreted compounds).

For isotope discrimination studies in which cells are cultivated in a chemostat and a single timepoint is taken, it is necessary to verify steady-state conditions. For example, one could measure whether biomass, substrate, and/or product concentrations are constant for a reasonable interval before and after sampling.

Another commonsense consideration for cultivation experiments, be they batch- or chemostat culture based, is that the vessel must be well stirred to prevent localized nutrient depletion as well as undue influence from the kinetic isotope effect associated with diffusion.

3.3. Analytical concerns

3.3.1. Headspace analyses

It is common, with gaseous substrates such as methane and CO_2, to measure headspace gas composition and calculate dissolved gas parameters from these analyses. However, this calculation assumes isotopic and chemical equilibrium between the two phases. This assumption is problematic, as diffusion is a slow process with a kinetic isotope effect; even at low cell densities, organisms can exert large multipermil effects on the dissolved substrate. Any assertion of chemical and isotopic equilibrium between the phases of a two-phase culture must be verified with measurements. Reliance on headspace analysis is likely a major contributor to the heterogeneity in isotope fractionation observed for culture studies.

3.3.2. Biomass

Cells that are to be used for isotopic analyses should be washed upon harvesting to remove compounds that could skew estimates of biomass isotopic composition (e.g., DIC). If possible, it is helpful to design the growth medium to minimize the presence of compounds that could compromise the interpretation of the biomass values. For example, if the $\delta^{13}C$ value of biomass in a methanogen is going to be compared to the $\delta^{13}C$ of the DIC in the growth medium, it is wise to use inorganic buffers to prevent the influence of the isotopic signal from an organic buffer (e.g., TRIS or HEPES) that may be difficult to completely remove from cells harvested from a culture.

Biomass must be combusted prior to mass spectrometry, and it is important to verify complete combustion, as early combustion products are likely to be ^{12}C-enriched due to differences in bond energy between ^{12}C and ^{13}C. Complete combustion can be verified either by comparing to an organic standard combusted in parallel with a sample, or detected gravimetrically.

4. Calculations

To calculate a kinetic isotope effect from a multiple-timepoint assay using the RDE, it is simplest to use a linearized form of Eq. (14.4). It has been common practice to simply natural log-transform the RDE

$$\ln\left(\frac{R_{st}}{R_{si}}\right) = \left(\frac{1}{\alpha} - 1\right) \ln\left(\frac{C_{st}}{C_{si}}\right) \tag{14.8}$$

However, this form of the equation places more emphasis on the initial measurement than on the other timepoints, since each timepoint is divided by the initial one (C_{si}, R_{si}), which can skew the estimate of α (McNevin *et al.*, 2006; Scott *et al.*, 2004a). A better approach is to regress the natural log of the substrate isotope ratio on the natural log of the substrate concentration:

$$\ln(R_{st}) = \left(\frac{1}{\alpha} - 1\right)\ln(C_{st}) + \ln\left[\frac{R_{si}}{C_{si}{}^{1/\alpha - 1}}\right] \quad (14.9)$$

If the product is monitored, it is not possible to linearize Eq. (14.5) to estimate α. Nonparametric bootstrapping could be used instead. To obtain an unbiased estimate of α, if N datapoints are taken, the R and C values from each datapoint could be used to calculate N estimates of α. One might be concerned about simply using the mean and sample standard error for such a derived quantity, because the sampling distribution may well not be normal. To derive a bootstrap estimate, a set of N-1 of these α values could be chosen at random (sampling with replacement), saving the mean of these (α'). If this procedure were repeated a large number (1000 or 5000) of times, the mean of these α' values would be an unbiased estimate of the mean α. A 95% confidence interval would be given by finding the smallest and largest 2.5% of the α' values.

If the substrate or product of interest is one of the components of the DIC system, it is necessary to modify these equations to include DIC equilibrium isotope effects. For example, if the substrate for an enzyme reaction is CO_2, and the concentration and isotopic composition of DIC are being monitored, Eq. (14.9) must be modified to calculate α. If

$$K = \frac{R_{HCO_3^-}}{R_{CO_2}} \quad (14.10)$$

and the reaction conditions are circumneutral so that HCO_3^- is the dominant form of DIC present, combining Eq. (14.9) and Eq. (14.10) results in

$$\ln(R_{DICt}) = \left[\frac{1}{\alpha K} - 1\right]\ln(DIC_t) + \ln\left[\frac{R_{DICi}}{DIC_i{}^{(1/\alpha K) - 1}}\right] \quad (14.11)$$

where R_{DICt} and DIC_t are the isotope ratio and concentration of DIC at a timepoint, R_{DICi} and DIC_i are the corresponding values for the DIC initially present in the reaction (Scott *et al.*, 2004b).

These linear forms of the RDE assume that α is constant over the course of the reaction. As mentioned above, some enzymes do not fractionate to the same degree at all concentrations of the substrate, and cells in a closed system often fractionate differently when growth conditions change. If this is the case, it is not appropriate to apply these equations.

The best estimate of α, and the error associated with it, will be obtained with multiple independent reactions or cultures. Often, different reactions will have different numbers of datapoints, and different amounts of error associated with their individual α values. Taking the average of the collected α values weighs more-informative reactions (more datapoints) equally with less-informative ones (fewer datapoints). To elucidate the best approach to take when substrate concentrations and isotope compositions are measured, simulated datasets with realistic measurement error in DIC quantification and R_{DIC} were used to find the method of combining datasets that had the best probability of generating α values and associated error estimates that covered the "true value" of α (Scott *et al.*, 2004a). If the error is the same for each individual dataset, combining the reactions into a single regression with dummy variables for each reaction reliably generated accurate estimates of α (Scott *et al.*, 2004a). For example, to combine data from two independent reactions, Eq. (14.9) would be modified to

$$\begin{aligned}\ln(R_{st}) = {} & \left(\frac{1}{\alpha} - 1\right)\ln(C_{st}) + \ln\left[\frac{R_{si1}}{C_{si1}{}^{1/\alpha-1}}\right] \\ & + \left(\ln\frac{R_{si1}}{C_{si1}{}^{1/\alpha-1}} - \ln\left[\frac{R_{si2}}{C_{si2}{}^{1/\alpha-1}}\right]\right)D2\end{aligned} \tag{14.12}$$

in which R_{si1} and C_{si1} are the isotope ratio and concentration of substrate initially present in reaction 1, R_{si2} and C_{si2} are the corresponding values for the second reaction, and $D2$ is a dummy variable, and $=0$ for reaction 1, and $=1$ for reaction 2. If the error is not equal between reactions, it is better to combine them using Pitman estimators (Scott *et al.*, 2004a). As Pitman estimators are a novelty beyond the discipline of statistics, and space does not permit a full description here, the reader is referred to (Scott *et al.*, 2004a), and a Matlab program to combine datasets using this method is available online at http://kmscott.myweb.usf.edu/.

If product concentrations and isotope compositions are being monitored instead, an extension of the approach suggested for a single experiment of this nature is suggested here. The populations of α' values calculated from each experiment could be compared via single-factor ANOVA. If these populations are statistically indistinguishable, they could be pooled into a single population from which the overall average α and 95% confidence interval could be determined.

ACKNOWLEDGMENTS

We gratefully acknowledge support from the National Science Foundation (MCB-0643713 to K. M. S., DEB-0614468 to G. F., and MCB-0702504 and OCE-0838107 to P. R. G.).

REFERENCES

Arnelle, D., and O'Leary, M. H. (1992). Binding of carbon dioxide to phosphoenolpyruvate carboxykinase deduced from carbon kinetic isotope effects. *Biochemistry* **31,** 4363–4368.

Baer, D. S., Paul, J. B., Gupta, J. B., and O'Keefe, A. (2002). Sensitive absorption measurements in the near-infrared region using off-axis integrated-cavity-output spectroscopy. *Appl. Phys. B Lasers Opt.* **75,** 261–265.

Bartlome, R., Baer, M., and Sigrist, M. W. (2007). High-temperature multipass cell for infrared spectroscopy of heated gases and vapors. *Rev. Sci. Instrum.* **78,** 013110.

Beckmann, M., and Lloyd, D. (2001). Mass spectrometric monitoring of gases (CO_2, CH_4, O_2) in a mesotrophic peat core from Kopparas Mire, Sweden. *Glob. Change Biol.* **7,** 171–180.

Behrens, M., Schmitt, J., Richter, K. U., Bock, M., Richter, U. C., Levin, I., and Fischer, H. (2008). A gas chromatography/combustion/isotope ratio mass spectrometry system for high-precision delta^{13}C measurements of atmospheric methane extracted from ice core samples. *Rapid Commun. Mass Spectrom.* **22,** 3261–3269.

Benstead, J., and Lloyd, D. (1994). Direct mass-spectrometric measurement of gases in peat cores. *FEMS Microbiol. Ecol.* **13,** 233–240.

Bock, M., Schmitt, J., Behrens, M., Moller, L., Schneider, R., Sapart, C., and Fischer, H. (2010). A gas chromatography/pyrolysis/isotope ratio mass spectrometry system for high-precision deltaD measurements of atmospheric methane extracted from ice cores. *Rapid Commun. Mass Spectrom.* **24,** 621–633.

Borjesson, G., Samuelsson, J., and Chanton, J. (2007). Methane oxidation in Swedish landfills quantified with the stable carbon isotope technique in combination with an optical method for emitted methane. *Environ. Sci. Technol.* **41,** 6684–6690.

Brass, M., and Röckmann, T. (2010). Continuous-flow isotope ratio mass spectrometry method for carbon and hydrogen isotope measurements on atmospheric methane. *Atmos. Meas. Tech. Discuss.* **3,** 433–2476.

Cook, P. F. (1998). Mechanism from isotope effects. *Isot. Environ. Health Stud.* **34,** 3–17.

Court, R. W., and Sephton, M. A. (2009). Quantitative flash pyrolysis Fourier transform infrared spectroscopy of organic materials. *Anal. Chim. Acta* **639,** 62–66.

Dobrinski, K. P., Longo, D. L., and Scott, K. M. (2005). A hydrothermal vent chemolithoautotroph with a carbon concentrating mechanism. *J. Bacteriol.* **187,** 5761–5766.

Fergus, J. W. (2007). Solid electrolyte based sensors for the measurement of CO and hydrocarbon gases. *Sens. Actuators B Chem.* **122,** 683–693.

Fergus, J. W. (2008). A review of electrolyte and electrode materials for high temperature electrochemical CO_2 and SO_2 gas sensors. *Sens. Actuators B Chem.* **134,** 1034–1041.

Fisher, R., Lowry, D., Wilkin, O., Sriskantharajah, S., and Nisbet, E. G. (2006). High-precision, automated stable isotope analysis of atmospheric methane and carbon dioxide using continuous-flow isotope-ratio mass spectrometry. *Rapid Commun. Mass Spectrom.* **20,** 200–208.

Fitzgerald, P. A. (1996). Comprehensive monitoring of a fluidised bed reactor for anaerobic treatment of high strength wastewater. *Chem. Eng. Sci.* **51,** 2829–2834.

Foger, K., and Ahmed, K. (2005). Catalysis in high-temperature fuel cells. *J. Phys. Chem. B* **109,** 2149–2154.

Fuchs, G., Thauer, R., Ziegler, H., and Stichler, W. (1979). Carbon isotope fractionation by *Methanobacterium thermoautotrophicum*. *Arch. Microbiol.* **120,** 135–139.

Girguis, P. R., Lee, R. L., Desaulniers, N., Childress, J. J., Pospesel, M., Felbeck, H., and Zal, F. (2000). Fate of nitrate acquired by the tubeworm *Riftia pachyptila*. *Appl. Environ. Microbiol.* **66,** 2783–2790.

Girguis, P. R., Childress, J. J., Freytag, J. K., Klose, K., and Stuber, R. (2002). Effects of metabolite uptake on proton-equivalent elimination by two species of deep-sea vestimentiferan tubeworm, *Riftia pachyptila* and *Lamellibrachia* cf. *luymesi*: Proton elimination is a necessary adaptation to sulfide-oxidizing chemoautotrophic symbionts. *J. Exp. Biol.* **205,** 3055–3066.

Girguis, P. R., Orphan, V. J., Hallam, S. J., and DeLong, E. F. (2003). Growth and methane oxidation rates of anaerobic methanotrophic archaea in a continuous-flow bioreactor. *Appl. Environ. Microbiol.* **69,** 5472–5482.

Girguis, P. R., Cozen, A. E., and DeLong, E. F. (2005). Growth and population dynamics of anaerobic methane-oxidizing archaea and sulfate-reducing bacteria in a continuous-flow bioreactor. *Appl. Environ. Microbiol.* **71,** 3725–3733.

Griffith, D. W., Bryant, G. R., Hsu, D., and Reisinger, A. R. (2008). Methane emissions from free-ranging cattle: Comparison of tracer and integrated horizontal flux techniques. *J. Environ. Qual.* **37,** 582–591.

Guy, R. D., Fogel, M. L., and Berry, J. A. (1993). Photosynthetic fractionation of the stable isotopes of oxygen and carbon. *Plant Physiol.* **101,** 37–47.

Harrison, D., Seakins, P. W., and Lewis, A. C. (2000). Simultaneous monitoring of atmospheric methane and speciated non-methane hydrocarbon concentrations using Peltier effect sub-ambient pre-concentration and gas chromatography. *J. Environ. Monit.* **2,** 59–63.

Hayes, J. M. (1993). Factors controlling ^{13}C contents of sedimentary organic compounds: Principles and evidence. *Mar. Geol.* **113,** 111–125.

Hemond, H. F., Mueller, A. V., and Hemond, M. (2008). Field testing of lake water chemistry with a portable and an AUV-based mass spectrometer. *J. Am. Soc. Mass Spectrom.* **19,** 1403–1410.

Hügler, M., Wirsen, C. O., Fuchs, G., Taylor, C. D., and Sievert, S. M. (2005). Evidence for autotrophic CO_2 fixation via the reductive tricarboxylic acid cycle by members of the epsilon subdivision of proteobacteria. *J. Bacteriol.* **187,** 3020–3027.

Jacq, K., Delaney, E., Teasdale, A., Eyley, S., Taylor-Worth, K., Lipczynski, A., Reif, V. D., Elder, D. P., Facchine, K. L., Golec, S., Oestrich, R. S., Sandra, P., *et al.* (2008). Development and validation of an automated static headspace gas chromatography-mass spectrometry (SHS-GC-MS) method for monitoring the formation of ethyl methane sulfonate from ethanol and methane sulfonic acid. *J. Pharm. Biomed. Anal.* **48,** 1339–1344.

Kampbell, D. H., and Vandegrift, S. A. (1998). Analysis of dissolved methane, ethane, and ethylene in ground water by a standard gas chromatographic technique. *J. Chromatogr. Sci.* **36,** 253–256.

Kassi, S., Gao, B., Romanini, D., and Campargue, A. (2008). The near-infrared (1.30-1.70 microm) absorption spectrum of methane down to 77 K. *Phys. Chem. Chem. Phys.* **10,** 4410–4419.

Kim, S. S., Menegazzo, N., Young, C., Chan, J., Carter, C., and Mizaikoff, B. (2009). Mid-infrared trace gas analysis with single-pass fourier transform infrared hollow waveguide gas sensors. *Appl. Spectrosc.* **63,** 331–337.

Lewis, E., and Wallace, D. W. R. (1998). *Program Developed for CO_2 System Calculations. ORNL/CDIAC-105* Oak Ridge National Laboratory, US Department of Energy, Oak Ridge.

Lloyd, D., Thomas, K. L., Cowie, G., Tammam, J. D., and Williams, A. G. (2002). Direct interface of chemistry to microbiological systems: Membrane inlet mass spectrometry. *J. Microbiol. Methods* **48,** 289–302.

Londry, K. L., Dawson, K. G., Grover, H. D., Summons, R. E., and Bradley, A. S. (2008). Stable carbon isotope fractionation between substrates and products of *Methanosarcina barkeri*. *Org. Geochem.* **39,** 608–621.

Madigan, M. T., Takigiku, R., Lee, R. G., Gest, H., and Hayes, J. M. (1989). Carbon isotope fractionation by thermophilic phototrophic sulfur bacteria: Evidence for autotrophic growth in natural populations. *Appl. Environ. Microbiol.* **55,** 639–644.

Mariotti, A., Germon, J. C., Hubert, P., Kaiser, P., Letolle, R., Tardieux, A., and Tardieux, P. (1981). Experimental determination of nitrogen kinetic isotope fractionation: Some principles; illustration for the denitrification and nitrification processes. *Plant Soil* **62,** 413–430.

Mastepanov, M., and Christensen, T. R. (2008). Bimembrane diffusion probe for continuous recording of dissolved and entrapped bubble gas concentrations in peat. *Soil Biol. Biochem.* **40,** 2992–3003.

McNevin, D. B., Badger, M. R., Kane, H. J., and Farquhar, G. D. (2006). Measurement of (carbon) kinetic isotope effect by Rayleigh fractionation using membrane inlet mass spectrometry for CO_2-consuming reactions. *Funct. Plant Biol.* **33,** 1115–1128.

Melander, L., and Saunders, W. H. (1980). Reaction Rates of Isotopic Molecules. Wiley-Interscience, New York.

Mihalcea, R. M., Baer, D. S., and Hanson, R. K. (1997). Diode laser sensor for measurements of CO, CO_2, and CH_4 in combustion flows. *Appl. Opt.* **36,** 8745–8752.

Mihalcea, R. M., Baer, D. S., and Hanson, R. K. (1998). A diode-laser absorption sensor system for combustion emission measurements. *Meas. Sci. Technol.* **9,** 327–338.

Mook, W. G., Bommerson, J. C., and Staverman, W. H. (1974). Carbon isotope fractionation between dissolved bicarbonate and gaseous carbon dioxide. *Earth Planet Sci. Lett.* **22,** 169–176.

Nagali, V., Chou, S. I., Baer, D. S., Hanson, R. K., and Segall, J. (1996). Tunable diode-laser absorption measurements of methane at elevated temperatures. *Appl. Opt.* **35,** 4026–4032.

O'Leary, M. H. (1984). Measurement of the isotopic fractionation associated with diffusion of carbon dioxide in aqueous solution. *J. Phys. Chem.* **88,** 823–825.

O'Leary, M. H., Rife, J. E., and Slater, J. D. (1981). Kinetic and isotope effect studies of maize phosphoenolpyruvate carboxylase. *Biochemistry* **20,** 7308–7314.

Owen, W. F., Stuckey, D. C., Healy, J. B., Young, L. Y., and McCarty, P. L. (1979). Bioassay for monitoring biochemical methane potential and anaerobic toxicity. *Water Res.* **13,** 485–492.

Panikov, N. S., Mastepanov, M. A., and Christensen, T. R. (2007). Membrane probe array: Technique development and observation of CO_2 and CH_4 diurnal oscillations in peat profile. *Soil Biol. Biochem.* **39,** 1712–1723.

Pardue, J. W., Scalan, R. S., VanBaalen, C., and Parker, P. L. (1976). Maximum carbon isotope fractionation in photosynthesis by blue-green algae and a green alga. *Geochim. Cosmochim. Acta* **40,** 309–312.

Park, S., Vohs, J. M., and Gorte, R. J. (2000). Direct oxidation of hydrocarbons in a solid-oxide fuel cell. *Nature* **404,** 265–267.

Pedersen, I. T., Holmen, K., and Hermansen, O. (2005). Atmospheric methane at Zeppelin Station in Ny-Alesund: Presentation and analysis of in situ measurements. *J. Environ. Monit.* **7,** 488–492.

Penning, H., Plugge, C. M., Galand, P. E., and Conrad, R. (2005). Variation of carbon isotope fractionation in hydrogenotrophic methanogenic microbial cultures and environmental samples at different energy status. *Glob. Change Biol.* **11,** 2103–2113.

Quandt, L., Gottschalk, G., Ziegler, H., and Stichler, W. (1977). Isotope discrimination by photosynthetic bacteria. *FEMS Microbiol. Lett.* **1,** 125–128.

Robinson, J. J., Scott, K. M., Swanson, S. T., O'Leary, M. H., Horken, K., Tabita, F. R., and Cavanaugh, C. M. (2003). Kinetic isotope effect and characterization of form II RubisCO from the chemoautotrophic endosymbionts of the hydrothermal vent tubeworm *Riftia pachyptila. Limnol. Oceanogr.* **48,** 48–54.

Roeske, C. A., and O'Leary, M. H. (1984). Carbon isotope effects on the enzyme-catalyzed carboxylation of ribulose bisphosphate. *Biochemistry* **23,** 6275–6284.

Roeske, C. A., and O'Leary, M. H. (1985). Carbon isotope effect on carboxylation of ribulose bisphosphate catalyzed by ribulosebisphosphate carboxylase from *Rhodospirillum rubrum*. *Biochemistry* **24,** 1603–1607.

Ruby, E. G., Jannasch, H. W., and Deuser, W. G. (1987). Fractionation of stable carbon isotopes during chemoautotrophic growth of sulfur-oxidizing bacteria. *Appl. Environ. Microbiol.* **53,** 1940–1943.

Sakata, S., Hayes, J. M., Rohmer, M., Hooper, A. B., and Seemann, M. (2008). Stable carbon-isotopic compositions of lipids isolated from the ammonia-oxidizing chemoautotroph *Nitrosomonas europaea*. *Org. Geochem.* **39,** 1725–1734.

Schluter, M., and Gentz, T. (2008). Application of membrane inlet mass spectrometry for online and in situ analysis of methane in aquatic environments. *J. Am. Soc. Mass Spectrom.* **19,** 1395–1402.

Schluter, M., Gentz, T., and Bussmann, I. (2009). Quantification of methane emissions from pockmarks (Lake Constance) by online and onsite membrane inlet mass spectrometry. *Geochim. Cosmochim. Acta* **73,** A1176.

Scott, K. M. (2003). A d13C-based carbon flux model for the hydrothermal vent chemoautotrophic symbiosis *Riftia pachyptila* predicts sizeable CO_2 gradients at the host-symbiont interface. *Environ. Microbiol.* **5,** 424–432.

Scott, K. M., Lu, X., Cavanaugh, C. M., and Liu, J. (2004a). Optimal methods for estimating kinetic isotope effects from different forms of the Rayleigh distillation equation. *Geochim. Cosmochim. Acta* **68,** 433–442.

Scott, K. M., Schwedock, J., Schrag, D. P., and Cavanaugh, C. M. (2004b). Influence of form IA RubisCO and environmental dissolved inorganic carbon on the $d^{13}C$ of the clam-bacterial chemoautotrophic symbiosis *Solemya velum*. *Environ. Microbiol.* **6,** 1210–1219.

Scott, K. M., Henn-Sax, M., Longo, D., and Cavanaugh, C. M. (2007). Kinetic isotope effect and biochemical characterization of form IA RubisCO from the marine cyanobacterium *Prochlorococcus marinus* MIT9313. *Limnol. Oceanogr.* **52,** 2199–2204.

Stetter, J. R., and Li, J. (2008). Amperometric gas sensors—A review. *Chem. Rev.* **108,** 352–366.

Tang, J. H., Bao, Z. Y., Xiang, W., Qiao, S. Y., and Li, B. (2006). On-line method for measurement of the carbon isotope ratio of atmospheric methane and its application to atmosphere of Yakela condensed gas field. *Huan Jing Ke Xue* **27,** 14–18.

Thomas, K. L., Price, D., and Lloyd, D. (1995). A comparison of different methods for the measurement of dissolved gas gradients in waterlogged peat cores. *J. Microbiol. Methods* **24,** 191–198.

Tortell, P. D., and Long, M. C. (2009). Spatial and temporal variability of biogenic gases during the Southern Ocean spring bloom. *Geophys. Res. Lett.* **36**: L01603.

Uotila, J., and Kauppinen, J. (2008). Fourier transform infrared measurement of solid-, liquid-, and gas-phase samples with a single photoacoustic cell. *Appl. Spectrosc.* **62,** 655–660.

Valentin, J. R., Carle, G. C., and Phillips, J. B. (1985). Determination of methane in ambient air by multiplex gas chromatography. *Anal. Chem.* **57,** 1035–1039.

Valentine, D. L., Chidthaisong, A., Rice, A., Reeburgh, W. S., and Tyler, S. C. (2004). Carbon and hydrogen isotope fractionation by moderately thermophilic methanogens. *Geochim. Cosmochim. Acta* **68,** 1571–1590.

vanderMeer, M. T. J. (2007). Impact of carbon metabolism on ^{13}C signatures of cyanobacteria and green non-sulfur-like bacteria inhabiting a microbial mat from an alkaline siliceous hot spring in Yellowstone National Park. *Environ. Microbiol.* **9,** 482–491.

vanderMeer, M. T. J., Schouten, S., Rijpstra, W. I. C., Fuchs, G., and Damste, J. S. S. (2001a). Stable carbon isotope fractionations of the hyperthermophilic crenarchaeon *Metallosphaera sedula*. *FEMS Microbiol.* **196,** 67–70.

vanderMeer, M. T. J., Schouten, S., vanDongen, B. E., Rijpstra, I. C., Fuchs, G., Damste, J. S. S., deLeeuw, J. W., and Ward, D. M. (2001b). Biosynthetic controls on the ^{13}C content of organic components in the photoautotrophic bacterium *Chloroflexus aurantiacus*. *J. Biol. Chem.* **276,** 10971–10976.

Webber, M. E., Wang, J., Sanders, S. T., Baer, D. S., and Hanson, R. K. (2000). In situ combustion measurements of CO, CO_2, H_2O and temperature using diode laser absorption sensors. *Proc. Combust. Inst.* **28,** 407–413.

Williams, T. J., Zhang, C. L., Scott, J. H., and Bazylinski, D. A. (2006). Evidence for autotrophy via the reverse tricarboxylic acid cycle in the marine magnetotactic coccus strain MC-1. *Appl. Environ. Microbiol.* **72,** 1322–1329.

Yokota, A., and Kitaoka, S. (1985). Correct pK values for dissociation constant of carbonic acid lower the reported Km values of ribulose bisphosphate carboxylase to half. Presentation of a nomograph and an equation for determining the pK values. *Biochem. Biophys. Res. Commun.* **131,** 1075–1079.

Zeebe, R. E., and Wolf-Gladrow, D. (2003). CO_2 in Seawater: Equilibrium, Kinetics, Isotopes. Elsevier, New York.

CHAPTER FIFTEEN

2-Oxoacid Metabolism in Methanogenic CoM and CoB Biosynthesis

David E. Graham

Contents

Abstract

Coenzyme M (CoM) and coenzyme B (CoB) are essential for methane production by the euryarchaea that employ this specialized anaerobic metabolism. Two pathways are known to produce CoM, 2-mercaptoethanesulfonate, and both converge on the 2-oxoacid sulfopyruvate. These cells have recruited the rich

Oak Ridge National Laboratory, Biosciences Division and University of Tennessee Knoxville, Microbiology Department, Oak Ridge, Tennessee, USA

Methods in Enzymology, Volume 494
ISSN 0076-6879, DOI: 10.1016/B978-0-12-385112-3.00015-9

biochemistry of amino acid and 2-oxoacid metabolizing enzymes to produce a compound that resembles oxaloacetate, but with a more stable and acidic sulfonate group. 7-Mercaptoheptanoylthreonine phosphate, CoB, likewise owes its carbon backbone to a 2-oxoacid. Three enzymes recruited from leucine biosynthesis have evolved to catalyze the elongation of 2-oxoglutarate to 2-oxosuberate in CoB biosynthesis. This chapter describes the enzymology, synthesis, and analytical techniques used to study 2-oxoacid metabolism in these pathways. Protein structure and mechanistic information from enzymes provide insight into the evolution of new enzymatic activity, and the evolution of substrate specificity from promiscuous enzyme scaffolds.

1. Introduction

Both hydrogenotrophic and aceticlastic methanogens use the methyl-coenzyme M reductase (MCR) enzyme to catalyze the terminal reduction of a methyl group to methane. This remarkable metalloenzyme uses a nickel-tetrapyrrole coenzyme F_{430} to catalyze the reductive cleavage of a methyl thioether form of coenzyme M (CoM) using the thioacyl chain of coenzyme B (CoB; Fig. 15.1). The products are methane and a hetero-disulfide compound (CoM–S–S–CoB), which is reduced to generate potential across the cellular membrane. This reaction was considered unique to methanogens, and the most common inhibitor of methanogenesis in the laboratory is bromoethanesulfonate, a CoM analog that inhibits MCR activity (Ellermann *et al.*, 1989; Goenrich *et al.*, 2004).

Almost two decades after its initial discovery in methanogenic archaea (McBride and Wolfe, 1971; Taylor and Wolfe, 1974), CoM was reisolated as an epoxide-opening cofactor from the α-proteobacterium *Xanthobacter autotrophicus* Py2 (Allen *et al.*, 1999). CoM's role in oxidative alkene metabolism has since been solidified by reports of its biosynthesis and activity in epoxide carboxylation reactions from diverse bacteria (Krishnakumar *et al.*, 2008). Recently, anaerobic methane-oxidizing archaea were reported to express the MCR enzyme to activate methane using CoM–S–S–CoB, that is, reverse methanogenesis (Scheller *et al.*, 2010).

2. Chemistry of 2-Oxoacids

The 2-oxoacids, also known as α-ketoacids, are key compounds in central metabolism. In metabolic networks, pyruvate, oxaloacetate, and 2-oxoglutarate form nodes with numerous connections (representing reactions) to amino acids, carbohydrates, and coenzymes (Table 15.1). These "hubs" in scale-free networks are natural starting points for the evolution of

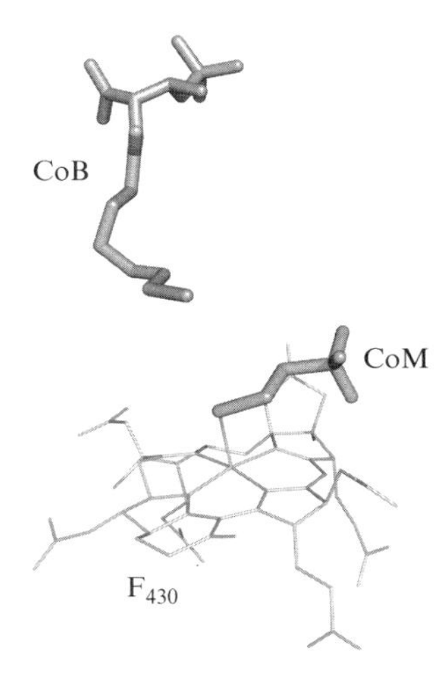

Figure 15.1 The methyl-coenzyme M reductase (MCR) active site contains a nickel-tetrapyrrole coenzyme F_{430}, which coordinates the thiol group of 2-mercaptoethane-sulfonate (Coenzyme M; CoM). The thioacyl chain of 7-mercaptoheptanoylthreonine phosphate (Coenzyme B; CoB or HS-HT*P*) is oriented to form a heterodisulfide with CoM, following the reductive cleavage of a methyl thioether. Figure drawn using PyMol, from the MCR crystal structure model with PDB identification number 1HBN (Grabarse *et al.*, 2001).

new pathways (Jeong *et al.*, 2000). Modern CoM biosynthetic pathways evolved by recruiting paralogs of genes encoding pyruvate decarboxylase, aspartate aminotransferase, or threonine synthase and adapting them to recognize a new 2-oxoacid, sulfopyruvate. CoB biosynthesis evolved by recruiting three successive enzymes in leucine biosynthesis that extend a 2-oxoacid by one methylene group and adapting them to recognize dicarboxylate substrates. Remarkably, these enzymes act as polymerases, extending 2-oxoglutarate (five carbon atoms) to 2-oxosuberate (eight carbons) in three rounds of catalysis.

With a carbonyl adjacent to a carboxylic acid group, the 2-oxoacids form both enol and keto tautomers: the enol/keto equilibrium constant for oxaloacetic acid is 0.12, which is substantially higher than 0.01% of pyruvate molecules found in the enol form at neutral pH (Burbaum and Knowles, 1989; Hess and Reed, 1972; Kozlowski and Zuman, 1987). On a faster timescale, the keto form can be hydrated to form the *gem*-diol, with keto/hydrate equilibrium constants of 0.34 for pyruvic acid, 0.46 for 2-oxoglutaric acid, and 0.14 for oxaloacetic acid (Fig. 15.2; Damitio *et al.*, 1992; Kerber and Fernando, 2010; Kozlowski and Zuman, 1987). The hydrated proportion of these molecules decreases to 5–8% at neutral pH, where they form dianions (Copper and Redfield, 1975); nevertheless, a substantial proportion of these molecules appear to proteins as hydrates rather than the textbook keto forms. The molecules' reported pK_a values are also composites of ionization constants for these interconverting forms.

Table 15.1 Common amino acids and their 2-oxoacid analogs

	2-Oxoacid	
Amino acid[a]	Common name	Systematic name
Alanine	Pyruvate	2-Oxopropanoate
Arginine	2-Oxo-δ-guanidinovalerate	5-Guanidino-2-oxopentanoate
Asparagine	Oxosuccinamate	4-Amino-2,4-dioxobutanoate
Aspartate	Oxaloacetate	Oxobutanedioate
Cysteine	3-Mercaptopyruvate	3-Mercapto-2-oxopropionate
Cysteate	3-Sulfopyruvate	2-Oxo-3-sulfopropanoate
Glutamate	2-Oxoglutarate	2-Oxopentanedioate
Glutamine	2-Oxoglutaramate	5-Amino-2,5-dioxopentanoate
Glycine	Glyoxylate	Oxoethanoate
Histidine	β-Imidazolepyruvate	3-(1*H*-imidazol-4-yl)-2-oxopropanoate
Isoleucine	2-Oxo-3-methylvalerate	3-Methyl-2-oxopentanoate
Leucine	2-Oxoisocaproate	4-Methyl-2-oxopentanoate
Lysine	2-Oxo-ε-aminocaproate	6-Amino-2-oxohexanoate
Methionine	2-Oxo-γ-methiolbutyrate	4-(Methylthio)-2-oxobutanoate
Phenylalanine	Phenylpyruvate	2-Oxo-3-phenylpropanoate
Serine	3-Hydroxypyruvate	3-Hydroxy-2-oxopropionate
Threonine	2-Oxo-β-hydroxybutyrate	3-Hydroxy-2-oxobutanoate
Tryptophan	β-Indoylpyruvate	3-(1*H*-indol-3-yl)-2-oxopropanoate
Tyrosine	4-Hydroxy-phenylpyruvate	3-(4-hydroxyphenyl)-2-oxopropanoate
Valine	2-Oxoisovalerate	3-Methyl-2-oxobutanoate

[a] See Meister (1957) for an extended list of corresponding amino acids and 2-oxoacids.

3. Biosynthesis of Coenzyme M

CoM can be readily synthesized using 2-bromoethanesulfonate in the laboratory (Romesser and Balch, 1980), or it can be purchased from chemical distributors: CoM or MESNA is used clinically to scavenge reactive oxygen species produced during chemotherapy. The small size of CoM (2-mercaptoethanesulfonate) and its simple chemical synthesis belie its complicated biosynthesis. Some ruminant methanogens such as

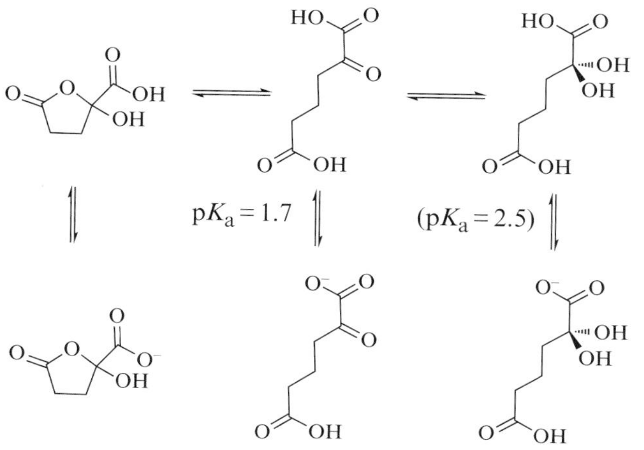

$pK_a = 1.7$ $(pK_a = 2.5)$

Figure 15.2 2-Oxoglutarate can cyclize to form the lactol 2-hydroxy-5-oxotetrahydrofuran-2-carboxylic acid (Kerber and Fernando, 2010). Alternatively, 2OG can be rapidly hydrated to form a ketal, 2,3-dihydroxypentanedioic acid. The pK_a of the oxo form was estimated to be 1.7–1.9, while a composite of ionization constants for the lactol and ketal forms was estimated as 2.5. The empirical pK_a' for 2-oxoglutarate, a composite of all three ionization constants, has been reported between 2.0 and 2.3 at 25 °C (Kerber and Fernando, 2010).

Methanobrevibacter ruminantium are auxotrophic for CoM (Balch and Wolfe, 1976). Isotope incorporation experiments showed that methanogenic rumen isolate 10–16b produced CoM containing both carbon atoms and two protons from acetate, in ratios consistent with acetate incorporation into phosphoenolpyruvate-derived amino acids (White, 1985). The biosynthetic intermediates sulfolactate, sulfopyruvate, and sulfoacetaldehyde were subsequently identified by GC–MS analysis of *Methanobacterium formicicum* extracts, consistent with the current pathway shown in Fig. 15.3A (White, 1986). The known enzymes of CoM biosynthesis were first identified in the model hydrogenotrophic archaeon *Methanocaldococcus jannaschii.*

In the first dedicated reaction for CoM biosynthesis in this organism, the phosphosulfolactate synthase enzyme (ComA; EC 4.4.1.19) catalyzes the nucleophilic addition of sulfite to phosphoenolpyruvate, forming (*R*)-phosphosulfolactate (Graham *et al.*, 2002). A molecular model produced from X-ray diffraction data of crystallized ComA protein identified a novel protein family with a triosephosphate isomerase $(\alpha/\beta)_8$ barrel fold (Wise *et al.*, 2003). Land plants have homologs of ComA, called Hsa32, that are important for acquired thermotolerance (Charng *et al.*, 2006). However, no catalytic activity has been reported for other members of this family.

The 2-phosphosulfolactate phosphatase enzyme (ComB; EC 3.1.3.71) catalyzes the metal ion-dependent hydrolysis of phosphosulfolactate

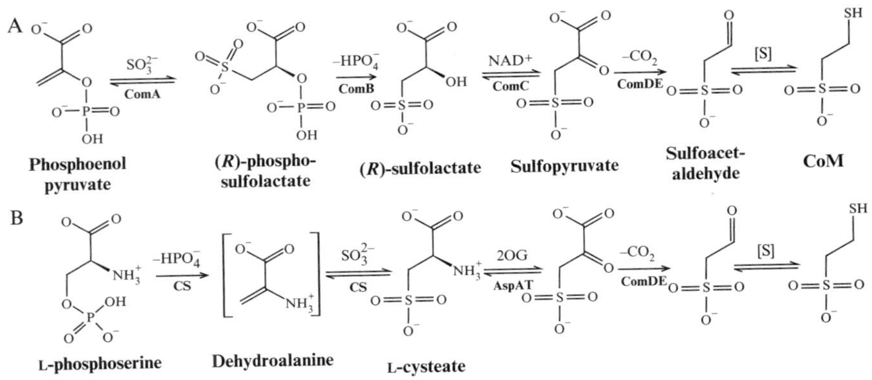

Figure 15.3 Methanogens use two different pathways to produce sulfopyruvate for coenzyme M biosynthesis (Graham *et al.*, 2009). These pathways use nonhomologous enzymes to catalyze the nucleophilic addition of sulfite to an unsaturated carboxylate. Both pathways converge on sulfopyruvate. Members of the Methanococcales, Methanobacteriales, and Methanopyrales are predicted to use pathway A. Members of the Methanosarcinales and Methanomicrobiales are predicted to use pathway B.

(Graham *et al.*, 2001). A structural model of a *Clostridium acetobutylicum* homolog cocrystallized with the sulfolactate product revealed a new protein fold, differentiating the ComB phosphatase superfamily from other phosphatases (DiDonato *et al.*, 2006). *C. acetobutylicum* lacks a ComA homolog to produce phosphosulfolactate. Due to the broad substrate specificity of *M. jannaschii* ComB, additional biochemical and physiological information will be required to determine the function of homologs in organisms that do not produce CoM.

A sulfolactate dehydrogenase enzyme (ComC; EC 1.1.1.272) catalyzes the NAD^+-dependent oxidation of (*R*)-sulfolactate to produce sulfopyruvate (Graupner *et al.*, 2000a). This unusual enzyme catalyzes hydride transfer from L-2-hydroxyacids to the Si-face of NAD^+ (Graupner and White, 2001). Although (*R*)-sulfolactate resembles (*S*)-malate and sulfopyruvate resembles oxaloacetate, the ComC enzyme has significantly higher activity reducing sulfopyruvate compared to oxaloacetate (Graupner *et al.*, 2000a). The protein's novel structure, a three-layer $\alpha\beta\alpha$ sandwich, lacks the Rossman fold topology typical of most NAD(P)-dependent oxidoreductases (Irimia *et al.*, 2004). A pair of basic amino acid residues, Arg^{48} and His^{116}, could bind the substrate's sulfonate group.

Sulfopyruvate decarboxylase (ComDE; EC 4.1.1.79) uses a thiamine diphosphate (ThDP) cofactor to catalyze sulfopyruvate decarboxylation, forming sulfoacetaldehyde (Graupner *et al.*, 2000b). Together, the two subunits of the *M. jannaschii* protein resemble the single-subunit sulfopyruvate decarboxylase from *Methanosarcina acetivorans* (Graham *et al.*, 2009) and the phosphonopyruvate decarboxylase enzyme (Zhang *et al.*, 2003). The chemistry of these decarboxylation reactions should follow the well-characterized pyruvate decarboxylase reaction (Jordan, 2003). However, the aggregation-prone sulfopyruvate decarboxylases appeared to be inactivated by air, an unprecedented sensitivity for ThDP-dependent proteins without metallocenters. This air-inactivation could be reversed by adding an unusually high concentration of 5 m*M* ThDP, suggesting that the coenzyme undergoes unique side reactions (Graham *et al.*, 2009).

While *Methanobrevibacter smithii* has homologs of the *comABCDE* genes, the CoM auxotroph *M. ruminantium* has only *comBC* homologs (Leahy *et al.*, 2010). However, genome sequences from members of the Methanosarcinales that produce CoM contain no *comABC* homologs. Therefore, these organisms were proposed to use a different pathway to make CoM, from the phosphoserine precursor of L-serine (Fig. 15.3B; Graham *et al.*, 2009).

In this second pathway, a paralog of the pyridoxalamine 5′-phosphate (PLP)-dependent threonine synthase enzyme catalyzes the β-elimination of phosphate, followed by the nucleophilic addition of sulfite to α,β-unsaturated dehydroalanine, forming L-cysteate in a reaction analogous to the one catalyzed by ComA (Graham *et al.*, 2009). The PLP-dependent aspartate aminotransferase enzyme can convert L-cysteate to sulfopyruvate (Helgadóttir *et al.*, 2007). This separate pathway could account for reduced deuterium incorporation reported for *Methanosarcina thermophila* TM-1 cells grown on [2,2,2-^{2}H]acetate (White, 1985). Each of the three PLP-dependent enzymes that forms a Schiff base intermediate presents an opportunity to exchange β-protons through a side reaction (Kimmich *et al.*, 2002). Subsequent steps in this pathway are presumed to be similar in all methanogens that make CoM.

4. Synthesis of 2-Sulfopyruvate

Sulfopyruvate can be readily prepared using a nucleophilic substitution reaction (Griffith and Weinstein, 1987; Weinstein and Griffith, 1986). In a representative synthesis, 2-bromopyruvate (10 mmol; TCI America) was dissolved in 10 ml of deionized water and heated at 75 °C under a stream of nitrogen gas. Sodium sulfite (22.5 mmol) dissolved in 12.5-ml deionized water was added dropwise to the stirred bromopyruvate solution over 3 min. The mixture was stirred for 45 min, although the substitution reaction was substantially complete after 5 min. Aliquots (0.5 μl) of reaction mixture were removed periodically and combined with 20 μl of a 2,4-dinitrophenylhydrazine (DNPH) solution to produce the hydrazone derivatives of substrate and product (Friedemann and Haugen, 1943; Yukawa *et al.*, 1993). The DNPH solution (Brady's reagent) was prepared by dissolving

3 g of 2,4-DNPH in 15 ml of concentrated sulfuric acid. This mixture was added, with stirring, to 90 ml of 74% ethanol (Shriner, 2004). Hydrazone derivatives were spotted on a silica gel TLC plate and developed in acetonitrile:water:formic acid (95:10:5 by volume). Under these conditions, the retention factors of hydrazone derivatives were 0.63 for sulfopyruvate and 0.87 for bromopyruvate.

To purify sulfopyruvate, a column of AG1-X8-Cl^- resin (Bio-Rad, 200–400 mesh, 15 mm by 16 cm) was washed with 1 *M* HCl, followed by deionized water until the pH of the eluate was 6.0. The reaction mixture was cooled and applied to the column, followed by 27 ml of deionized water. A linear gradient maker was used with 50-ml deionized water and 50-ml 2N HCl, followed by 25 ml of 4N HCl to wash and elute the product. Fractions (5 ml) were collected and screened for carbonyl compounds using a spot test with DNPH reagent and then sodium hydroxide on a white porcelain multiwell plate to produce the characteristic red, *aci*-nitro quinoid. Maximum hydrazone formation was noted in fractions corresponding to ~1.8–4N HCl eluate. These fractions were concentrated by rotary evaporation under vacuum and adjusted to pH 6.5 with lithium hydroxide. The neutralized product was concentrated under vacuum to form tan oil, which was dried over P_2O_5 to yield 4.6 mmol lithium sulfopyruvate (46% yield). ^{1}H NMR (400 MHz, ^2H_2O + TSP, δ) 4.43 (s, 2H). ^{13}C NMR (^2H_2O, 100 MHz, δ) 195.2 (C1), 167.5 (C2), 58.1 (C3; Graham *et al.*, 2009).

The oil was dissolved in deionized water, and the sulfopyruvate concentrations in stock solutions were determined by colorimetric carbonyl analysis using DNPH reagent. Sample or standard solutions (500 μl) were mixed with 100 μl of a 50 m*M* DNPH solution and incubated for 15 min at 37 °C. An aliquot (400 μl) of 2N NaOH was mixed with each sample, and the absorbance at 415 nm was measured by spectrophotometry. A standard curve of sodium pyruvate (0–50 nmol) was used to estimate 2-oxoacid content of the samples.

5. Discussion of CoM Biosynthesis

CoM biosynthesis has not been fully understood yet. Sulfopyruvate decarboxylase's requirement for high levels of ThDP or dithionite suggests that we have more to learn about that enzyme's reaction mechanism. And the final, reductive thiolation reaction has no precedent in biochemistry. Depending on the reduction state of the transferred sulfur species, additional electrons may be required to reduce a thio-hemiacetal intermediate to CoM. Finally, the bacteria known to produce CoM do not appear to use either of the two known methanogen pathways. The genome sequence of *X. autotrophicus* Py2 encodes three phosphosulfolactate synthase homologs, but lacks phosphatase, dehydrogenase, and decarboxylase homologs, presaging a third pathway that might skip the sulfopyruvate intermediate.

6. Coenzyme B Biosynthesis

The structure of CoB, isolated from *Methanothermobacter thermautotrophicus*, was determined to be 7-mercaptoheptanoylthreonine phosphate (HS-HT*P*; Noll *et al.*, 1986). Noll *et al.* (1987) chemically synthesized this cofactor and showed that it was equivalent to CoB isolated from *M. thermautotrophicus* in its role as a substrate for MCR. The same form of CoB was copurified and cocrystallized with MCR (Ermler *et al.*, 1997; Noll and Wolfe, 1986). However, a much larger form of CoB, containing a UDP-disaccharide moiety reminiscent of bacterial peptidoglycan precursors, was later reported to be present in the same organism (Marsden *et al.*, 1989; Sauer *et al.*, 1990). It is possible that the larger form is a biosynthetic precursor of HS-HT*P*, or that the labile phosphoanhydride bond between phosphothreonine and *N*-acetylmannosaminuronate is rapidly hydrolyzed to produce HS-HT*P*. Methanogens can produce both UDP-*N*-acetylmannosaminuronate and UDP-*N*-acetylglucosamine precursors (Namboori and Graham, 2008), and complex *N*- and *O*-linked glycosylation pathways have been described in the archaea (Jarrell *et al.*, 2010). Further studies on the biosynthesis and structure of CoB will be required to resolve the role of this extended headgroup.

Deuterium-labeling studies using *Methanococcus voltae* and *M. thermophila* concluded that methanogens produce HS-HT*P* from 7-mercaptoheptanoate and threonine (Fig. 15.4; White, 1989a,c). Amide bond formation between L-threonine and 7-mercaptoheptanoate requires ATP in *M. thermophila*, although it is not clear whether the reaction proceeds through an adenylate or phosphoanhydride intermediate (Solow and White, 1997; White, 1994). *M. thermophila* and *M. jannaschii* cells produce the carbon backbone of 7-mercaptoheptanoate using reactions that are analogous to the Krebs cycle and the isopropylmalate pathway in leucine biosynthesis (Howell *et al.*, 1998; White, 1989b). These pathways extend a 2-oxoacid by one methylene group, forming one carbon–carbon bond and cleaving another to release CO_2. Some methanogens, including *M. jannaschii,* use a similar citramalate pathway to make 2-oxobutyrate for isoleucine biosynthesis as well. The first reactions for CoB biosynthesis convert 2-oxoglutarate to 2-oxoadipate, the same reaction series found in the α-aminoadipate pathway for lysine biosynthesis in some bacteria and yeast (Bhattacharjee, 1985). But unlike amino acid biosynthetic reactions, the methanogen homocitrate pathway recycles its products, extending 2-oxoadipate to 2-oxopimelate and 2-oxopimelate to 2-oxosuberate (Fig. 15.5). With at least four 2-oxoacid

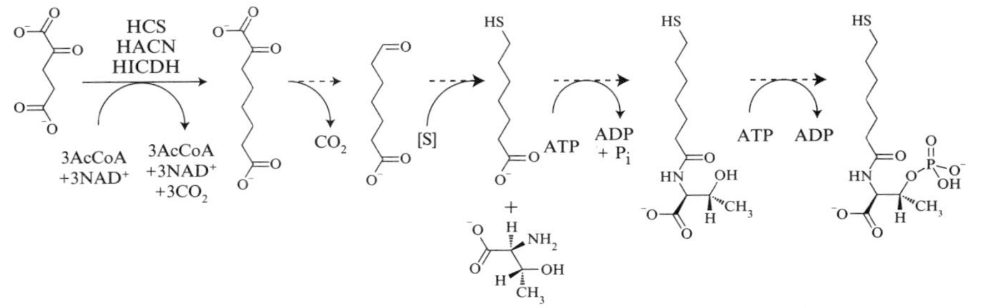

Figure 15.4 In the proposed biosynthetic pathway to CoB, 2-oxoglutarate is extended to 2-oxosuberate through three iterations of the 2-oxoacid elongation process. Reactions in that process include homocitrate synthase (HCS), homoaconitase (HACN), and homoisocitrate dehydrogenase (HICDH), which are shown in full in Fig. 15.5. Subsequent enzymes have not yet been identified for the decarboxylation of 2-oxosuberate to produce 7-oxoheptanoate; the formation of an amide bond with L-threonine; or the phosphorylation of the threonine headgroup (White, 1989c).

elongation pathways operating in *M. jannaschii*, it was essential to characterize the specificity of all paralogs of the three requisite enzymes.

All the three enzymes in the 2-oxoacid elongation pathway rely on metal ions to coordinate carboxylates and 2-hydroxy or 2-keto groups. In each case, the substrates and proteins provide oxygen-rich, hard ligands that coordinate these cations, while in some enzymes, the metal ions also act as Lewis acids. Comparative structural and functional analysis has illustrated how a small number of active site amino acid replacements enabled paralogous proteins to recognize substrate γ-moieties of varying chain length and hydrophobicity.

7. Homocitrate Synthase

One might expect the first enzyme in the 2-oxoglutarate elongation pathway to have evolved from the canonical Si-citrate synthase enzyme in the Krebs cycle that catalyzes the aldol addition of acetyl-CoA to oxaloacetate (Spector, 1972), followed by the hydrolysis of citryl-CoA. Instead, homocitrate synthase (HCS; EC 2.3.3.14) shares a common ancestor with *Re*-citrate synthase and isopropylmalate synthase (Higgins *et al.*, 1972; Howell *et al.*, 1998; Li *et al.*, 2007). HCS enzymes function in several bacterial, archaeal, and fungal biosynthetic pathways. In diazotrophic bacteria like *Azotobacter vinelandii*, HCS produces the tricarboxylate ligand of the FeMo nitrogenase cofactor (Zheng *et al.*, 1997). HCS also catalyzes the first reaction in the α-aminoadipate pathway to make lysine (Xu *et al.*, 2006). At least, one legume plant expresses its own HCS to support

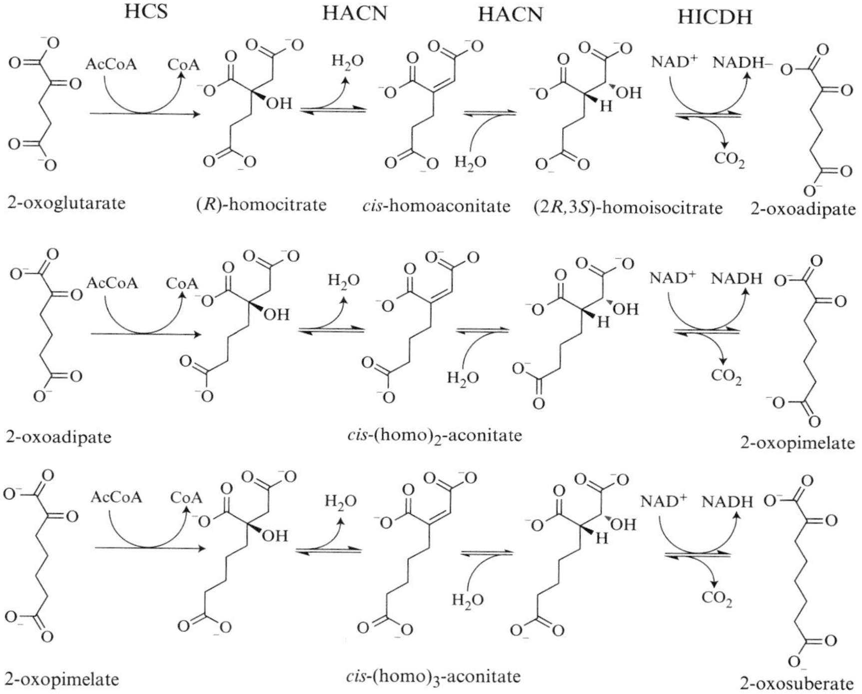

Figure 15.5 Three enzymes catalyze the elongation of 2-oxoacids for CoB biosynthesis. The homocitrate synthase (HCS) enzyme catalyzes the aldol addition of an acetyl group to the ketone of 2-oxoglutarate and the hydrolysis of a CoA thioester, producing (*R*)-homocitrate. The homoaconitase (HACN) enzyme catalyzes a dehydration reaction, producing *cis*-homoaconitate. HACN also catalyzes the hydration of the flipped *cis*-homoaconitate intermediate, producing (2*R*, 3*S*)-homoisocitrate. The homocitrate dehydrogenase enzyme (HICDH) catalyzes hydride transfer to NAD^+ and a subsequent β-decarboxylation reaction, producing 2-oxoadipate. As a result of these reactions, the substrate is extended by one methylene group, although both acetyl carbons are incorporated into the 2-oxoacid. The same three enzymes are capable of extending 2-oxoadipate to 2-oxopimelate and 2-oxosuberate, in successive reactions.

diazotrophic bacteria growing in its rhizosphere (Hakoyama *et al.*, 2009). Phylogenetic analysis suggests that HCS activity evolved several times within this acetyltransferase family.

8. Homocitrate Synthase Enzymology

The methanogen HCS is the only family member reported to catalyze acetyl addition to 2-oxoacids of varying chain length (Howell *et al.*, 1998). The low solubility and activity of heterologously expressed methanogen

HCS proteins has hindered their investigation. Therefore, the stabilized yeast and bacterial HCS proteins have become model systems for mechanistic and structural analyses, leveraging numerous studies of the homologous isopropylmalate synthase enzyme (Andi *et al.*, 2004b; de Carvalho and Blanchard, 2006; Koon *et al.*, 2004). 2-Oxoglutarate binds HCS, coordinating a divalent cation, followed by acetyl-CoA binding and a conformational change in the C-terminal HCS domain (Andi *et al.*, 2004a). A general base abstracts a proton from the acetyl group acetyl-CoA, forming an enolate intermediate (Bulfer *et al.*, 2009; Qian *et al.*, 2008). Nucleophilic attack by the enolate on the keto group of 2-oxoglutarate produces an alkoxide of (*R*)-homocitryl-CoA, which is protonated by a general acid. Base-catalyzed hydrolysis of the CoA thioester results in the release of CoA, followed by the release of (*R*)-homocitrate. This catalytic mechanism appears conserved in all the acetyltransferase homologs, but the proteins differ significantly in their 2-oxoacid substrate specificities. This reaction is expected to be exergonic due to the substantial free energy of CoA thioester hydrolysis.

9. Substrate Specificity Determinants in HCS

While HCS probably evolved from an isopropylmalate synthase ancestral gene, there are few experimental reports characterizing the proteins' specificity determinants for different 2-oxoacid substrates. The *Serratia marcescens* bacterial isopropylmalate synthase catalyzes acetyl-CoA addition to pyruvate, 2-oxobutyrate, 2-oxovalerate, and 2-oxoisovalerate with similar specificity constants (Kisumi *et al.*, 1976). A crystal structure model of the *Leptospira interrogans* citramalate synthase bound to pyruvate shows that Tyr and two Leu side chains make hydrophobic contacts with the pyruvate methyl group, occluding larger 2-oxoacids from binding (Fig. 15.6; Ma *et al.*, 2008). In a *Mycobacterium tuberculosis* isopropylmalate synthase complex, Ser, Tyr, His, and Leu side chains form a hydrophobic pocket that accommodates the isopropyl group of 2-oxoisovalerate (Koon *et al.*, 2004). In contrast, Arg, His, and Ser residues form hydrogen bonds with the C-5 carboxylate of 2-oxoglutarate in *Schizosaccharomyces pombe* HCS (Bulfer *et al.*, 2009) and *Thermus thermophilus* HCS (Okada *et al.*, 2010). This HCS active site has convergently evolved to resemble the Si-citrate synthase active site (Karpusas *et al.*, 1990).

Compared to *T. thermophilus* HCS, the *M. jannaschii* MJ0503 protein shares 33% amino acid identity and 68% similarity. Further studies of the methanogen HCS will be needed to understand how that enzyme's active site can accommodate 2-oxoacid substrates differing by three methylene groups in length. One might expect this enzyme to regulate γ-chain length

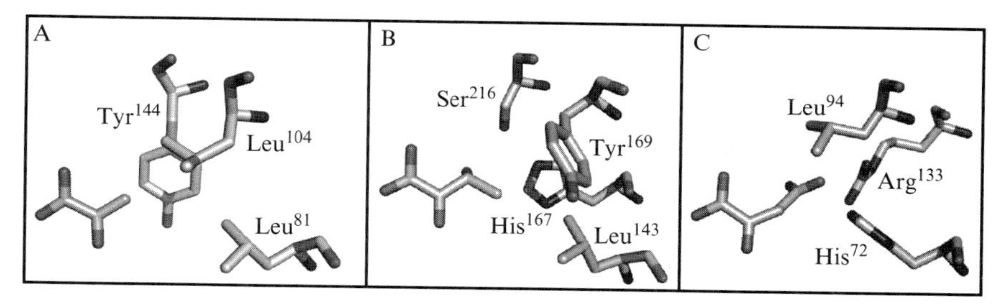

Figure 15.6 Side chains of active site residues affect 2-oxoacid ligand binding in crystal structure models of acetyltransferases. Part A shows pyruvate bound in the active site of *L. interrogans* citramalate synthase (PDB identifier 3BLI), where the hydroxyphenyl ring of Tyr^{144} and the isobutyl group of Leu^{104} constrain substrate binding (Ma *et al.*, 2008). Part B shows 2-oxoisovalerate in the active site of *M. tuberculosis* isopropylmalate synthase (PDB identifier 1SR9; Koon *et al.*, 2004). Part C shows 2-oxoglutarate in the active site of homocitrate synthase from *T. thermophilus* (PDB identifier 2ZYF; Okada *et al.*, 2010). The γ-chain carboxylate forms hydrogen bonds with the guandinium group of Arg^{133} and the imidazolium group of His^{72}. Both hydrogen bonding residues are conserved in the homologous *Schizosaccharomyces pombe* HCS structure (Bulfer *et al.*, 2009). In all the three structures, the C-1 carboxylate and C-2 carbonyl coordinate a divalent metal ion (not shown).

to avoid producing excessively long 2-oxoacids, yet there is no evidence for a molecular ruler in HCS. Residues that bind the metal ion, C-1 carboxylate, and C-2 ketone are conserved. However, corresponding residues that interact with the γ-chain carboxylate are less polar in the methanogen HCS: His^{72} is replaced by Leu, Leu^{94} is replaced by Phe, and Arg^{133} is replaced by Ala in most methanogen homologs. Therefore, the mode of 2-oxoacid binding may be very different in the methanogen HCS.

10. Chemical Synthesis of 2-Oxoacids

Due to their biological significance, 2-oxoacids have been synthesized by a variety of chemical and enzymatic schemes (Gault, 1912; Meister, 1957). 2-Oxoglutaric acid and 2-oxoadipic acid are currently available from commercial suppliers (e.g., Sigma-Aldrich). The longer chain acids, 2-oxopimelic and 2-oxosuberic acids, must be synthesized for biochemical analysis, from diethyl adipate and diethyl pimelate, respectively (Cooper *et al.*, 1983; Waters, 1947). A Claisen condensation of diethyl oxalate with the diethyl ester in the presence of base, followed by acid-catalyzed decarboxylation and deprotection reactions readily produces the free 2-oxoacid (Drevland *et al.*, 2008; Nelson and Gribble, 1973).

In a typical synthesis of 2-oxosuberic acid, 3.3 g (49 mmol) of sodium ethoxide was added with stirring to 30 ml of dry diethyl ether in a dry

100-ml round bottom Schlenk flask under N_2. Diethyl pimelate (6.5 g; 30 mmol) was added and the mixture was stirred for 15 min under N_2 at room temperature. Diethyl oxalate (5.3 g; 36 mmol) was added dropwise with stirring under N_2. The mixture gradually changed from an orange to a dark red color, coincident with the formation of α-oxalyldiethylpimelate. Reaction progress was monitored by TLC, using a silica-coated plate with fluorescent indicator developed in 2:1 (v:v) hexanes:ethyl acetate with UV detection. The reactants and products had the following R_f values: α-oxalyldiethylpimelate (0.71), diethyl oxalate (0.82), and diethyl pimelate (0.85).

After stirring under nitrogen for 4 h at room temperature, the mixture was poured into 25 ml of water. HCl was then added with mixing until the solution was clarified. This solution was extracted with diethyl ether, and the combined organic phase was dried over sodium sulfate. The solvent was removed by rotary evaporation under vacuum to afford 8.5 g of red–orange oil (~ 85% yield). The α-oxalyldiethylpimelate mixture was placed in a 100-ml round bottom flask with 36 ml 4 *M* HCl, and the mixture was heated with stirring at 80 °C using a reflux condenser. After 6 h reflux, the homogeneous solution was cooled and extracted with diethyl ether. The organic phase was evaporated under vacuum to produce a slurry (5.4 g), containing 2-oxosuberic acid as well as excess reagents and side-products. Recrystallization from diethyl ether yielded the free acid as white or tan crystals. The unoptimized recrystallization of 2-oxosuberic acid from diethyl ether was relatively inefficient (yields were as low as 6%), compared with the other 2-oxodicarboxylic acids. Therefore, it would be worthwhile to screen other solvents. LRMS-CI (m/z): 143 (100%, $[M-CHO_2]^+$), 125 (53%, $[M-CH_3O_3]^+$), 91 (52%), 171 (42%, $[M-OH]^+$), 229 (36%), 211 (14%, $[M + Na^+]$), 115 (8%, $[M-C_2HO_3]^+$), and 189 (3%, MH^+). ^{1}H NMR and ^{13}C NMR spectra for the 2-oxoacids were reported previously, as supplementary material (Drevland *et al.*, 2008).

11. Chemical Synthesis of Homocitrate and Its Analogs

Several chemical or enzymatic syntheses of homocitrate have been reported, and a racemic homocitrate mixture is available commercially (Sigma-Aldrich; Howell *et al.*, 1998). However, only the (*R*)-(–)-homocitrate enantiomer appears naturally (Thomas *et al.*, 1966). Homocitric acid lactones have been produced by several syntheses, either as a racemic mixture (Li and Xu, 1998) or as an enantiopure compound (Paju *et al.*, 2004). Ma *et al.* (2008) used a diastereoselective alkylation from D-malate to produce (*R*)-homocitrate in 12% yield (Ma and Palmer, 2000). A subsequent report simplified that synthesis and increased the yield of (*R*)-homocitric acid lactone to more than 30% (Xu *et al.*, 2005).

12. Homoaconitase

The homoaconitase enzyme was assumed to function like aconitase in the Krebs cycle, which uses a [4Fe–4 S] cluster to catalyze the dehydration of citrate, forming *cis*-aconitate, and the hydration of this intermediate to produce isocitrate (Beinert *et al.*, 1996). Although the enzyme from the α-aminoadipate pathway to make lysine has been called *cis*-homoaconitase in prior literature, both the yeast and *T. thermophilus* enzymes were reported to catalyze only one half-reaction—the hydration of *cis*-homoaconitate to produce homoisocitrate (Jia *et al.*, 2006; Strassman and Ceci, 1966). Therefore, this enzyme should be called homoaconitate hydratase (EC 4.2.1.36). Aconitase itself has been proposed to catalyze the dehydration of homocitrate to produce *cis*-homoaconitate (Jia *et al.*, 2006).

Methanogens have no aconitase enzyme: they reduce oxaloacetate to produce succinate and 2-oxoglutarate for biosynthesis. Instead, they have two paralogs of the large and small subunit genes whose proteins comprise the [4Fe–4S]-dependent isopropylmalate isomerase. In *M. jannaschii*, the MJ0499 and MJ1277 proteins form the isopropylmalate/citramalate isomerase (IPMI; EC 4.2.1.33) that is involved in leucine and isoleucine biosyntheses (Drevland *et al.*, 2007). The other set of subunits, MJ1003 and MJ1271, forms homoaconitase (HACN; EC 4.2.1.114; Drevland *et al.*, 2008). This phylogeny parallels the evolution of HCS. HACN evolved from IPMI rather than aconitase.

13. Homoaconitase Enzymology

Heterologously expressed HACN protein lacks the requisite [4Fe–4S] cofactor. Therefore, the MJ1003 and MJ1271 proteins were copurified aerobically from an *Escherichia coli* strain that expressed both subunits (Drevland *et al.*, 2008). Chemical reconstitution of the iron–sulfur cluster using $Fe(NH_4)_2(SO_4)_2$ and sodium sulfide could be performed by stirring the purified, chilled protein solution in a septum-sealed vial under argon for 3–4 h (Drevland *et al.*, 2007). Although chemical analysis, EPR, or Mössbauer spectroscopy are the most direct methods to analyze cluster formation, enzyme activity measurements using a generic substrate such as maleate are more practical for routine quality control of small-scale reconstitutions. HACN proteins with reconstituted clusters rapidly lose activity when oxygen permeates septa.

The methanogen HACN enzyme catalyzes the complete isomerization of (*R*)-homocitrate to produce (2*R*, 3*S*)-homoisocitrate. Neither (*S*)-homocitrate nor *trans*-homoaconitate is a substrate for the enzyme, confirming the pathway shown in Fig. 15.5. The equilibrium for the hydration of homoaconitate favors homocitrate formation (the reverse reaction), making *cis*-homoaconitate the preferred substrate for activity assays. It is convenient to monitor reaction progress in continuous assays by measuring the decrease in UV absorbance at 235–250 nm (depending on substrate concentration). The specificity constants for HACN-catalyzed hydration of *cis*-(homo)$_{1-4}$aconitate substrates are all similar (2.2–6.8 × 10^4 M^{-1} s^{-1}). The enzyme has similar activity for maleate hydration, producing D-malate (k_{cat}/K_M = 1.8 × 10^4 M^{-1} s^{-1}), due to a proportional increase in both turnover and Michaelis constant (Drevland *et al.*, 2008).

The complete isomerization of (*R*)-homocitrate can be analyzed using a coupled assay with the homoisocitrate dehydrogenase enzyme, described below. This effectively irreversible reaction forms a 2-oxoadipate and NADH, which can be measured by chromatography and UV absorbance spectroscopy, respectively.

As discussed for HCS and IPMS, the specificities of HACN and IPMI depend on the active site amino acid residues that interact with the substrate's γ-moiety. No crystal structure model of the heterodimeric HACN has been reported. Therefore, specificity determinants have been inferred from a crystal structure model of the small HACN subunit from *M. jannaschii*, steady-state kinetic characterization of altered specificity proteins, and aconitase crystal structures (Jeyakanthan *et al.*, 2010). Two residues in a flexible loop of the *M. jannaschii* HACN small subunit, Arg^{26} and Thr^{27}, are conserved in most methanogen HACN proteins but are replaced by hydrophobic and bulky residues in IPMI subunits. The Arg^{26}Val, Thr^{27}Ala, and Arg^{26}Val + Thr^{27}Tyr variants of the *M. jannaschii* HACN small subunit have broad substrate specificities: they can catalyze both IPMI and HACN reactions, at reduced efficiency (Jeyakanthan *et al.*, 2010). Based on these experiments, the modern HACN enzyme is remarkable in its specificity for *cis*-unsaturated tricarboxylates and lack of promiscuity in catalyzing IPMI reactions.

14. Chemical Synthesis of Homoaconitate and Its Analogs

Neither *cis*-homoaconitate nor its longer chain analogs is commercially available. A synthesis of both *cis*- and *trans*-homoaconitate was reported by Massoudi *et al.*, using the di-*t*-butyl ester of 2-oxoglutarate and a phosphonoacetate compound in a Wittig–Horner reaction (Massoudi *et al.*, 1983). A modified procedure was developed using stable dimethyl esters of 2-oxoacids and commercially available trimethyl phosphonoacetate to preferentially form *cis*-(homo)$_{1-4}$aconitate trimethyl esters (Drevland *et al.*, 2008; Howell *et al.*, 1998). Flash chromatography resolved the *cis*- and *trans* isomers, and subsequent saponification and extraction resulted in limited isomerization. The *trans* isomers inhibit HACN activity; so careful preparation, storage, and analysis of the tricarboxylate salts is required to assure substrate purity (Krebs and Eggleston, 1944).

15. Homoisocitrate Dehydrogenase

The final enzyme in the 2-oxoacid elongation pathway catalyzes the oxidation and decarboxylation of (2*R*, 3*S*)-homoisocitrate and its analogs (Strassman and Ceci, 1965). Homoisocitrate dehydrogenase (HICDH; EC 1.1.1.87) shares a common ancestor with 3-isopropylmalate dehydrogenase (IPMDH; EC 1.1.1.85), in the pyridine nucleotide-dependent β-hydroxyacid oxidative dehydrogenase family (Aktas and Cook, 2009). Based on the standard free energy of the similar isocitrate dehydrogenase reaction (−8.5 kJ/mol), the HICDH reaction probably drives the 2-oxoacid elongation process (Londesborough and Dalziel, 1968).

16. Enzymology of Homoisocitrate Dehydrogenase

The proposed mechanism of HICDH resembles those of other family members. The *Saccharomyces cerevisiae* HICDH protein binds NAD^+ and then an Mg^{2+}-homoisocitrate complex (Lin, 2009). Lys^{206} assists in abstracting a proton from the substrate's hydroxyl, while hydride is transferred to form NADH. Decarboxylation subsequently produces an Mg^{2+}-stabilized enol that is transiently protonated by Lys^{206}. The Tyr^{150} hydroxyphenol transfers a proton to the C-3 position, and Lys^{206} deprotonates the hydroxyl group, stabilizing the keto form of the Mg^{2+}-2-oxoadipate product.

The *T. thermophilus* HICDH has detectable isocitrate and 3-isopropylmalate dehydrogenase activities (Miyazaki *et al.*, 2005), although IPMDH activity is too low to rescue growth of an *E. coli leuB* mutant by complementation. The *S. cerevisiae* HICDH is highly specific for homoisocitrate. Directed evolution experiments substantially increased the *T. thermophilus* and *S. cerevisiae* enzymes' specificity constants for IPMDH activity, by replacing a handful of amino acids (mostly away from the isopropylmalate binding site) and by displacing the α4 helix to interact appropriately with the isopropylmalate substrate (Suzuki *et al.*, 2010). Genetic experiments in *T. thermophilus* demonstrated that the mutated *T. thermophilus* genes could still complement an *hicdh* mutation. Therefore, these protein variants appear more promiscuous than the wild-type proteins, as observed for the homoaconitase variants. It is unclear whether mutations in the yeast HICDH gene that introduced IPMDH activity had the same effect. A crystal structure of either HICDH in the presence of tricarboxylate ligand would clarify the roles of active site amino acids in determining substrate specificity.

The *M. jannaschii* HICDH catalyzes the oxidative decarboxylation of $(\text{homo})_{1-4}$isocitrate with similar specificity, in both direct and coupled assays with homoaconitase (Drevland *et al.*, 2008; Howell *et al.*, 2000). This protein had no detectable activity with 3-isopropylmalate or isocitrate, despite the 48% amino acid identity it shares with the paralogous *M. jannaschii* IPMDH. The *T. thermophilus* active site residues Ile^{84}, Arg^{96}, Arg^{118}, Asp^{232}, and Glu^{122} are conserved, but one of the key residues for homoisocitrate recognition, Arg^{85}, is replaced by isoleucine in the *M. jannaschii* HICDH. An Arg^{85}Val substitution in the *T. thermophilus* enzyme caused it to gain IPMDH activity, lose isocitrate dehydrogenase activity, and reduce HICDH activity (Miyazaki *et al.*, 2003). Further experiments will be required to determine how the methanogen HICDH accommodates these variable-length γ-chains.

17. Synthesis of Homoisocitrate

The two chiral centers of homoisocitrate present a challenge to synthesize and isolate a single diastereomer. One synthesis used dimethyl D-malate and an alkynylsilane to form the (−)-*threo*-homoisocitrate trimethyl ester (Schmitz *et al.*, 1996). Subsequent syntheses prepared this isomer by the condensation of dimethyl D-malate with methyl-3-bromopropionate or allyl bromide, in yields up to 30% (Howell *et al.*, 1998; Ma and Palmer, 2000).

18. Chromatographic Analysis of 2-Oxoacids

Carboxylic acids are routinely separated and analyzed by HPLC, TLC, or GC. However, multiple carboxylate groups and fully ionized sulfonate groups complicate the analyses. Underivatized sulfopyruvate is poorly retained on typical reversed-phase octadecylsilyl HPLC columns. Retention can be improved using ion-pair chromatography or a modified reversed-phase column with polar functionality (Table 15.2). Anion exchange chromatography often produces broader peaks and poor

Table 15.2 Reversed-phase HPLC analysis of organic acids

Compound	Retention factor[a]	Absorbance maximum (nm)	Limit of quantification (μ*M*)[b]
Sulfopyruvate	0.11	199	6
2-Oxoadipate	0.16	201	2.2
Oxaloacetate	0.43	193	46
Pyruvate	0.85	195	12
Citrate	1.0	215	11
(*R*)-Homocitrate	1.1	203	127
(*S*)-Homocitrate	1.1	203	133
Homoisocitrate	1.3	216	0.9
2-Oxoglutarate	1.33	196	14
cis-Aconitate	1.7	215	1.5
trans-Aconitate	1.7	215	2.1
cis-Homoaconitate	3.6	213	N.D.[c]

[a] Compounds (10 μl injection volume) were separated on a reversed-phase synergy 4μ Hydro-RP column (80 Å; 250 by 4.6 mm; Phenomenex) at 35 °C, with isocratic elution in mobile phase containing 20 m*M* KH_2PO_4, 5% acetonitrile, adjusted to pH 2.5 with H_3PO_4 ($t_0 = 2.23$ min).

[b] The limit of quantification by absorbance at 215 nm was calculated based on the concentration of analyte required to produce a peak with 10 times the signal/noise ratio of the baseline.

[c] N.D., Not determined.

separation between similarly charged analytes. Detection by measuring the carbonyl group absorption at 215 nm identifies most organic acids, but the low molar absorption coefficients make the method's limits of detection and quantification unsuitably high for many biochemical analyses.

Several classes of derivatization reactions exploit the electrophilic carbonyl group of 2-oxoacids to produce imine, hydrazone, or oxime derivatives (Fig. 15.7). Many of these derivatives have enhanced volatility, increased chromatographic retention, or characteristic fragmentation patterns in mass spectrometry. Nevertheless, derivatization reagents must be chosen carefully to permit the resolution of multiple 2-oxoacids and their separation from excess reagent or by-products. 2,4-DNPH reacts rapidly with 2-oxoacids to form a yellow Schiff base (Friedemann and Haugen, 1943). Dansyl hydrazine reacts similarly to produce fluorescent hydrazones (Chayen *et al.*, 1971); however, their chromatographic properties were unsuitable for sulfopyruvate analysis. *N*-Methyl-4-hydrazine-7-nitrobenzofurazan (MNBDH) reacts with 2-oxoacids to form stable, red-colored hydrazones with high molar absorption coefficients at 500 nm, few by-products, and good chromatographic retention and peak symmetry (Büldt and Karst, 1999). Although the dicarboxylate MNBDH derivatives are not well resolved by HPLC, this method is well suited for the analysis of

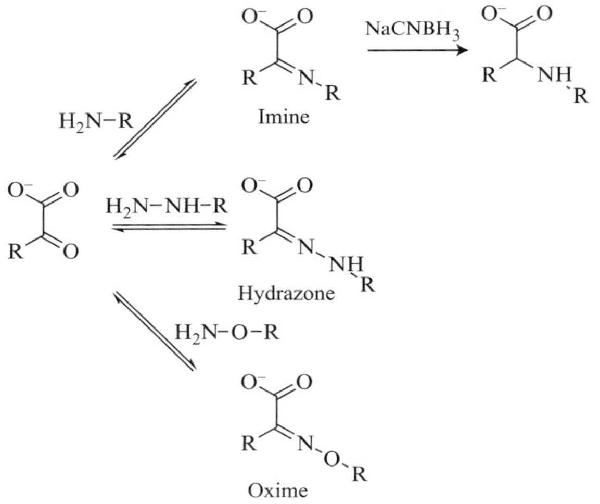

Figure 15.7 2-Oxoacid derivatives can be readily prepared by nucleophilic substitution reactions under basic conditions by reaction with a primary amine to form an imine (top), reaction with a hydrazine to form a hydrazone (middle), or reaction with a hydroxylamine to form an oxime (bottom). Reducing the imine with sodium cyanoborohydride forms a stable secondary amine. The imine, hydrazone, and oxime derivatives can form both *syn* and *anti*stereoisomers.

individual 2-oxoacids in complex biological matrices. A standard derivatization procedure was used for samples dissolved in ethanol, and the derivatives were adequately separated by reversed-phase HPLC on an octadecylsilyl column, described previously (Graham *et al.*, 2009). Retention factors and limits of quantification (10 × signal/noise ratio) were as follows (t_0 = 4.5 min): sulfopyruvate (0.11, 17 μ*M*), oxaloacetate (0.13, 26 μ*M*), 2-oxoglutarate (0.15, 53 μ*M*), pyruvate (0.92, 16 μ*M*), and bromopyruvate (1.2, 76 μ*M*).

2-Oxoacids must be derivatized for GC–MS analysis, to enhance their volatility (Knapp, 1979). Sulfonates are particularly difficult to modify due to their low pK_a values and low reactivities. Howell *et al.* (1998, 2000) reported the methylation of relevant carboxylic acids using 1 *M* HCl in anhydrous methanol at 50 °C for 12 h. In addition to the dimethyl ester of 2-oxoacids, the dimethyl ketal dimethyl ester was obtained. Hydroxyl groups in homocitrate were derivatized using trifluoroacetic anhydride in methylene chloride (1:1). In the electron ionization method used for these analyses, the molecular ion was rarely detected and base peaks were often identical among analogous 2-oxoacid esters. These derivatization reactions have low efficiencies, presumably due to intramolecular reactions and hydrolysis caused by residual water. Softer ionization methods now available are preferable and more likely to produce stable molecular ions.

Despite the simplicity of new LC–MS–ESI methods for the analysis of polar compounds, the 2-oxoacid dicarboxylates will often pass through a reversed-phase microcapillary column unretained, causing them to coelute with salts that suppress their detection. The derivatives described above and in Fig. 15.7 can improve retention, eliminate hydrated ketal forms, and enhance fragmentation and detection in positive ion mode (Drevland *et al.*, 2008).

19. Conclusions

Once tied to amino acid metabolism, 2-oxoacids have emerged as biosynthetic nodes for a variety of natural product biosyntheses, including CoM and CoB. Their keto/enol tautomers support both carbon bond formation and cleavage reactions. The 2-oxoacid elongation pathway for CoB biosynthesis is reminiscent of Wächtershäuser's proposed archaic cycle, fueled by pyrite formation, for prebiotic polymer synthesis (Wächtershäuser, 1990). The isopropylmalate pathway has been adapted to produce 5–8 carbon alcohols in *E. coli*, for potential biofuel production (Zhang *et al.*, 2008). Compared to fatty acid or polyketide syntheses, this pathway uses carbon inefficiently, but preserves dicarboxylic acid functionality. Therefore, the dicarboxylate elongation pathway is better suited for the synthesis of specialty chemicals or high-value polymer precursors. Given the large number of homologs of these genes found in diverse microbial genomes, it is likely that more 2-oxoacid biosynthetic pathways have evolved from the isopropylmalate pathway. Moreover, nature has shown us that coenzyme biosynthesis can be reinvented with new reactions and new proteins, challenging us to identify this biosynthetic diversity and understand the context under which it evolved.

ACKNOWLEDGMENTS

The author's research on coenzyme biosynthesis is supported by National Science Foundation grant MCB-0817903. This chapter was sponsored by the Laboratory Directed Research and Development Program of Oak Ridge National Laboratory, managed by UT-Battelle, LLC, for the U. S. Department of Energy under contract DE-AC05-00OR22725.

This chapter has been authored by UT-Battelle LLC under Contract No. DE-AC05-00OR22725 with the U.S. Department of Energy. The publisher, by accepting the chapter for publication, acknowledges that the United States Government retains a nonexclusive paid-up irrevocable worldwide license to publish or reproduce the published form of this chapter, or allow others to do so for United States government purposes.

REFERENCES

Aktas, D. F., and Cook, P. F. (2009). A lysine-tyrosine pair carries out acid-base chemistry in the metal ion-dependent pyridine dinucleotide-linked α-hydroxyacid oxidative decarboxylases. *Biochemistry* **48,** 3565–3577.

Allen, J. R., Clark, D. D., Krum, J. G., and Ensign, S. A. (1999). A role for coenzyme M (2-mercaptoethanesulfonic acid) in a bacterial pathway of aliphatic epoxide carboxylation. *Proc. Natl. Acad. Sci. USA* **96,** 8432–8437.

Andi, B., West, A. H., and Cook, P. F. (2004a). Kinetic mechanism of histidine-tagged homocitrate synthase from *Saccharomyces cerevisiae. Biochemistry* **43,** 11790–11795.

Andi, B., West, A. H., and Cook, P. F. (2004b). Stabilization and characterization of histidine-tagged homocitrate synthase from *Saccharomyces cerevisiae. Arch. Biochem. Biophys.* **421,** 243–254.

Balch, W. E., and Wolfe, R. S. (1976). New approach to the cultivation of methanogenic bacteria: 2-mercaptoethanesulfonic acid (HS-CoM)-dependent growth of *Methanobacterium ruminantium* in a pressureized atmosphere. *Appl. Environ. Microbiol.* **32,** 781–791.

Beinert, H., Kennedy, M. C., and Stout, C. D. (1996). Aconitase as iron-sulfur protein, enzyme, and iron-regulatory protein. *Chem. Rev.* **96,** 2335–2374.

Bhattacharjee, J. K. (1985). α-Aminoadipate pathway for the biosynthesis of lysine in lower eukaryotes. *CRC Crit. Rev. Microbiol.* **12,** 131–151.

Büldt, A., and Karst, U. (1999). *N*-Methyl-4-hydrazino-7-nitrobenzofurazan as a new reagent for air monitoring of aldehydes and ketones. *Anal. Chem.* **71,** 1893–1898.

Bulfer, S. L., *et al.* (2009). Crystal structure and functional analysis of homocitrate synthase, an essential enzyme in lysine biosynthesis. *J. Biol. Chem.* **284,** 35769–35780.

Burbaum, J. J., and Knowles, J. R. (1989). The nature of pyruvate bound to pyruvate kinase as determined by ^{13}C NMR spectroscopy. *Bioorg. Chem.* **17,** 359–371.

Charng, Y.-Y., *et al.* (2006). Arabidopsis Hsa32, a novel heat shock protein, is essential for acquired thermotolerance during long recovery after acclimation. *Plant Physiol.* **140,** 1297–1305.

Chayen, R., Dvir, R., Gould, S., and Harell, A. (1971). 1-Dimethylaminonaphthalene-5-sulfonyl hydrazine (dansyl hydrazine): A fluorometric reagent for carbonyl compounds. *Anal. Biochem.* **42,** 283–286.

Cooper, A. J. L., Ginos, J. Z., and Meister, A. (1983). Synthesis and properties of the α-keto acids. *Chem. Rev.* **83,** 321–358.

Copper, A. J., and Redfield, A. G. (1975). Proton magnetic resonance studies of α-keto acids. *J. Biol. Chem.* **250,** 527–532.

Damitio, J., Smith, G., Meany, J. E., and Pocker, Y. (1992). A comparative study of the enolization of pyruvate and the reversible dehydration of pyruvate hydrate. *J. Am. Chem. Soc.* **114,** 3081–3087.

de Carvalho, L. P. S., and Blanchard, J. S. (2006). Kinetic and chemical mechanism of α-isopropylmalate synthase from *Mycobacterium tuberculosis. Biochemistry* **45,** 8988–8999.

DiDonato, M., *et al.* (2006). Crystal structure of 2-phosphosulfolactate phosphatase (ComB) from *Clostridium acetobutylicum* at 2.6 Å resolution reveals a new fold with a novel active site. *Proteins* **65,** 771–776.

Drevland, R. M., Waheed, A., and Graham, D. E. (2007). Enzymology and evolution of the pyruvate pathway to 2-oxobutyrate in *Methanocaldococcus jannaschii. J. Bacteriol.* **189,** 4391–4400.

Drevland, R. M., Jia, Y., Palmer, D. R. J., and Graham, D. E. (2008). Methanogen homoaconitase catalyzes both hydrolyase reactions in coenzyme B biosynthesis. *J. Biol. Chem.* **283,** 28888–28896.

Ellermann, J., *et al.* (1989). Methyl-coenzyme-M reductase from *Methanobacterium thermoautotrophicum* (strain Marburg). Purity, activity and novel inhibitors. *Eur. J. Biochem.* **184,** 63–68.

Ermler, U., *et al.* (1997). Crystal structure of methyl-coenzyme M reductase: The key enzyme of biological methane formation. *Science* **278,** 1457–1462.

Friedemann, T. E., and Haugen, G. E. (1943). Pyruvic acid: II determination keto acids blood urine. *J. Biol. Chem.* **147,** 415–442.

Gault, M. H. (1912). Recherches sur les α-cétodiacides acycliques. Acide α cétoadipique. *Bull. Soc. Chim. Fr.* **11,** 382–389.

Goenrich, M., *et al.* (2004). Probing the reactivity of Ni in the active site of methyl-coenzyme M reductase with substrate analogues. *J. Biol. Inorg. Chem.* **9,** 691–705.

Grabarse, W., *et al.* (2001). On the mechanism of biological methane formation: Structural evidence for conformational changes in methyl-coenzyme M reductase upon substrate binding. *J. Mol. Biol.* **309,** 315–330.

Graham, D. E., Graupner, M., Xu, H., and White, R. H. (2001). Identification of coenzyme M biosynthetic 2-phosphosulfolactate phosphatase: a member of a new class of Mg^{2+}-dependent acid phosphatases. *Eur. J. Biochem.* **268,** 5176–5188.

Graham, D. E., Xu, H., and White, R. H. (2002). Identification of coenzyme M biosynthetic phosphosulfolactate synthase new family sulfonate biosynthesizing enzymes. *J. Biol. Chem.* **277,** 13421–13429.

Graham, D. E., Taylor, S. M., Wolf, R. Z., and Namboori, S. C. (2009). Convergent evolution of coenzyme M biosynthesis in the Methanosarcinales: Cysteate synthase evolved from an ancestral threonine synthase. *Biochem. J.* **424,** 467–478.

Graupner, M., and White, R. H. (2001). The first examples of (*S*)-2-hydroxyacid dehydrogenases catalyzing the transfer of the pro-4 *S* hydrogen of NADH are found in the archaea. *Biochim. Biophys. Acta* **1548,** 169–173.

Graupner, M., Xu, H., and White, R. H. (2000a). Identification of an archaeal 2-hydroxy acid dehydrogenase catalyzing reactions involved in coenzyme biosynthesis in methanoarchaea. *J. Bacteriol.* **182,** 3688–3692.

Graupner, M., Xu, H., and White, R. H. (2000b). Identification of the gene encoding sulfopyruvate decarboxylase, an enzyme involved in biosynthesis of coenzyme M. *J. Bacteriol.* **182,** 4862–4867.

Griffith, O. W., and Weinstein, C. L. (1987). β-Sulfopyruvate. *Methods Enzymol.* **143,** 221–223.

Hakoyama, T., *et al.* (2009). Host plant genome overcomes the lack of a bacterial gene for symbiotic nitrogen fixation. *Nature* **462,** 514–517.

Helgadóttir, S., Rosas-Sandoval, G., Söll, D., and Graham, D. E. (2007). Biosynthesis of phosphoserine in the *Methanococcales*. *J. Bacteriol.* **189,** 575–582.

Hess, J. L., and Reed, R. E. (1972). Considerations of oxalacetic acid keto-enol equilibria in various solvents. *Arch. Biochem. Biophys.* **153,** 226–232.

Higgins, M. J. P., Kornblatt, J. A., and Rudney, H. (1972). Acyl-CoA ligases. *In* "The Enzymes," (P. D. Boyer, ed.), pp. 407–434. Academic Press, New York.

Howell, D. M., Harich, K., Xu, H., and White, R. H. (1998). α-Keto acid chain elongation reactions involved in the biosynthesis of coenzyme B (7-mercaptoheptanoyl threonine phosphate) in methanogenic Archaea. *Biochemistry* **37,** 10108–10117.

Howell, D. M., Graupner, M., Xu, H., and White, R. H. (2000). Identification of enzymes homologous to isocitrate dehydrogenase that are involved in coenzyme B and leucine biosynthesis in methanoarchaea. *J. Bacteriol.* **182,** 5013–5016.

Irimia, A., Madern, D., Zaccaï, G., and Vellieux, F. M. (2004). Methanoarchaeal sulfolactate dehydrogenase: Prototype of a new family of NADH-dependent enzymes. *EMBO J.* **23,** 1234–1244.

Jarrell, K. F., Jones, G. M., Kandiba, L., Nair, D. B., and Eichler, J. (2010). S-Layer glycoproteins and flagellins: Reporters of archaeal posttranslational modifications. *Archaea*, Article ID 612948, 13 pages. doi:10.1155/2010/612948.
Jeong, H., *et al.* (2000). The large-scale organization of metabolic networks. *Nature* **407,** 651–654.
Jeyakanthan, J., *et al.* (2010). Substrate specificity determinants of the methanogen homoaconitase enzyme: Structure and function of the small subunit. *Biochemistry* **49,** 2687–2696.
Jia, Y., *et al.* (2006). Kinetics and product analysis of the reaction catalysed by recombinant homoaconitase from *Thermus thermophilus*. *Biochem. J.* **396,** 479–485.
Jordan, F. (2003). Current mechanistic understanding of thiamin diphosphate-dependent enzymatic reactions. *Nat. Prod. Rep.* **20,** 184–201.
Karpusas, M., Branchaud, B., and Remington, S. J. (1990). Proposed mechanism for the condensation reaction of citrate synthase: 1.9-Å structure of the ternary complex with oxaloacetate and carboxymethyl coenzyme A. *Biochemistry* **29,** 2213–2219.
Kerber, R. C., and Fernando, M. S. (2010). α-Oxocarboxylic acids. *J. Chem. Educ.* **87,** 1079–1084.
Kimmich, G. A., Roussie, J. A., and Randles, J. (2002). Aspartate aminotransferase isotope exchange reactions: Implications for glutamate/glutamine shuttle hypothesis. *Am. J. Physiol. Cell Physiol.* **282,** C1404–C1413.
Kisumi, M., Sugiura, M., and Chibata, I. (1976). Biosynthesis of norvaline, norleucine, and homoisoleucine in *Serratia marcescens*. *J. Biochem.* **80,** 333–339.
Knapp, D. R. (1979). Carboxylic acids. *In* "Handbook of Analytical Derivatization Reactions,". pp. 146–224. Wiley, New York.
Koon, N., Squire, C. J., and Baker, E. N. (2004). Crystal structure of LeuA from *Mycobacterium tuberculosis*, a key enzyme in leucine biosynthesis. *Proc. Natl. Acad. Sci. USA* **101,** 8295–8300.
Kozlowski, J., and Zuman, P. (1987). Polarographic reduction of aldehydes and ketones: Part XXX. Effects of acid-base, hydration-dehydration and keto-enol equilibria on reduction of α-ketoglutaric and oxalacetic acid and their esters. *J. Electroanal. Chem.* **226,** 69–102.
Krebs, H. A., and Eggleston, L. V. (1944). Micro-determination of isocitric and *cis*-aconitic acids in biological material. *Biochem. J.* **38,** 426–437.
Krishnakumar, A. M., *et al.* (2008). Getting a handle on the role of coenzyme M in alkene metabolism. *Microbiol. Mol. Biol. Rev.* **72,** 445–456.
Leahy, S. C., *et al.* (2010). The Genome sequence of the rumen methanogen *Methanobrevibacter ruminantium* reveals new possibilities for controlling ruminant methane emissions. *PLoS ONE* **5,** e8926.
Li, Z.-C., and Xu, J.-Q. (1998). An improved synthesis of homocitrate. *Molecules* **3,** 31–34.
Li, F., *et al.* (2007). *Re*-Citrate synthase from *Clostridium kluyveri* Is phylogenetically related to homocitrate synthase and isopropylmalate synthase rather than to *si*-citrate synthase. *J. Bacteriol.* **189,** 4299–4304.
Lin, Y., West, A.H., Cook, P.F. (2009). Site-directed mutagenesis as a probe of the acid-base catalytic mechanism of homoisocitrate dehydrogenase from *Saccharomyces cerevisiae*. *Biochemistry* **48,** 7305–7312.
Londesborough, J. C., and Dalziel, K. (1968). The equilibrium constant of the isocitrate dehydrogenase reaction. *Biochem. J.* **110,** 217–222.
Ma, G., and Palmer, D. R. J. (2000). Improved asymmetric syntheses of (R)-(-)-homocitrate and (2R, 3S)-(-)-homoisocitrate, intermediates in the α-aminoadipate pathway of fungi. *Tetrahedron Lett.* **41,** 9209–9212.
Ma, J., *et al.* (2008). Molecular basis of the substrate specificity and the catalytic mechanism of citramalate synthase from *Leptospira interrogans*. *Biochem. J.* **415,** 45–56.
Marsden, B. J., Sauer, F. D., Blackwell, B. A., and Kramer, J. K. G. (1989). Structure determination of the UDP-disaccharide fragment of cytoplasmic cofactor isolated from *Methanobacterium thermoautotrophicum*. *Biochem. Biophys. Res. Commun.* **159,** 1404–1410.

Massoudi, H. H., Cantacuzene, D., Wakselman, C., and de la Tour, C. B. (1983). Synthesis of the *cis*- and *trans*-isomers of homoaconitic and fluorohomoaconitic acid. *Synthesis* **12,** 1010–1012.

McBride, B. C., and Wolfe, R. S. (1971). A new coenzyme of methyl transfer, coenzyme M. *Biochemistry* **10,** 2317–2324.

Meister, A. (1957). Preparation of α-keto acids. *Methods Enzymol.* **3,** 404–414.

Miyazaki, J., Kobashi, N., Nishiyama, M., and Yamane, H. (2003). Characterization of homoisocitrate dehydrogenase involved in lysine biosynthesis of an extremely thermophilic bacterium, *Thermus thermophilus* HB27, and evolutionary implication of beta-decarboxylating dehydrogenase. *J. Biol. Chem.* **278,** 1864–1871.

Miyazaki, J., *et al.* (2005). Crystal structure of tetrameric homoisocitrate dehydrogenase from an extreme thermophile, *Thermus thermophilus*: Involvement of hydrophobic dimer-dimer interaction in extremely high thermotolerance. *J. Bacteriol.* **187,** 6779–6788.

Namboori, S. C., and Graham, D. E. (2008). Acetamido sugar biosynthesis in the Euryarchaea. *J. Bacteriol.* **190,** 2987–2996.

Nelson, R. B., and Gribble, G. W. (1973). On the preparation of α-ketoadipic acid. *Org. Prep. Proced. Int.* **5,** 55–58.

Noll, K. M., and Wolfe, R. S. (1986). Component C of the methylcoenzyme M methylreductase system contains bound 7-mercaptoheptanoylthreonine phosphate (HS-HTP). *Biochem. Biophys. Res. Commun.* **139,** 889–895.

Noll, K. M., Rinehart, K. L., Jr., Tanner, R. S., and Wolfe, R. S. (1986). Structure of component B (7-mercaptoheptanoylthreonine phosphate) of the methylcoenzyme M methylreductase system of *Methanobacterium thermoautotrophicum*. *Proc. Natl. Acad. Sci. USA* **83,** 4238–4242.

Noll, K. M., Donnelly, M. I., and Wolfe, R. S. (1987). Synthesis of 7-mercaptoheptanoylthreonine phosphate and its activity in the methylcoenzyme M methylreductase system. *J. Biol. Chem.* **262,** 513–515.

Okada, T., *et al.* (2010). Mechanism of substrate recognition and insight into feedback inhibition of homocitrate synthase from *Thermus thermophilus*. *J. Biol. Chem.* **285,** 4195–4205.

Paju, A., *et al.* (2004). A short enantioselective synthesis of homocitric acid-[gamma]-lactone and 4-hydroxy-homocitric acid-[gamma]-lactones. *Tetrahedron* **60,** 9081–9084.

Qian, J., Khandogin, J., West, A. H., and Cook, P. F. (2008). Evidence for a catalytic dyad in the active site of homocitrate synthase from *Saccharomyces cerevisiae*. *Biochemistry* **47,** 6851–6858.

Romesser, J. A., and Balch, W. E. (1980). Coenzyme M: Preparation and assay. *Methods Enzymol.* **67,** 545–552.

Sauer, F. D., Blackwell, B. A., Kramer, J. K. G., and Marsden, B. J. (1990). Structure of a novel cofactor containing *N*-(7-mercaptoheptanoyl)-*O*-3-phosphothreonine. *Biochemistry* **29,** 7593–7600.

Scheller, S., *et al.* (2010). The key nickel enzyme of methanogenesis catalyses the anaerobic oxidation of methane. *Nature* **465,** 606–608.

Schmitz, C., Rouanet-Dreyfuss, A.-C., Tueni, M., and Biellmann, J.-F. (1996). Syntheses of (-)-isocitric acid lactone and (-)-homoisocitric acid. A new method of conversion of alkynylsilanes into the alkynyl thioether and corresponding carboxylic acids. *J. Org. Chem.* **61,** 1817–1821.

Shriner, R. L. (2004). The Systematic Identification of Organic Compounds. 8th edn. Wiley, Hoboken, NJ.

Solow, B., and White, R. H. (1997). Biosynthesis of the peptide bond in the coenzyme *N*-(7-mercaptoheptanoyl)-L-threonine phosphate. *Arch. Biochem. Biophys.* **345,** 299–304.

Spector, L. B. (1972). Citrate cleavage and related enzymes. *In* "The Enzymes," (P. D. Boyer, ed.), pp. 357–389. Academic Press, New York.
Strassman, M., and Ceci, L. N. (1965). Enzymatic formation of α-ketoadipic acid from homoisocitric acid. *J. Biol. Chem.* **240,** 4357–4361.
Strassman, M., and Ceci, L. N. (1966). Enzymatic formation of *cis*-homoaconitic acid, an intermediate in lysine biosynthesis in yeast. *J. Biol. Chem.* **241,** 5401–5407.
Suzuki, Y., *et al.* (2010). Enhancement of the latent 3-isopropylmalate dehydrogenase activity of promiscuous homoisocitrate dehydrogenase by directed evolution. *Biochem. J.* **431,** 401–410.
Taylor, C. D., and Wolfe, R. S. (1974). Structure and methylation of coenzyme M ($HSCH_2CH_2SO_3$). *J. Biol. Chem.* **249,** 4879–4885.
Thomas, U., Kalyanpur, M. G., and Stevens, C. M. (1966). The absolute configuration of homocitric acid (2-hydroxy-1, 2, 4-butanetricarboxylic acid), an intermediate in lysine biosynthesis. *Biochemistry* **5,** 2513–2516.
Wächtershäuser, G. (1990). Evolution of the first metabolic cycles. *Proc. Natl. Acad. Sci. USA* **87,** 200–204.
Waters, K. L. (1947). The α-keto acids. *Chem. Rev.* **41,** 585–598.
Weinstein, C. L., and Griffith, O. W. (1986). β-Sulfopyruvate: Chemical and enzymatic syntheses and enzymatic assay. *Anal. Biochem.* **156,** 154–160.
White, R. H. (1985). Biosynthesis of coenzyme M (2-mercaptoethanesulfonic acid). *Biochemistry* **24,** 6487–6493.
White, R. H. (1986). Intermediates in the biosynthesis of coenzyme M (2-mercaptoethanesulfonic acid). *Biochemistry* **25,** 5304–5308.
White, R. H. (1989a). Biosynthesis of the 7-mercaptoheptanoic acid subunit of component B [(7-mercaptoheptanoyl)threonine phosphate] of methanogenic bacteria. *Biochemistry* **28,** 860–865.
White, R. H. (1989b). A novel biosynthesis of medium chain length a-ketodicarboxylic acids in methanogenic archaebacteria. *Arch. Biochem. Biophys.* **270,** 691–697.
White, R. H. (1989c). Steps in the conversion of α-ketosuberate to 7-mercaptoheptanoic acid in methanogenic bacteria. *Biochemistry* **28,** 9417–9423.
White, R. H. (1994). Biosynthesis of (7-mercaptoheptanoyl)threonine phosphate. *Biochemistry* **33,** 7077–7081.
Wise, E. L., Graham, D. E., White, R. H., and Rayment, I. (2003). The structural determination of phosphosulfolactate synthase from *Methanococcus jannaschii* at 1.7 Å resolution: An enolase that is not an enolase. *J. Biol. Chem.* **278,** 45858–45863.
Xu, P.-F., Matsumoto, T., Ohki, Y., and Tatsumi, K. (2005). A facile method for synthesis of (*R*)-(−)- and (*S*)-(+)-homocitric acid lactones and related [alpha]-hydroxy dicarboxylic acids from d- or l-malic acid. *Tetrahedron Lett.* **46,** 3815–3818.
Xu, H., *et al.* (2006). The α-aminoadipate pathway for lysine biosynthesis in fungi. *Cell Biochem. Biophys.* **46,** 43–64.
Yukawa, N., Takamura, H., and Matoba, T. (1993). Determination of total carbonyl compounds in aqueous media. *J. Am. Oil Chem. Soc.* **70,** 881–884.
Zhang, G., Dai, J., Lu, Z., and Dunaway-Mariano, D. (2003). The phosphonopyruvate decarboxylase from *Bacteroides fragilis*. *J. Biol. Chem.* **278,** 41302–41308.
Zhang, K., Sawaya, M. R., Eisenberg, D. S., and Liao, J. C. (2008). Expanding metabolism for biosynthesis of nonnatural alcohols. *Proc. Natl. Acad. Sci. USA* **105,** 20653–20658.
Zheng, L., White, R., and Dean, D. (1997). Purification of the *Azotobacter vinelandii* nifV-encoded homocitrate synthase. *J. Bacteriol.* **179,** 5963–5966.

CHAPTER SIXTEEN

Biomethanation and Its Potential

Irini Angelidaki,* Dimitar Karakashev,* Damien J. Batstone,† Caroline M. Plugge,‡ and Alfons J. M. Stams‡

Contents

Abstract

Biomethanation is a process by which organic material is microbiologically converted under anaerobic conditions to biogas. Three main physiological groups of microorganisms are involved: fermenting bacteria, organic acid oxidizing bacteria, and methanogenic archaea. Microorganisms degrade organic matter via cascades of biochemical conversions to methane and carbon dioxide. Syntrophic relationships between hydrogen producers (acetogens) and hydrogen scavengers (homoacetogens, hydrogenotrophic methanogens, etc.) are critical to the process. Determination of practical and theoretical methane potential is very important for design for optimal process design, configuration, and effective evaluation of economic feasibility. A wide variety of process applications for biomethanation of wastewaters, slurries, and solid waste

* Department of Environmental Engineering, Technical University of Denmark, Lyngby, Denmark
† Advanced Water Management Centre, The University of Queensland, Brisbane, Queensland, Australia
‡ Laboratory of Microbiology, Wageningen University, Wageningen, The Netherlands

Methods in Enzymology, Volume 494
ISSN 0076-6879, DOI: 10.1016/B978-0-12-385112-3.00016-0

have been developed. They utilize different reactor types (fully mixed, plug-flow, biofilm, UASB, etc.) and process conditions (retention times, loading rates, temperatures, etc.) in order to maximize the energy output from the waste and also to decrease retention time and enhance process stability. Biomethanation has strong potential for the production of energy from organic residues and wastes. It will help to reduce the use of fossil fuels and thus reduce CO_2 emission.

1. Introduction

Biomethanation is a natural process of anaerobic degradation of organic materials resulting in production of biogas (a mixture of methane and carbon dioxide). It occurs in natural environments, such as landfills, rice fields, sediments, and intestinal tracts of animals, where light and inorganic electron acceptors (oxygen, nitrate, sulfate, iron, etc.) are not present or limiting (Hattori, 2008). The anaerobic degradation process is a multistep complex process performed by the combined action of three major physiological groups of microorganisms (Hattori, 2008; Thauer *et al.*, 2008): primary fermenting (hydrolytic–acidogenic) bacteria, anaerobic oxidizing (syntrophic–acetogenic) bacteria, and methanogenic archaea. Fermenting microorganisms decompose the biopolymers (lipids, proteins, nucleic acids, carbohydrates, etc.) to soluble monomers (long-chain fatty acids, glycerol, amino acids, purines, pyrimidines, monosugars, etc.) that are further converted to short chain fatty acids (butyrate, propionate, acetate, etc.), alcohols (ethanol and methanol), hydrogen, and carbon dioxide by the same microbes. Short chain fatty acids and also alcohols are oxidized by proton-reducing syntrophic acetogens to hydrogen, acetate, formate, and carbon dioxide. These end products are ultimately transformed to methane and carbon dioxide by the methanogenic archaea.

The ability of anaerobic microorganisms to decompose complex organic matter and to produce biogas has been utilized for centuries in artificial (man-made) systems (biogas reactors) for production of energy. Methane derived from anaerobic digestion (AD) of different organic wastes and residues has strong potential as an alternative to fossil fuels. It combines production of renewable energy with environmental benefits, such as reduction of greenhouse emissions, reduction of odors, and controlled waste disposal. Moreover, AD process offers the possibility for recycling of nutrients (nitrogen and phosphorous), as the digested material can be applied on agricultural land as biofertilizer, and thereby replace artificial fertilizers.

Anaerobic microorganisms obtain little energy from the conversions that they catalyze and consequently, low cell yields are obtained. The cell yields of anaerobes are in the range of 0.02–0.2 g cells g^{-1} substrate, contrary to aerobic microorganisms that have much higher yields (0.4–0.7 g cells g^{-1} substrate; Zinder, 1993). The remaining energy is conserved in reduced products like methane, ethanol, H_2, and volatile fatty acids. Besides production of these valuable energy storing carriers, this low cell biomass production is another major advantage of the anaerobic process, compared to aerobic processes. In addition, anaerobic wastewater treatment has low power requirements, as there is no need for aeration.

In this chapter, the biomethanation process is described in terms of microorganisms (phylogeny, morphology, physiology, ecology, etc.) involved in different stages (hydrolysis, acidogenesis, acetogenesis, methanogenesis, etc.) and the biochemical pathways that are used by the different microorganisms. Methods for determination of theoretical and practical methane potential are also given. The most important practical applications of the process for anaerobic treatment of different residues (slurries, solid wastes, wastewaters, etc.) are summarized with respect to operating conditions (temperature, retention time, organic loading, reactor configuration, etc.) and yields of the final products (biogas).

2. The AD Process: Microbiology and Metabolic Pathways

AD is a multistep process, with a number of microbial interrelationships and dependencies (Fig. 16.1). Individual processes are kinetically nonlinear with respect to substrate concentration and inhibitors, and under most circumstances, either hydrolysis or aceticlastic methanogenesis is the rate-limiting processes. This is not a fixed rule, and under certain conditions (e.g., highly loaded glucose-fed systems), a buildup of hydrogen can prevent acetogenesis from occurring.

2.1. Hydrolysis

In AD of complex materials, the term hydrolysis is used to describe a wide range of depolymerization and solubilization processes by which complex polymeric organic compounds are broken down into soluble monomers. Most of these reactions, such as carbohydrate, polypeptide, triglyceride, and nucleic acid hydrolysis, are true hydrolysis processes, while others (e.g., scission of disulfide bonds) are reductive or oxidative biotransformations (Sevier and Kaiser, 2002).

The three main primary substrates (biopolymers) for hydrolysis are carbohydrates, lipids, and proteins, which hydrolyze to monosaccharides, long-chain fatty acids plus glycerol, and amino acids, respectively. In waste or feed

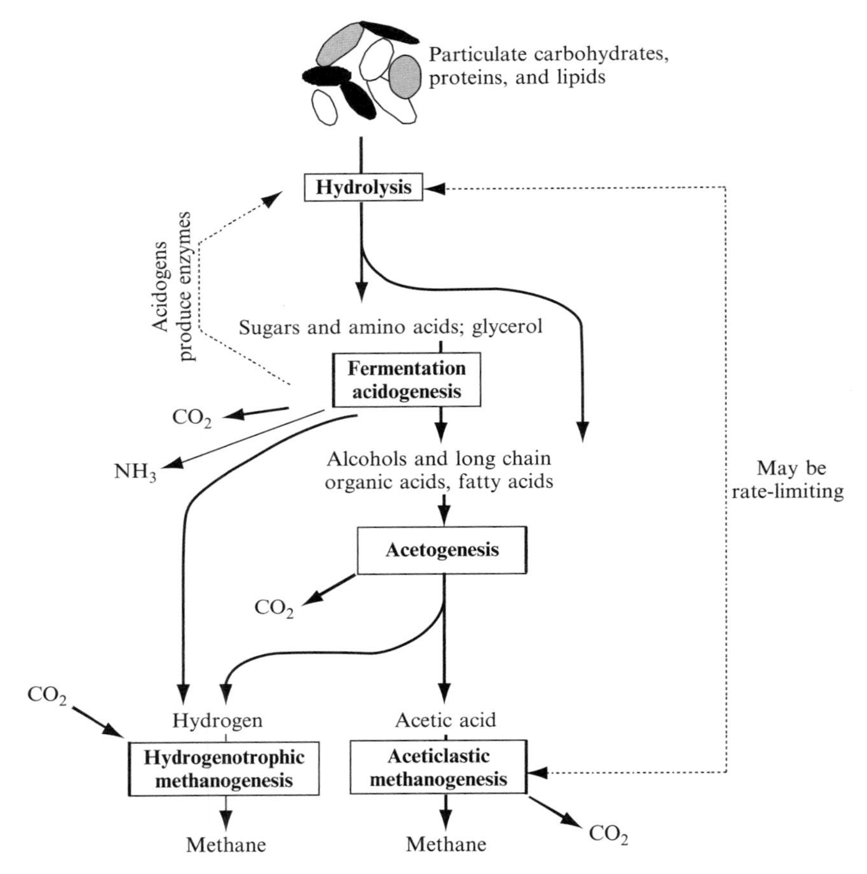

Figure 16.1 Key steps in the anaerobic digestion process. Modified from Batstone and Jensen (2010).

streams to anaerobic digesters, materials can either be a mixture of the three substrates (e.g., primary sludge, manure, agricultural wastes, etc.), or a complex composite with lumped properties that will be degraded to the monomers (Batstone *et al.*, 2002; Nopens *et al.*, 2009). An example of the latter is activated sludge. In all cases, hydrolysis is a complex multistep process, generally mediated by extracellular enzymes. These enzymes may be attached to the microbial cell (Tong and McCarty, 1991) or secreted to the bulk solution.

Hydrolysis is widely regarded as the rate-limiting step of degradation of particulate organic matter (e.g., manure, sewage sludge, crop residues, etc.; Pavlostathis and Giraldo-Gomez, 1991). Therefore, the overall rate of the process is determined by the hydrolysis rate of this complex substrate. For the purposes of process design, this determines the required retention time in a stirred methanogenic bioreactor. Despite being a very complicated process, hydrolysis is most commonly represented by a first-order process (Batstone *et al.*, 2002). This is justified as a cumulative effect of a wide range of processes that are difficult to characterize or identify (Eastman and Ferguson, 1981). Hydrolysis rate coefficients (constants) are normally in the order of 0.1–0.3 d^{-1} (Batstone *et al.*, 2002).

There are also particular considerations for different primary materials, classified as biofibers, carbohydrates, proteins, and fats. The most common composite feed is waste-activated sludge from sewage treatment plants. This is a mixture of microbial material, decay products, and inert materials originating from the feed. The degradation of decay products and inerts is generally limited in anaerobic processes, and hence the degradability of waste-activated sludge depends heavily on upstream properties (Ekama *et al.*, 2007).

Biofibers are a mixture of cellulose, hemicellulose, and lignins (Tong and McCarty, 1991). Degradation of biofibers largely depends on their structure and composition, with dense materials high in lignin such as woody material, being practically nondegradable, and relatively accessible materials such as straw being degradable (Tong and McCarty, 1991; Yang *et al.*, 2009).

Proteins and lipids are generally found together in streams such as originating from meat processing. Protein degradation depends on protein structure, with semisoluble globular proteins being highly degradable, and fibrous proteins being relatively difficult to hydrolyze (McInerney, 1988). Lipids are normally triglycerides, which are hydrolyzed by lipases. Hydrolysis rates of lipids depends less on the chemical properties of the substrate and more on particle size and environmental conditions such as pH and surface tension.

2.2. Fermentation

Fermentation is anaerobic conversion of organic materials in the absence of inorganic electron acceptors such as sulfate, nitrate, or oxygen. Reduction of protons to form hydrogen may take place, but this is generally facultative. This is in contrast to degradation of propionate or butyrate to acetate and hydrogen, a process more properly referred to as anaerobic oxidation (see Section 2.2.1 below).

A wide range of substrates can be fermented, including monosaccharides, amino acids, unsaturated fatty acids, glycerol, and halogenated organics (Madigan *et al.*, 2009). However, the most abundant substrates for fermentation, and a primary route for carbon flux, are monosaccharides and amino acids. These two fermentation processes are completely different, though capability to utilize these substrates is widespread, and a wide range of microorganisms, mostly *Clostridia* and other low GC Gram-positive microorganisms, can use both substrates (Ramsay and Pullammanappallil, 2001). Fermentation of amino acids and monosaccharides has common elements, in that both fermentative processes are relatively energy rich and rapid, and have a wide range of operating conditions in terms of pH and oxidation/reduction potential (Batstone *et al.*, 2002; Madigan *et al.*, 2009; Ramsay and Pullammanappallil, 2001). However, pathways, current level of knowledge, and regulating factors are very different for these two primary substrates.

2.2.1. Fermentation of monosaccharides

Fermentation of monosaccharides is probably the most widely applied biotechnology process worldwide to produce a wide range of products. In pure culture, regulation and product spectrum have been extensively characterized, particularly for organisms commonly used in industrial biotechnology. The way in which these well-characterized pure cultures are regulated by environmental conditions is precisely understood. However, where multiple organisms exist in a competitive environment (mixed culture fermentation), the basic pathways are known, but factors regulating product mixture are less understood (Batstone *et al.*, 2002; Rodríguez *et al.*, 2006; Temudo *et al.*, 2008). Monosaccharides ferment via the Emben–Meyerhof–Parnas (EMP) or Entner Doudoroff (ED) pathway, and subsequently to C3 products (lactate and propionate), or C2/C4/C6 products (acetate/butyrate/caproate) via acetyl-CoA. C3 products are uncommon, except under overload conditions, and the most common products are acetate, butyrate, and ethanol, with waste carbon going to carbon dioxide, and excess electrons to hydrogen gas (Rodríguez *et al.*, 2006). There is some dispute as to whether ethanol production is enhanced at low (Ren *et al.*, 1997) or high pH (Temudo *et al.*, 2008).

2.2.2. Fermentation of amino acids

Whereas glucose uses itself as an electron acceptor to produce oxidized (e.g., acetate) and reduced (e.g., propionate, ethanol, etc.) products, mixed amino acids commonly degrade in pairs in a coupled oxidation/reduction reaction. This is called the Stickland reaction (Winter *et al.*, 1987). In this pathway, one amino acid is oxidized (becoming one carbon shorter), while the other is reduced. An example of the Stickland reaction is the fermentation of alanine plus glycine:

$$\text{Alanine (oxidation)} : CH_3CHNH_2COO^- + 2H_2O \rightarrow CH_3COO^- + CO_2 + NH_3 + 4H^+ + 4e^-$$

$$\text{Glycine (reduction)} : 2CH_2NH_2COO^- + 4H^+ + 4e^- \rightarrow 2CH_3COO^- + 2NH_3$$

Amino acids can act as electron acceptor, donor, or in some cases as both (Ramsay and Pullammanappallil, 2001). Glutamate fermentation is an example of uncoupled amino acid degradation (Buckel, 2001). For

Stickland fermentation, there is normally a shortfall in acceptor amino acids, and the remainder needs to be oxidized by uncoupled degradation. In conditions where hydrogen concentrations are low and the energetics are suitable, uncoupled oxidation can also occur (Stams, 1994). In the case of uncoupled oxidation, alanine would then produce two molecules of hydrogen gas instead of electrons that are used to convert glycine to acetate.

2.3. Acetogenesis

Acetogenesis refers to the synthesis of acetate, which includes the formation of acetate by the reduction of CO_2 and the formation of acetate from organic acids. Hydrogen-utilizing acetogens, previously also termed homoacetogens, are strict anaerobic bacteria that can use the acetyl-CoA pathway as (i) their predominant mechanism for the reductive synthesis of acetyl-CoA from CO_2, (ii) terminal electron-accepting, energy-conserving process, and (iii) mechanism for the synthesis of cell carbon from CO_2 (Drake, 1994). These bacteria compete with methanogens for substrates like hydrogen, formate, and methanol.

Organic acids (such as propionate and butyrate) and alcohols (such as ethanol) produced during the fermentation step are oxidized to acetate by hydrogen-producing acetogens. Electrons produced from this oxidation reaction are transferred to protons (H^+) to produce H_2 or bicarbonate to generate formate. The term obligate means that the primary substrate cannot be used as alternative electron acceptor, and electrons must be wasted to hydrogen ions or carbon dioxide, with generally unfavorable energetics (Stams and Plugge, 2009).

Acetogens that oxidize organic acids obligately use hydrogen ions and carbon dioxide as electron acceptor. Acetogenic bacteria are limited by the unfavorable energetics of the conversion processes (Schink and Stams, 2006). Table 16.1 illustrates the conversion of propionate and butyrate, important intermediates in anaerobic fermentations of complex organic matter, to the methanogenic substrates, acetate and hydrogen. It is evident that bacteria can only derive energy for growth from these conversions if the concentration of the products is kept low. This results in an obligate dependence of acetogenic bacteria on methanogenic archaea or other hydrogen scavengers (e.g., sulfate reducers) for product removal (McInerney *et al.*, 2008; Sousa *et al.*, 2009; Stams and Plugge, 2009). The coupling or syntrophic relationship of hydrogen producers and hydrogen consumers is called interspecies hydrogen transfer (Fig. 16.2). Obligately syntrophic communities of acetogenic bacteria and methanogenic archaea have several unique features: (i) they degrade fatty acids coupled to growth, while neither the methanogen nor the acetogen alone is able to degrade these compounds, (ii) intermicrobial distances between acetogens and hydrogen-scavenging microorganisms influence specific growth rates (Batstone *et al.*,

Table 16.1 Standard Gibbs free energy changes for some of the reactions involved in syntrophic fatty acid degradation

Reaction	ΔGo′ (kJ/reaction)
Fatty-acid oxidation	
$Acetate^- + 4\,H_2O \rightarrow H_2 + 2HCO_3^- + H^+$	+105
$Propionate^- + 3H_2O \rightarrow acetate^- + HCO_3^- + H^+ + 3H_2$	+76
$Butyrate^- + 2H_2O \rightarrow 2\,acetate^- + H^+ + 2H_2$	+48
Hydrogen utilization by methanogens and acetogens	
$4\,H_2 + HCO_3^- + H^+ \rightarrow CH_4 + 3H_2O$	−136
$4\,H_2 + 2\,HCO_3^- + H^+ \rightarrow acetate^- + 4H_2O$	−105
Syntrophic acetate oxidation coupled with hydrogenotrophic methanogenesis	
$CH_3COO^- + H_2O \rightarrow CH_4 + HCO_3^-$	−31

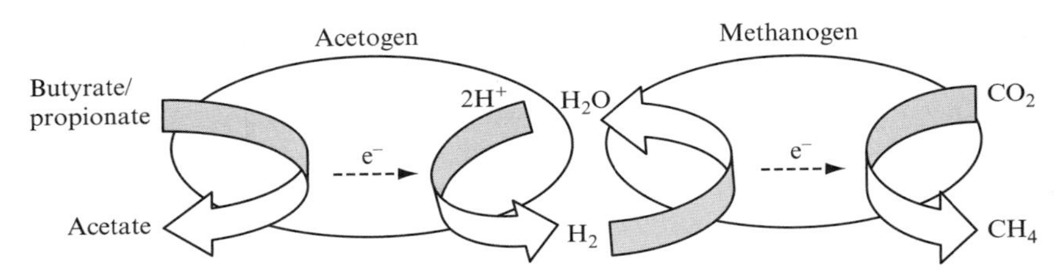

Figure 16.2 Scheme of interspecies hydrogen transfer.

2006), (iii) the communities grow in conditions that are close to thermodynamic equilibrium, and (iv) the communities have evolved biochemical mechanisms that allow sharing of chemical energy.

Acetate formed by acetogenic bacteria can be either used directly by aceticlastic methanogens (*Methanosarcina* spp. and *Methanosaeta* spp.), or it can be degraded by syntrophic associations of bacteria (syntrophic acetate oxidizers) and hydrogen-consuming methanogenic archaea. The syntrophic acetate-oxidation phenomenon was observed first in a thermophilic reactor system (Zinder and Koch, 1984) and later also at lower temperature in sludges with enhanced ammonia content (Schnürer *et al.*, 1996). The overall reaction (Table 16.1) is relatively low in energy, but this increases with rising temperature (Schink and Stams, 2006). A number of thermophilic (*Thermacetogeneum phaeum*, *Thermotoga lettingae*; Hattori, 2008) and mesophilic (*Clostridium ultunense*, *Syntrophaceticus schinkii*; Schnürer *et al.*, 1996; Westerholm *et al.*, 2010) acetate oxidizers have been isolated and described. The biochemistry of syntrophic acetate oxidation appears to be basically a reversal of the homoacetogenic acetate formation pathway (so-called

Wood–Ljungdahl pathway or CO-dehydrogenase pathway). Acetate is activated to acetyl-CoA and cleaved by a CO-dehydrogenase/acetyl-CoA synthase enzyme complex to a methyl and a carbonyl residue, which are oxidized separately by well-described pathways (Hattori, 2008). The question remains how the bacteria couple this pathway to ATP synthesis, especially since some of these acetate-degrading bacteria can also grow using the reverse reaction to form acetate.

2.4. Methanogenesis

Methanogenic microorganisms belong to the *Archaea* domain, phylum *Euryarchaeota*. Five phylogenetic orders of methanogens have been identified: *Methanosarcinales*, *Methanobacteriales*, *Methanomicrobiales*, *Methanococcales*, and *Methanopyrales*. Very recently, a new, sixth order *Methanocellales* was described (Sakai *et al.*, 2008) and phylogenetically placed between orders *Methanosarcinales* and *Methanomicrobiales*.

Methanogens are very diverse with respect to cell morphology—from regular and irregular coccoidal cell shape (*Methanococcales*, *Methanomicrobiales* sp.) to short rods (*Methanobacteriales*, *Methanopyrales*), and long filaments (*Methanosaetacea* sp. within *Methanosarcinales*).

Methanogenic archaea are responsible for the final step in AD process for methane formation from acetate and/or from carbon dioxide and hydrogen, formate, alcohols, and methylated C_1 compounds (Thauer *et al.*, 2008). They are strictly anaerobic microorganisms abundant in environments where external electron acceptors such as O_2, NO_3^-, Fe^{3+}, and SO_4^{-2} are limited. Common habitats for those archaea are anoxic marine and freshwater sediments, gastrointestinal tracts of ruminants and insects, anaerobic digesters, hot springs, and flooded soils.

Methanogens have a unique metabolism involving a number of unique enzymes and coenzymes (Deppenmeier, 2002). The most interesting feature is that none of the methanogenic archaea can utilize energy from substrate level phosphorylation, and ATP is probably generated by a proton motive force and, for hydrogenotrophic methanogens, by a sodium motive force (Boone *et al.*, 1993). The main physiological characteristics (carbon sources utilized, temperature- and pH growth range) of the methanogens are summarized in Table 16.2. It can be pointed out that some methanogenic orders such as *Methanosarcinales* (except the Methanosaetaceae) and *Methanomicrobiales* are versatile regarding carbon sources utilized, and the others (*Methanobacteriales*, *Methanococcales*, *Methanopyrales*, and *Methanocellales*) are more specialized. With respect to temperature preferences, most known methanogens are mesophilic and moderate to extreme thermophilic archaea. An exception is the order *Methanopyrales*, containing one strict hyperthermophilic strain, *Methanopyrus kandleri*, growing up to 110 °C (Kurr *et al.*, 1991). Some psychrophiles, such as *Methanosarcina lacustris*

Table 16.2 Main physiological features of the methanogenic orders

Methanogenic order	Carbon sources	Temperature range (°C)	pH range	References
Methanosarcinales	Acetate, $H_2 + CO_2$, CO, methanol, methylamines, methylmercaptopropionate, dimethylsulfide	1.0–70	4.0–10.0	Kendall and Boone (2006), Liu and Whitman (2008), Deppenmeier (2002), Cheng *et al.* (2007), Thauer *et al.* (2008)
Methanomicrobiales	$H_2 + CO_2$, formate, ethanol,[a] 2-propanol,[b] 2-butanol,[b] cyclopentanol[b]	15–60	6.1–8.0	Garcia *et al.* (2006), Liu and Whitman (2008), Dianou *et al.* (2001), Thauer *et al.* (2008)
Methanobacteriales	$H_2 + CO_2$, CO, formate, C_1-methylated compounds[c]	20–88	5.0–8.8	Bonin and Boone (2006), Liu and Whitman (2008), Thauer *et al.* (2008)
Methanococcales	$H_2 + CO_2$, formate	$<$20–88	4.5–9.8	Whitman and Jeanthon (2006), Liu and Whitman (2008), Thauer *et al.* (2008)
Methanopyrales	$H_2 + CO_2$	84–110	5.5–7.0	Kurr *et al.* (1991), Huber and Stetter (2001), Liu (2010)
Methanocellales	$H_2 + CO_2$, formate	25–40	6.5–7.8	Sakai *et al.* (2008)

[a] Only for *Methanogenium* sp.
[b] Only for *Methanoculleus* sp.
[c] Only for *Methanosphaera* sp.

(Kendall and Boone, 2006) and *Methanogenium frigidum* (Garcia *et al.*, 2006), have also been reported. Most methanogens are neutrophiles growing optimally at a pH around 7. Some species, such as *Methanosarcina baltica* (Kendall and Boone, 2006) and *Methanothermococcus okinawensis* (Whitman and Jeanthon, 2006), are acidotolerant (growth occurs at pH down to 4–4.5) and species such as *Methanosalsum zhilinae* (Kendall and Boone, 2006) and *Methanothermococcus thermolitotrophicus* (Whitman and Jeanthon, 2006) are alkalotolerant (growth occurs under pH up to 9.8–10).

Three main pathways for methane formation in anaerobic environments are known (Conrad *et al.*, 2010; Deppenmeier, 2002; Liu and Whitman, 2008): aceticlastic methanogenesis where acetate is cleaved to methane and carbon dioxide (Eq. 16.1), hydrogenotrophic methanogenesis where carbon dioxide is reduced to methane (Eq. 16.2), and methylotrophic methanogenesis where methylated C_1 compounds (methanol, methylamines, methylmercaptopropionate, dimethylsulfide, etc.) are converted to methane (Eq. 16.3).

$$CH_3COO^- + H^+ \rightarrow CH_4 + CO_2 \quad (16.1)$$

$$4H_2 + CO_2 \rightarrow CH_4 + 2H_2O \quad (16.2)$$

$$4CH_3OH \rightarrow 3CH_4 + CO_2 + 2H_2O \quad (16.3)$$

Methylotrophic methanogenesis is limited to species from order *Methanosarcinales* with one exception, *Methanosphaera* sp. which belongs to order *Methanobacteriales*. Acetate is the most important source of methane in the anaerobic environment, giving rise to approx. In all, 70% of the total global methane output, while the remaining 30% is formed from hydrogen and carbon dioxide, or formate (Conrad *et al.*, 2010). *Methanosarcinales* is the only order where aceticlastic methanogens are found so far: Methanosaetaceae, with members that only use acetate, and *Methanosarcinacea*, with members that are more versatile, being able to utilize other carbon sources (carbon monoxide, carbon dioxide, and methylated C_1 compounds) in addition to acetate (Kendall and Boone, 2006). Recently, Cheng *et al.* (2007) discovered a new family Methermicoccaceae within the order *Methanosarcinales*, comprised only by strict methyltrophic (utilizing methanol, methylamine, and trimethylamine as carbon sources) archaea, not able to grow on acetate or carbon dioxide/hydrogen.

Methanosaetacea is pH- and ammonia-sensitive and dominate at low acetate concentrations (below 10^{-3} *M*; Zinder, 1993), while *Methanosarcinacea* generally dominate in environments with high ammonia and organic acid content (Karakashev *et al.*, 2005). Previous investigations (Karakashev *et al.*, 2006) led to the hypothesis that some *Methanosarcinacea* sp. (instead of conventional acetate cleavage to methane and carbon dioxide) might be able to oxidize acetate to hydrogen and carbon dioxide (acetate oxidation) and subsequently perform hydrogenotrophic methanogenesis, mediating the entire process of nonaceticlastic conversion of acetate to methane. However, more detailed biochemical investigations are required to confirm and clarify the biochemical mechanisms involved in this process.

The ability to utilize hydrogen as an electron donor for reduction of carbon dioxide (hydrogenotrophic activity) is widespread among methanogens. An interesting feature is that hydrogenotrophic methanogens from the order *Methanomicrobiales* can also grow mixotrophically; for example, *Methanoculeus* sp. can also utilize cyclopentanol and secondary alcohols, such as 2-propanol and 2-butanol (Dianou *et al.*, 2001), and some *Methanogenium* species can oxidize ethanol (Liu and Whitman, 2008).

Hydrogenotrophic methanogenesis performed by *Methanobacteriales*, *Methanomicrobiales*, *Methanococcales*, *Methanocellales*, and *Methanopyrales* is one of the major sinks for electrons in electron acceptor-depleted anaerobic environments through interspecies electron transfer where reducing equivalents are transferred between hydrogen-forming acetogenic bacteria and hydrogen-consuming archaea (Stams and Plugge, 2009). Methanogens maintain a low dissolved hydrogen concentration by close contact with hydrogen producers. In practice, the majority of hydrogen-dependent methane formation occurs in flocks or biofilms, making a direct transfer of hydrogen from the hydrogen-producing microorganism to the hydrogen-consuming methanogens possible.

3. Biochemical Methane Potential

Different biomass feedstocks (organic wastes and residues) have different methanogenic potential, depending on the inherent degradability, as well as the carbon-oxidation state. As the economy of the existing biogas plant is strongly dependent on the methane potential of the biomass used, prediction of the biogas (methane) potential is a key issue in AD. Biogas consists mainly of methane and carbon dioxide, and minor amounts of other gases such as ammonia, hydrogen sulfide, hydrogen, and nitrogen (Zinder, 1993). The prime interest of the biogas is the methane, because methane is the energy-rich component of the gas. The amount of biogas produced and the content of methane in the gas phase depend on the waste being degraded, both its degradability and its oxidation state. The better degradability and the lower the oxidation state, the more methane will be produced.

When considering the biogas process for a specific application, the characteristics of the substrate/waste are naturally of prime interest. Waste and wastewater have a complex composition which is difficult to describe in detail, but can readily be analyzed by bulk chemical processes as described below.

The most common single measure used to describe the concentration of waste or waste water is the chemical oxygen demand (COD) expressed as g-O_2/liter, or the Volatile Solids (VS) content expressed as g-VS/liter or % W/V VS (APHA, 2001). The COD content describes the amount of oxygen needed to completely oxidize the organics in the waste, and is determined experimentally by measuring the amount of a chemical oxidizing agent needed to fully oxidize a sample of the waste. The VS content describes the content of organic material in the waste, and is defined as the amount of matter of a dried sample, lost after 1 h at a temperature of approx. 550 °C. Both methods are relatively easy to perform, and give a good first impression of the "strength" of a waste. However, determination of COD for very concentrated samples is inaccurate, as large dilutions of the sample are required.

If the atomic composition of the organic material is known, the relation between COD and VS content can be calculated from the stoichiometric equation for the complete oxidation of the organic matter. For many types of organic waste, the oxidation state of carbon is close to zero (as for glucose) and in these cases, the COD/VS ratio will be close to unity. However, in more reduced compounds such as long-chain fatty acids and lipids, the oxidation state of carbon is negative and the COD/VS ratio is significantly higher than 1. This means that 1 g of lipids produce a higher COD wastewater than 1 g of sugars. COD can be balanced across a process for a given removal efficiency to give either aeration required (in the case of aerobic process), or methane produced (in the case of anaerobic processes). Methane has a molecular weight of 16 g mol^{-1}, and COD of 64 g mol^{-1}, and therefore has a theoretical COD:VS ratio of 4 (with the lowest possible carbon-oxidation state), though this is not measured by the standard potassium dichromate COD test method.

In a more generalized form, the oxidation reaction for organic material might be written as (Eq. 16.4)

$$C_nH_aO_b + \left(n + \frac{a}{4} - \frac{b}{2}\right) O_2 \rightarrow +nCO_2 + \frac{a}{2}H_2O \qquad (16.4)$$

and the COD/VS ratio becomes

$$\frac{\text{COD}}{\text{VS}} = \frac{(n + \frac{a}{4} - \frac{b}{2})32}{12n + a + 16b} \qquad (16.5)$$

3.1. Theoretical potential

The theoretical methane potential (or methane yield) is widely recognized to give an indication of the maximum methane production expected from a specific waste.

If the exact waste composition is known, methane production can be predicted by a stoichiometric equation, balancing the total conversion of the organic material, to CH_4 and CO_2 with H_2O, as the only external source, that is, under anaerobic conditions (Eq. (16.6)). The ratio between CH_4 and CO_2 depends on the oxidation state of the carbon present in the organic material, that is, the more reduced the organic carbon content is, the more CH_4 will be produced.

$$C_nH_aO_b + \left(n - \frac{a}{4} - \frac{b}{2}\right)H_2O \rightarrow \left(\frac{n}{2} + \frac{a}{8} - \frac{b}{4}\right) CH_4 + \left(\frac{n}{2} - \frac{a}{8} + \frac{b}{4}\right)CO_2 \tag{16.6}$$

The specific theoretical methane potential ($B_{o,th}$), usually expressed as l CH_4/g-VS, may then be calculated as (Eq. 16.7): assuming 22.4 as the volume of 1 mole of gas under standard conditions (i.e., 273 K and 1 atm. pressure).

Alternatively, the specific methane potential ($B_o{}'{}_{,th}$) can also be expressed as l-CH_4/g-COD according to Eq. (16.8):

$$B_{o,th} = \frac{(\frac{n}{2} + \frac{a}{8} - \frac{b}{4})22.4}{12n + a + 16b} \quad \left(\mathrm{STP}\frac{l\mathrm{CH_4}}{\mathrm{g-VS}}\right) \tag{16.7}$$

$$B_{o,th} = \frac{(\frac{n}{2} + \frac{a}{8} - \frac{b}{4})22.4}{(n + \frac{a}{4} - \frac{b}{2})32} \quad \left(\mathrm{STP}\frac{l\mathrm{CH_4}}{\mathrm{g-COD}}\right) \tag{16.8}$$

In Table 16.3, the characteristics of a number of typical organic materials suitable for anaerobic degradation are listed.

However, often the exact atomic composition of the feedstock is unknown, and only the approximate content of the main organic groups is known. In this case, an alternative and easier way to get an estimation of the theoretical methane potential is, based on data from Table 16.1, to calculate the theoretical methane potential as Eq. (16.9):

$$\begin{aligned} B_{o,th} = {} & 0.415\,\mathrm{carbohydrates} + 0.496\,\mathrm{proteins} + 1.014\,\mathrm{lipids} \\ & + 0.373\,\mathrm{acetate} + 0.530\,\mathrm{propionate} \end{aligned} \tag{16.9}$$

3.2. Practical potential

Although the theoretical methane potential gives a rough indication of the amount of biogas produced from any waste, the practical potential obtained will always be lower due to a number of factors:

Table 16.3 Characteristics of organic materials

Substrate type	Composition	COD/VS (g.COD/ g.VS)	CH_4 yield $B_{o,th}$ (STP l/g.VS)	CH_4 yield $B_{o,th}$ (STP l/g.COD)	CH_4 (%)
Carbohydrate[a]	$(C_6H_{10}O_5)_n$	1.19	0.415	0.35	50
Protein[b]	$C_5H_7NO_2$	1.42	0.496	0.35	50
Lipids[c]	$C_{57}H_{104}O_6$	2.90	1.014	0.35	70
Ethanol	C_2H_6O	2.09	0.730	0.35	75
Acetate	$C_2H_4O_2$	1.07	0.373	0.35	50
Propionate	$C_3H_6O_2$	1.51	0.530	0.35	58

[a] Cellulose is used as a model carbohydrate.
[b] Gelatine is used as a model protein and nitrogen is converted to ammonia.
[c] Glycerol trioleate is used as a model lipid.

- A fraction of the substrate is utilized to synthesize bacterial mass, typically 2–5% of the organic material degraded.
- There is a substantial fraction of inert material, which may include lignins, cellular debris, and humics.
- Some material is not accessible, due to its recalcitrant organic structure.
- Limitation of other nutrient factors.

Under favorable conditions with mainly water-soluble matter, degrees of conversion up to 100% can be achieved. If the organic matter is highly particulate, 30–60% conversion is achievable. Manure, for instance, has a typical conversion of the VS of approximately 40–50% for cattle manure and 55–65% for swine manure (Møller, 2003). In continuously operated biogas reactors, the methane yield is even lower, as the effluent COD concentration is never zero. In well-operated reactors, this lost potential can be as low as 5%, while in other reactors, operated suboptimally, losses up to 25% have been observed (Angelidaki *et al.*, 2005).

Hashimoto *et al.* (1981) defined the practical methane potential (B_o and B_o') as the ultimate volume of methane produced from a specific amount of waste (in either weight or COD) in a batch experiment for indefinite degradation time, that is, until the methane production ceases.

In order to estimate practical methane potential, it is very important to predict the composition of the biogas produced, which depends first of all on the amount of CH_4 and CO_2 produced, and also on the pH of the reactor content. The CH_4 produced is mainly released to the gas phase, but CO_2 is partly dissolved in the liquid phase of the reactor or is converted to bicarbonate dependent on the pH. Consequently, the CH_4 percentage in the biogas produced will generally be higher than predicted by the stoichiometric ratio. Methane content in the gas phase is typically 55–65% and increases when the pH is high. As such, methane content in the biogas can reflect the pH in the reactor.

Determination of the practical methane potential is not always easy, as there are many factors that can lead to false results. Sampling of heterogeneous wastes, inoculum activity, presence of toxicants, inoculum to substrate ratio, and lack of nutrients, can be some of the sources for underestimation of the methane potential of a sample (Angelidaki *et al.*, 2009; Chen and Hashimoto, 1996; Gungor-Demirci and Demirer, 2004; Hansen *et al.*, 2004; Hashimoto 1989; Raposo *et al.*, 2006; Zhang *et al.*, 2007). Methane potential can also be overestimated, such as in the case of wastes containing easily volatilized compounds (ethanol, lactate, etc.). In this case, VS determination by the standard analytical methods (APHA), where drying at 105 °C is recommended, leads to evaporation of these compounds, which will result in underestimation of the organic matter (VS) content and, thereby, overestimating the methane potential.

Biogas production is quantified by measurement of the volume of the produced gas under constant pressure (volumetric), measurement of pressure increase in constant volume (manometric), or measurement of methane formation by gas chromatography (Angelidaki and Sanders, 2004; Angelidaki *et al.*, 2005, 2007, 2009). The chromatographic method directly measures methane formation, and is the most reliable method, although it requires more complex equipment. In contrast, the manometric and volumetric methods measure total biogas production (CH_4 plus CO_2), and are strongly dependent on the pH influencing CO_2 (HCO_3^-) distribution between gas and liquid phase. Critical to all gasometrical methods is the accurate measurement of biogas production at low headspace pressure to prevent errors associated with CO_2 solubility.

4. Biogas Applications

Many different technologies have emerged ranging from simple unmixed semicontinuous flow tanks, such as septics, to more advanced, immobilized, and multicompartment systems. Technologies applied are dependent on the total solids (TS) content of the influent to be treated. Wastewaters with TS ranging from 1% to 2%, are often utilizing immobilized biomass technology, such as the upflow anaerobic sludge blanket (UASB) digester. Slurries with TS contents ranging from 3–4% to 7–10% are most often applying fully mixed tank reactors, while with solid wastes with TS higher than 10%, special technology has to be applied taking into account difficulties with technical handling of the wastes, such as mixing and transport.

4.1. AD of slurries

Manure together with primary and activated sludge constitutes the largest groups of slurries. Manure is an abundant biomass in rural areas, especially in countries with intensive agriculture, and constitutes the main biomass resource available for biogas production (Raven and Gregersen, 2007). Due to the particulate consistence of manure, and the relatively high TS concentration, high-rate reactors (see Section 4.2) cannot be applied, and continuously stirred tank reactors (CSTR) are the most common configurations (Speece, 2008). In a CSTR reactor, there will be a continuous loss of active microbial biomass. It is therefore necessary to operate these reactors with a retention time that exceeds the doubling time of the slowest growing organisms present ($\mu > D$). As a consequence, CSTR hydraulic retention times (HRTs) are typically in the range 15–30 days at mesophilic conditions and 10–20 days at thermophilic conditions. Long retention times also benefit hydrolysis of the particulate matter of complex structure such as lignocellulose in manure. Doubling times of microorganisms as determined under optimal conditions in the laboratory, are often significantly shorter than retention times applied in biogas plants, due to suboptimal growth conditions in biogas reactors, such as high ammonia concentrations and the presence of toxicants.

Depending on the composition, it may be possible to separate and return solids from the effluent to reduce loss of active biomass and thus intensify the process. Slurries, together with small amounts of solid wastes, can be mixed together and codigested in biogas reactors. The codigestion concept is used to a large extent in Denmark in centralized biogas plants. In these plants, a mixture of manure slurry (60–80%) with various types of industrial wastes (Angelidaki *et al.*, 2005) is used. The economy of the biogas plants is dependent on effective utilization of the substrates treated in order to maximize energy yield in relation to treatment costs. In manure-based biogas plants, typically only 50–70% of the organic matter is converted to biogas despite rather long average retention times. Some of the residual organic matter is recalcitrant and cannot be digested. However, some degradable material is also lost with the effluent from the CSTR reactor. The reason for this loss of degradable matter is either the basic characteristics of a fully mixed reactor or, in the worst cases, the direct short-circuiting from influent to effluent.

4.2. Anaerobic digestion of solid waste

Solid waste is defined as organic material with solid content 10–40% TS, which is not fluid. The most important types of solid wastes with considerable biomethanation potential are municipal solid waste (MSW), kitchen waste, garden waste, energy crops (maize, grass, sugarcane, etc.), etc.

The main obstacle in the processing of MSW is the separation of the organic biodegradable matter from other parts of MSW such as plastics and metals. Separation includes magnetic treatment, screening, pulping, gravity separation, or pasteurization. Also, source-sorting of biodegradable material has been practiced in many cases. Biological treatment of solid wastes includes technical challenges, such as feedstock pumping, homogenization, and mixing, which needs special consideration. Digestion can take place either under mesophilic or thermophilic conditions, and the HRT is 10–30 days, depending on the process temperature, the technology used, and the waste composition.

Currently, both batch and continuous processes for treatment of biodegradable fraction of MSW are used (Monson *et al.*, 2007) according to the feeding mode of the wastes through the reactors (Fig. 16.3). Batch systems are basically engineered landfills (either confined or covered), with enhanced performance through recirculation of leachate. Three different batch configurations have been considered so far, including single batch (one tank with recirculation of the reactor effluent to the top of the same reactor), sequential batch (two tanks with recirculation of the effluent from first, acidogenic to the second, methanogenic reactor), and hybrid batch-UASB (combination of an acidification tank and methanogenic biofilm reactor where effluent from the tank is recirculated toward the methanogenic reactor). Although very cheap, batch systems have not gained substantial market share so far, due to clogging of the bed late in the digestion stage, and manual requirements of loading and unloading. However due to their simple design and process control, they are very attractive for commercial applications in developing countries (Quedraogo, 1999).

Continuous systems can be divided into one- or multistage (often two-stage) processes. The one-stage process utilizes either fully mixed reactors, such as the Valorga process, or plug-flow reactors such as the Dranco process (with vertical plug-low reactor) or Compogas (with horizontal

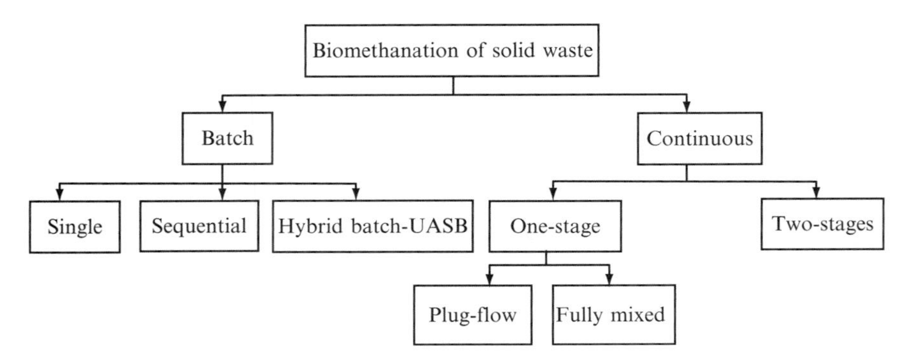

Figure 16.3 Process configurations for biomethanation of solid wastes.

plug-flow reactor). For the fully mixed reactors, process water is separated from the effluent and is recycled in order to dilute the high total solid wastes, and thus make pumping and mixing possible. Moreover, lack of mixing inside plug-flow reactors, is not facilitating contact of substrate and microorganisms and therefore, recycling of digested effluent is applied to ensure effective inoculation of the influent substrate. With respect to two-stage process, many different process configurations were developed so far, the most simple one being applied includes two completely mixed or plug-flow reactors in series (Monson *et al.*, 2007).

The AD process is feasible and efficient; however, technical difficulties concerning collection, separation, and handling of the wastes constitute a major problem in implementation of solid waste technologies. Therefore, codigestion with more liquid waste streams might be an attractive and economical way to treat these types of wastes.

4.3. Anaerobic treatment of wastewater

The major issue involving anaerobic treatment of wastewater has typically been the extended retention times required to grow methanogens (> 10 days). This is not a problem for digestion of solids, where the high strength means that reactor sizes are relatively small, and growth rates are high, but is a major issue for wastewater, where organic concentrations are far lower. To address these systems, reactors need to separate and recirculate, or retain biological solids. This was originally done by separate separation units (anaerobic contact reactors), but these have been since replaced by more integrated designs.

The major process configuration in current use is the UASB reactor. This integrates solids retention into the main digester by feeding at the base through a manifold, and collecting at the top in a gas/liquid/solid separation system, often referred to as high-rate AD (van Lier, 2008). Optimal operation depends on formation of a stable granular sludge bed in the base of the reactor. Granules are very dense biological aggregates of 0.3–1 mm in diameter and settle very rapidly. While reactors operate best with granules, it is possible to operate a UASB with floccular sludge. UASB reactors are now a highly mature and validated technology and, for optimal applications (highly soluble and nonnitrogenous organic feeds), are almost the default technology (van Lier, 2008). On these types of feeds, design is normally based on a loading rate of 10 kg-COD $m^{-3}d^{-1}$, with retention times of 8–24 h. Even higher loading rates (20 kg COD $m^{-3}d^{-1}$) are possible with newer low footprint designs, such as the internal circulation (IC), and expanded granular sludge bed (EGSB) reactors.

Thus, for highly soluble, low nitrogen wastewaters, high-rate AD is ideal. However, high-rate AD has a number of limitations in other applications, especially domestic wastewater (Seghezzo *et al.*, 1998; van Haandel *et al.*, 2006):

(a) Nitrogen is not removed catabolically in AD and is not substantially removed by anabolic processes.
(b) High-rate AD is sensitive to lipids and solids, which interfere with granulation and may cause granular sludge to be lost.
(c) Domestic wastewater strength (<1 g COD L^{-1}) is lower than optimal for high-rate AD (>5 g COD L^{-1}).
(d) Methane cannot be fully captured and may be lost in the effluent. This is particularly a problem for low-strength wastewater.

Many of these problems are being addressed by a number of investigations, including postdigestion aerobic treatment, phosphorous recovery, and nutrient removal by anaerobic oxidation (Foresti *et al.*, 2006), and it is possible that high-rate AD will play a major role in future domestic wastewater treatment.

5. Biomethanation—Global Aspects

Probably, the most complete projection of worldwide consumption of energy is published by the US Energy Information Administration (DOE/EIA-0484, 2010). Energy demand from the OECD (Organization for Economic Co-operation and Development) countries is predicted to be relatively stable at 220 EJ/a (exa-joules = 10^{18} J 1 EJ ~ 1 quadrillion British thermal units (BTU)) over the next 30 years, with net growth to be balanced by improved efficiency. However, non-OECD energy consumption is set to increase dramatically, driving an increase in overall consumption from 500 to 750 EJ in the next 30 years. This is almost entirely driven by an increase in consumption in non-OECD Asian countries (largely China and India). Renewable energy is predicted to increase at approximately the same rate as nonrenewable sources (50% over the next 35 years; DOE/EIA-0484, 2010). However, in the long term (up to 2035), this increase is predicted to be mostly due to a fourfold increase (0.7–3.0 EJ/a) in biomass-derived energy, while solar, wind, and geothermal-derived energy production is predicted to rise only in the short term (up to 2015).

The current paradigm for energy from biogas has been developed in Western Europe, and is based on centralized biogas plants. Biogas-derived energy in Europe contributes with 0.25 EJ/a (2007; Weiland, 2010), representing some 0.3% of present energy consumption in the EU. However, the share of biogas energy is expected to increase in the future. As an example, the Agency for Renewable Resources in Germany (FNR) has estimated a total biogas potential in Germany of 0.42 EJ/a, representing 3% of Germany's total energy consumption, of which 50% is crop residue or energy crop, 25% is animal manure, and 5% is wastewater-derived (FNR, 2010). Biogas applications in Asian countries (China, India, etc.) are more diverse, with domestic small biogas plant installations for heating, light, cooking, but also medium- and large-scale installations for electricity production (Fang, 2010). This means that the predicted increase in energy consumption in these regions (see above) can be partially offset by more diverse biogas usage patterns.

While utilization of arable land for cultivation of energy crops is generally not justified economically, biomethane formation from agricultural residues and waste streams is an effective way to produce renewable energy (Shilton and Guieysse, 2010). Recycling of organics is a key component in the global carbon cycle. By aerobic microbial decomposition, carbon dioxide is released without energy conservation. By AD, carbon dioxide formation is retarded by the formation of methane, which can serve as an alternative for natural gas. In this way, fossil fuels are preserved and net less carbon dioxide is emitted.

Although biomethanation through AD is already commercially applied, there are still technological bottlenecks that can be improved. Emerging reactor technologies, development of advanced monitoring and control systems, as well as methods for increasing biodegradability of relatively recalcitrant feedstocks are making biogas production more economically feasible. Although, biogas still needs subsidizing, it is foreseen that the biogas will become a significant biofuel in the coming years.

ACKNOWLEDGMENTS

This chapter was financially supported by Project No. 2010-1-10537, EnergiNet (ForskEL programme). Research was also funded by grants of the Chemical Science (CW) and the Earth and Life Sciences (ALW) divisions and the Technical Science Foundation (STW) of the Netherlands Science Foundation (NWO).

REFERENCES

Angelidaki, I., and Sanders, W. (2004). Assessment of the anaerobic biodegradability of macropollutants. *Rev. Environ. Sci. Biotechnol.* **3**(2), 117–129.

Angelidaki, I., Boe, K., and Ellegaard, L. (2005). Effect of operating conditions and reactor configuration on efficiency of full-scale biogas plants. *Water Sci. Technol.* **52**(1–2), 189–194.

Angelidaki, I., Alves, M., Bolzonella, D., Borzacconi, L., Campos, L., Guwy, A., Jenicek, P., Kalyuzhnui, S., and van Lier, J. (2007). Anaerobic Biodegradation. Activity and Inhibition (ABAI) Task Group Meeting 9th to 10th October 2006, Prague, Institute of Environment Engineering, Technical University of Denmark, Denmark. http://www.er.dtu.dk/publications/fulltext/2007/ MR2007-147.pdf.

Angelidaki, I., Alves, M., Bolzonella, D., Borzacconi, L., Campos, L., Guwy, A., Kalyuzhnyi, S., Jenicek, P., and van Lier, J. B. (2009). Defining the biomethane potential (BMP) of solid organic wastes and energy crops: A proposed protocol for batch assays. *Water Sci. Technol.* **59**(5), 927–934.

American Public Health Association (APHA), American Water Works Association (AWWA) & Water Environment Federation (WEF): *Standard Methods for the Examination of Water and Wastewater*, 21st Edition, 2005.

Batstone, D. J., and Jensen, P. D. (2010). Anaerobic processes. *In* "Treatise on Water Science," (K. Hanaki, ed.). Elsevier, London, Chapter 97.

Batstone, D. J., Keller, J., Angelidaki, I., Kalyuzhnyi, S. V., Pavlostathis, S. G., Rozzi, A., Sanders, W. T. M., Siegrist, H., and Vavilin, V. A. (2002). Anaerobic Digestion Model No. 1 (ADM1), IWA Task Group for Mathematical Modelling of Anaerobic Digestion Processes. IWA Publishing, London.

Batstone, D. J., Picioreanu, C., and van Loosdrecht, M. C. M. (2006). Multidimensional modelling to investigate interspecies hydrogen transfer in anaerobic biofilms. *Water Res.* **40**(16), 3099–3108.

Bonin, A. S., and Boone, D. R. (2006). The order methanobacteriales. *In* "The Prokaryotes, Volume 3: Archaea. Bacteria: Firmicutes, Actinomycetes," (M. Dworkin, S. Falkow, E. Rosenberg, K.-H. Schleifer, and E. Stackebrandt, eds.), pp. 231–243. Springer-Verlag, Singapore.

Boone, D. R., Whitman, W. B., and Rouviere, P. (1993). Diversity and taxonomy of methanogens. *In* "Methanogenesis: Ecology, Physiology, Biochemistry and Genetics," (J. G. Ferry, ed.). Chapman and Hall, New York.

Buckel, W. (2001). Unusual enzymes involved in five pathways of glutamate fermentation. *Appl. Microbiol. Biotechnol.* **57**(3), 263–273.

Chen, T. H., and Hashimoto, A. G. (1996). Effects of pH and substrate:inoculum ratio on batch methane fermentation. *Bioresour. Technol.* **56,** 179–186.

Cheng, L., Qiu, T.-L., Yin, X.-B., Wu, X.-L., Hu, G.-Q., Deng, Y., and Zhang, H. (2007). *Methermicoccus shengliensis* gen. nov., sp. nov., a thermophilic, methylotrophic methanogen isolated from oil-production water, and proposal of *Methermicoccaceae* fam. nov. *Int. J. Syst. Evol. Microbiol.* **57,** 2964–2969.

Conrad, R., Klose, M., Claus, P., and Enrich-Prast, A. (2010). Methanogenic pathway, ^{13}C isotope fractionation, and archaeal community composition in the sediment of two clear-water lakes of Amazonia. *Limnol. Oceanogr.* **55**(2), 689–702.

Deppenmeier, U. (2002). The unique biochemistry of methanogenesis. *Prog. Nucleic Acid Res. Mol. Biol.* **71,** 223–283.

Dianou, D., Miyaki, T., Asakawa, S., Morii, H., Nagaoka, K., Oyaizu, H., and Matsumoto, S. (2001). *Methanoculleus chikugoensis* sp. nov., a novel methanogenic archaeon isolated from paddy field soil in Japan, and DNA-DNA hybridization among *Methanoculleus* species. *Int. J. Syst. Evol. Microbiol.* **51,** 1663–1669.

DOE/EIA-0484 (2010). U.S. Energy Information Administration International Energy Outlook 2010 U.S. Energy Information Administration Office of Integrated Analysis and Forecasting U.S. Department of Energy, Washington, DC. http://www.eia.gov/oiaf/ieo/index.html, (Accessed 15 November 2010).

Drake, H. L. (1994). Acetogenesis. Chapman & Hall, New York.

Eastman, J. A., and Ferguson, J. F. (1981). Solubilization of particulate organic carbon during the acid phase of anaerobic digestion. *J. Water Pollut. Contin. Fed.* **53,** 352–366.

Ekama, G. A., Sötemann, S. W., and Wentzel, M. C. (2007). Biodegradability of activated sludge organics under anaerobic conditions. *Water Res.* **41**(1), 244–252.

Fang, H. P. P. (2010). ApplicaPons of Anaerobic Technology in East Asia, Keynote Presentation. Presentation, Proceedings of the 12th World Congress on Anaerobic Digestion, Guadalajara, Mexico October 31—November 4, 2010.

FNR (2010). Biogas Basisdaten Deutschland, Fachagentur Nachwachsende Rohstoffe. http://www.fnr-server.de/ftp/pdf/literatur/pdf_185-v8-basisdaten_biogas_2010_finale-fassung.pdf, (Accessed 15 November 2010).

Foresti, E., Zaiat, M., and Vallero, M. (2006). Anaerobic processes as the core technology for sustainable domestic wastewater treatment: Consolidated applications, new trends, perspectives, and challenges. *Rev. Environ. Sci. Biotechnol.* **5**(1), 3–19.

Garcia, J.-L., Ollivier, B., and Whitman, W. B. (2006). The order methanomicrobiales. *In* "The Prokaryotes, Vol. 3," (M. Dworkin, ed.), pp. 208–230. Springer, New York.

Gungor-Demirci, G., and Demirer, G. N. (2004). Effect of initial COD concentration, nutrient addition, temperature and microbial acclimation on anaerobic treatability of broiler and cattle manure. *Bioresour. Technol.* **93**(2), 109–117.

Hansen, T. L., Schmidt, J. E., Angelidaki, I., Marca, E., Jansen, J. C., Mosbæk, H., and Christensen, T. H. (2004). Measurement of methane potentials of solid organic waste. *Waste Manag.* **24**(4), 393–400.

Hashimoto, A. G. (1989). Effect of inoculum/substrate ratio on methane yield and production rate from straw. *Biol. Waste* **28,** 247–255.

Hashimoto, A. G., Varel, V. H., and Chen, Y. R. (1981). Ultimate methane yield from beef cattle manure: Effect of temperature, ration constituents, antibiotics and manure age. *Agric. Waste* **3,** 241–256.

Hattori, A. (2008). Syntrophic acetate-oxidizing microbes in methanogenic environments. *Microbes Environ.* **23**(2), 118–127.

Huber, H., and Stetter, K. O. (2001). Order I. *Methanopyrales* ord. nov. *In* "Bergey's Manual of Systematic Bacteriology, Vol. 1," (D. R. Boone, R. W. Castenholz, and G. M. Garrity, eds.), second ed., pp. 353–355. Springer, New York. (The Archaea and the Deeply Branching and Phototrophic Bacteria).

Karakashev, D., Batstone, D. J., and Angelidaki, I. (2005). Influence of environmental conditions on methanogenic compositions in anaerobic biogas reactors. *Appl. Environ. Microbiol.* **71**(1), 331–338.

Karakashev, D., Batstone, D. J., Trably, E., and Angelidaki, I. (2006). Acetate oxidation is the dominant methanogenic pathway from acetate in the absence of *Methanosaetaceae*. *Appl. Environ. Microbiol.* **72**(7), 5138–5141.

Kendall, M. M., and Boone, D. R. (2006). The order Methanosarcinales. *In* "The Prokaryotes. A Handbook on the Biology of Bacteria, Vol. 3," (M. Dworkin, S. Falkow, E. Rosenberg, K.-H. Schleifer, and E. Stackebrandt, eds.), pp. 244–256. Springer, New York, NY.

Kurr, M., Huber, R. K., König, H., Jannasch, H. W., Fricke, H., Trincone, A., Kristjansson, J. K., and Stetter, K. O. (1991). *Methanopyrus kandleri*, gen. and sp. nov. represents a novel group of hyperthermophilic methanogens, growing at 110 °C. *Arch. Microbiol.* **156**(4), 239–247.

Liu, Y. (2010). Methanopyrales. *In* "Handbook of Hydrocarbon and Lipid Microbiology. Part **7**," (K. N. Timmis, ed.), pp. 605–607. Springer-Verlag, Berlin Heidelberg.

Liu, Y., and Whitman, B. (2008). Metabolic, phylogenetic, and ecological diversity of the methanogenic archaea. *Ann. N. Y. Acad. Sci.* **1125,** 171–189.

Madigan, M. T., Martinko, J. M., Dunlap, P. V., and Clark, D. P. (2009). *Brock Biology of Microorganisms*. 12th edition. Pearson Benjamin Cummings. San Francisco.

McInerney, M. J. (1988). Anaerobic hydrolysis and fermentation of fats and proteins. *In* "Biology of Anaerobic Microorganisms," (A. J. B. Zehnder, ed.), first ed. pp. 373–416. John Wiley & Sons, New York.

McInerney, M. J., Struchtemeyer, C. G., Sieber, J., Mouttaki, H., Stams, A. J. M., Schink, B., Rohlin, L., and Gunsalus, R. P. (2008). Physiology, ecology, phylogeny and genomics of microorganisms capable of syntrophic metabolism. *Ann. N. Y. Acad. Sci.* **1125,** 58–72.

Møller, H. B. (2003). Methane productivity and nutrient recovery from manure. Technical University of Denmark, PhD thesis, BioCentrum-DTU.

Monson, K. D., Esteves, S. R., Guwy, A. J., and Dinsdale, R. M. (2007). Anaerobic digestion of biodegradable municipal wastes. A Review. Report ISBN: 978-1-84054-156-5.

Nopens, I., Batstone, D. J., Copp, J. B., Jeppsson, U., Volcke, E., Alex, J., and Vanrolleghem, P. A. (2009). An ASM/ADM model interface for dynamic plant-wide simulation. *Water Res.* **43**(7), 1913–1923.

Pavlostathis, S. G., and Giraldo-Gomez, E. (1991). Kinetics of anaerobic treatment: A critical review. *Crit. Rev. Environ. Control* **21**(5–6), 411–490.

Quedraogo, A. (1999). Pilot scale two-phase anaerobic digestion of the biodegradable organic fraction of Bamako district municipal solid waste. *In* "II Int. Symp. Anaerobic Dig. Solid Waste, Vol. 2," (J. Mata-Alvarez, A. Tilche, and F. Cecchi, eds.), pp. 73–76. Int. Assoc. Wat. Qual, Barcelona. June 15–17, 1999.

Ramsay, I. R., and Pullammanappallil, P. C. (2001). Protein degradation during anaerobic wastewater treatment: Derivation of stoichiometry. *Biodegradation* **12**(4), 247–257.

Raposo, F., Banks, C. J., Siegert, I., Heaven, S., and Borja, R. (2006). Influence of inoculum to substrate ratio on the biochemical methane potential of maize in batch tests. *Process Biochem.* **41**(6), 1444–1450.

Raven, R. P. J. M., and Gregersen, K. H. (2007). Biogas plants in Denmark: Successes and setbacks. *Ren. Sust. Energy Rev.* **11,** 116–132.

Ren, N., Wan, B., and JuChang, H. (1997). Ethanol-type fermentation from carbohydrate in high rate acidogenic reactor. *Biotechnol. Bioeng.* **54**(5), 428–433.

Rodríguez, J., Kleerebezem, R., Lema, J. M., and van Loosdrecht, M. C. M. (2006). Modeling product formation in anaerobic mixed culture fermentations. *Biotechnnol. Bioeng.* **93**(3), 592–606.

Sakai, S., Imachi, H., Hanada, S., Ohashi, A., Harada, H., and Kamagata, Y. (2008). *Methanocella paludicola* gen. nov., sp. nov., a methane-producing archaeon, the first isolate of the lineage 'Rice Cluster I', and proposal of the new archaeal order *Methanocellales* ord. nov. *Int. J. Syst. Evol. Microbiol.* **58,** 929–936.

Schink, B., and Stams, A. J. M. (2006). Syntrophism among prokaryotes. *In* "The Prokaryotes: An Evolving Electronic Resource for the Microbiological Community," (M. Dworkin, S. Falkow, E. Rosenberg, K.-H. Schleifer, and E. Stackebrandt, eds.), pp. 309–335. Springer-Verlag, New York.

Schnürer, A., Schink, B., and Svensson, B. H. (1996). *Clostridium ultunense* sp. nov., a mesophilic bacterium oxidizing acetate in syntrophic association with a hydrogenotrophic methanogenic bacterium. *Int. J. Syst. Bacteriol.* **46,** 1145–1152.

Seghezzo, L., Zeeman, G., van Lier, J. B., Hamelers, H. V. M., and Lettinga, G. (1998). A review: The anaerobic treatment of sewage in UASB and EGSB reactors. *Bioresour. Technol.* **65**(3), 175–190.

Sevier, C. S., and Kaiser, C. A. (2002). Formation and transfer of disulphide bonds in living cells. *Nat. Rev. Mol. Cell Biol.* **3**(11), 836–847.

Shilton, A., and Guieysse, B. (2010). Sustainable sunlight to biogas is via marginal organics. *Curr. Opin. Biotechnol.* **21**(3), 287–291.

Sousa, D. Z., Smidt, H., Alves, M. M., and Stams, A. J. M. (2009). Degradation of saturated and unsaturated long-chain fatty acids by syntrophic methanogenic communities. *FEMS Microbiol. Ecol.* **68,** 257–272.

Speece, R. E. (2008). Anaerobic Biotechnology and Odor/Corrosion Control for Municipalities and Industries. Archaeology Press, Nashville, Tennessee.

Stams, A. J. M. (1994). Metabolic interactions between anaerobic bacteria in methanogenic environments. *Antonie Van Leeuwenhoek Int. J. Genet. Mol. Microbiol.* **66**(1–3), 271–294.

Stams, A. J. M., and Plugge, C. M. (2009). Electron transfer in syntrophic communities of anaerobic bacteria and archaea. *Nat. Rev. Microbiol.* **7,** 568–577.

Temudo, M. F., Muyzer, G., Kleerebezem, R., and van Loosdrecht, M. C. M. (2008). Diversity of microbial communities in open mixed culture fermentations: Impact of the pH and carbon source. *Appl. Microbiol. Biotechnol.* **80**(6), 1121–1130.

Thauer, R. K., Kaster, A.-N., Seedorf, H., Buckel, W., and Hedderich, R. (2008). Methanogenic archae: Ecologically relevant differences in energy conservation. *Nat. Rev. Microbiol.* **6,** 579–591.

Tong, Z., and McCarty, P. (1991). Microbial hydrolysis of lignocellulosic materials. *In* "Methane from Community Wastes," (R. Isaacson, ed.), pp. 61–100. Elsevier Applied Science, London.

van Haandel, A., Kato, M., Cavalcanti, P., and Florencio, L. (2006). Anaerobic reactor design concepts for the treatment of domestic wastewater. *Rev. Environ. Sci. Biotechnol.* **5**(1), 21–38.

van Lier, J. B. (2008). High-rate anaerobic wastewater treatment: Diversifying from end-of-the-pipe treatment to resource-oriented conversion techniques. *Water Sci. Technol.* **57**(8), 1137–1148.

Weiland, P. (2010). Biogas production: Current state and perspectives. *Appl. Microbiol. Biotechnol.* **85,** 849–860.

Westerholm, M., Roos, S., and Schnürer, A. (2010). *Syntrophaceticus schinkii* gen. nov., sp. nov., an anaerobic, syntrophic acetate-oxidizing bacterium isolated from a mesophilic anaerobic filter. *FEMS Microbiol. Lett.* **309,** 100–104.

Whitman, W. B., and Jeanthon, C. (2006). Methanococcales. *In* "The Prokaryotes, Vol. 3," (S. Falkow, E. Rosenberg, K.-H. Schleifer, and E. Stackebrandt, eds.), 3rd edn. pp. 257–273. Springer, New York.

Winter, J., Schindler, F., and Wildenauer, F. X. (1987). Fermentation of alanine and glycine by pure and syntrophic cultures of *Clostridium sporogenes. FEMS Microbiol. Ecol.* **45,** 153–161.

Yang, S. J., Kataeva, I., Hamilton-Brehm, S. D., Engle, N. L., Tschaplinski, T. J., Doeppke, C., Davis, M., Westpheling, J., and Adams, M. W. W. (2009). Efficient degradation of lignocellulosic plant biomass, without pretreatment, by the thermophilic anaerobe "*Anaerocellum thermophilum*" DSM 6725. *Appl. Environ. Microbiol.* **75**(14), 4762–4769.

Zhang, R., El-Mashad, H. M., Hartman, K., Wang, F., Liu, G., Choate, C., and Gamble, P. (2007). Characterization of food waste as feedstock for anaerobic digestion. *Bioresour Technol.* **98**(4), 929–935.

Zinder, S. H. (1993). Physiological Ecology of Methanogens. *In* "Methanogenesis. Ecology, Physiology, Biochemistry and Genetics," (J. G. Ferry, ed.). Chapman and Hall, New York.

Zinder, S. H., and Koch, M. (1984). Non-acetoclastic methanogenesis form acetate: Acetate oxidation by a thermophilic syntrophic coculture. *Arch. Microbiol.* **138,** 263–273.

Author Index

D

E

F

G

H

I

J

K

L

M

Q

R

S

T

U

V

W

X

Y

Subject Index